VISCERAL SENSORY NEUROSCIENCE

VISCERAL SENSORY NEUROSCIENCE

Interoception

OLIVER G. CAMERON
Professor, Department of Psychiatry
The University of Michigan

UNIVERSITY PRESS
2002

OXFORD
UNIVERSITY PRESS

Oxford New York
Athens Auckland Bangkok Bogotá Buenos Aires Cape Town
Chennai Dar es Salaam Delhi Florence Hong Kong Istanbul
Kolkata Kuala Lumpur Madrid Melbourne Mexico City Mumbai Nairobi
Paris São Paulo Shanghai Singapore Taipei Tokyo Toronto Warsaw

and associated companies in
Berlin Ibadan

Published by Oxford University Press, Inc.,
198 Madison Avenue, New York, New York, 10016
http://www.oup-usa.org
1-800-334-4249

Library of Congress Cataloging-in-Publication Data

Cameron, Oliver G.
Visceral sensory neuroscience : interoception / Oliver G. Cameron.
p. ; cm.
Includes bibliographical references and index.
ISBN 0-19-513601-2
1. Senses and sensation. 2. Viscera–Innervation. I. Title.
[DNLM: 1. Visceral Afferents–physiology.
2. Psychophysiology. WL 102 C182v 2002]
QP435 .C35 2002
612.8'8–dc21
2001037056

2 4 6 8 9 7 5 3 1

Printed in the United States of America
on acid-free paper

In Memoriam: Richard Quinn LaPiana

PREFACE

What is this book about? It is concerned with the effects that afferent sensory impulses arising from the viscera, the organs and tissues within the trunk of the body, have on the higher nervous system processes and the behavior of the organism. This set of functions is called interoception, that is, perception of the functions and physiological activities of the interior of the body. The physiology of these afferent processes is essential to interoception, but the definition of interoception as used here is more than physiology and more than biology. It is psychobiology.

Why write a book on such a topic at this time? There are two reasons. First, interoception is an important phenomenon that is closely related to other topics and themes of great current interest, for example, the scientific study of emotion, and methods to study it more rigorously and in greater detail are becoming more readily available. Second, few up-to-date, in-depth reviews of this topic exist.

Is this book for you? It has been written for and is intended primarily for a professional audience, including scientists and clinicians working in all the fields related to psychobiology—neuroscience, neurophysiology, neuroanatomy, neuropharmacology, behavioral sciences, psychology, brain imaging, and related clinical fields such as psychiatry, neurology, internal medicine (especially cardiology and gastroenterology), and clinical psychology—and students in each of these fields. Individuals in the relatively new field of neurophilosophy should also find the material of relevance. In addition, any interested person who has studied basic biology and psychology, including information on human physiology and brain function, should be able to understand much of the information without undue difficulty. If you have ever used terms like "gut reaction" or "heartache" and wished to know more about how the gut and the heart are connected to cognitive and emotional events in the brain and mind, I believe you will be rewarded by reading this volume.

The thought and research that led up to and culminated in the writing of this book have been shaped and influenced by many forces and, especially, many individuals over a great number of years. There are too many people to mention separately, but those that affected me the most include *(1)* the faculty of the Department of Psychology at the University of Notre Dame in the late 1960s; *(2)* the faculty and fellow graduate students of the Department of Biopsychology and the Medical School at the University of Chicago in the early 1970s; *(3)* since the late 1970s, my colleagues in the Departments of Psychiatry, Nuclear Medicine (PET facility), and Pharmacology at the University of Michigan Medical Center, especially the faculty of the Department of Psychiatry Anxiety Disorders Program; *(4)* colleagues who are members and officers of the American Psychosomatic Society; and *(5)* my editor and the editorial staff at Oxford University Press. I thank the University of Michigan Medical School, which granted me a sabbatical during which time most of the writing was completed. And I thank my wife and children, Susan, Leah, and Peter, who constantly had to step over multiple piles of books and papers throughout our house for a year while I was writing.

Even though the chapters have different free-standing topics, the book is structured with the assumption that it will be read from beginning to end as a coherent whole. The first six chapters could be considered background. Chapters 2, 5, and 6 review mainly older research that may be considered the direct historical roots from which this topic grew. The early historical issues are *(1)* the emerging importance of the physical, the body including the brain, in philosophical and early psychological speculation; *(2)* the influence of the emerging biological sciences in the nineteenth century in further reinforcing this interest in the body per se; and *(3)* the focus on the study of sensation and perception in early experimental psychology. Chapters 7 through 9 contain the bulk of the recent areas of primary interest in interoception research—alimentary (gastrointestinal), respiratory, vascular, and (especially) cardiac systems, along with the underlying neural science. Chapters 10 and 11 discuss related areas that are not always considered interoceptive processes. Chapter 12 presents data concerning interoceptive processes in psychiatric disorders. Chapter 13 briefly touches on the relationship between interoception and consciousness. Chapter 14 presents a summary and some recommendations for future directions this field might take.

Ann Arbor, Michigan O. G. C.

CONTENTS

PART I. DEFINITIONS, HISTORY, BACKGROUND

PART II. THE ESSENTIAL RECENT SCIENCE

PART ONE

Definitions, History, Background

one

DEFINITIONS

> Interoceptor: Any sensory receptor which receives stimuli arising within the body, or specifically within the viscera.
>
> *Oxford English Dictionary*

> A specialized cell or end organ that responds to and transmits stimuli from the internal organs, muscles, blood vessels, and the ear labyrinth.
>
> *Webster's New World Dictionary of the American Language*

This book is about interoception. Interoception is a concept with which many, even in the professional psychobiology community, are not familiar despite the fact that investigation of interoceptors (defined in the quotes above) and interoceptive processes goes back more than a century. Basically, interoception is synonymous with sensory-perceptual processes for events occurring inside the body, including visceral perception (i.e., conscious awareness of visceral function). In the past, visceral pain was usually included as part of interoception. Proprioceptive sensations and labyrinthine functions often were not. Proprioception is included, as are other processes, in the broader definition of interoception to be proposed below. Before proceeding to this broader definition and perspective, a brief historical review of the early study of interoceptive processes is provided. The chapter concludes with a brief discussion of some additional issues and an outline of the book.

HISTORY

Charles Sherrington, the eminent physiologist, apparently was the first to use the word interoceptor approximately 100 years ago to apply to the sensory nerve receptors that reacted to stimuli originating within the body. But he was not the first physiologist to study the phenomenon. Before Sherrington, for example, in the nineteenth century the Hering–Breuer reflex, which is involved in the nervous system control of respiratory movements, was described. In the 1860s nerves were identified running with the carotid artery, and in the 1880s the existence of receptors that responded to changes in body chemistry were demonstrated in the cardiovascular system.

The phenomenon of interoception was of interest not only to physiologists in the 1800s but also to investigators interested in behavior. In the mid-nineteenth century Ivan Sechanov, the great Russian physiologist, in his writings about bodily sensations referred to "dim feelings," "faint sensations," and an "obscure muscular sense" at the border of consciousness. He was aware of the fact that individuals had a sense of their bodies, but that this was very difficult to describe with any precision. Furthermore, later investigators were aware that there was a distinction between a somatic sense, referable to muscles and to the body generally, in contrast to a specific sense of the visceral organs and their functions.

In discussing interoceptors, some of the early investigators focused very specifically on sensation presumably originating from the viscera. This usually included the cardiovascular, respiratory, alimentary, and genitourinary systems, but sometimes also included the hematologic–lymphatic system, the reticulo–endothelial system, and the endocrine system. More broadly still, chemical, osmotic, and volume changes were considered. What was, nonetheless, left out of this narrower definition was the somatic sense, and more specifically, sensory information from the muscles, joints, and connective tissue and skin, including the proprioceptive sense. Some later physiologists, such as John Fulton, did include these. Lastly, pain must be considered, especially since it is not always clear when pain is due to stimulation of different nerve endings and when it is due to a more intense (or in other ways different) stimulation of the same receptors. The first actual demonstration of conditioned interoception apparently was reported by K. M. Bykov and his colleagues in Ivan Pavlov's laboratory in 1926.

A BROADER DEFINITION FOR INTEROCEPTION

Although the term interoceptor was first used in the context of physiology, its use in this volume will refer not just to physiology but also to psychology and behavior, that is, to psychobiology. In this sense it includes the ability of visceral afferent information to either reach awareness and/or to directly or indirectly affect behavior. For example, a common traditional use of the term includes the implication that interoception can occur in humans if and only if a subject can report awareness of some visceral sensation. A broader definition would make no such assumption but would define the occurrence of interoception if molar organismic behavior was demonstrably influenced by visceral sensory impulses, with or without awareness. The broadest definition, in the physiology arena, would define the occurrence of interoception if visceral sensory nervous system impulses occurred, irrespective of any effect on either awareness or behavior.

This volume, focusing primarily on the realm of psychobiology, adopts the broader definition. Further, it is recommended and advocated that the broader definition be generally adopted in favor of the narrower, being more consistent with the hypothesis that visceral sensory impulses are important in the control of behavior even in situations in which conscious awareness does not occur. Adopting such a definition implies that lack of awareness (ability to report a sensation) is not evidence against the occurrence of an interoceptive event, and it further implies that methods to determine the occurrence of visceral sensory impulses that do not depend on indicators of awareness (such as verbal report in humans) will need to be developed. Several specific topic areas of relevance to this broader definition are reviewed in the chapters of this book, as indicated by the chapter titles.

RELATED ISSUES

Beyond the question of which putative interoceptors can be identified and which are to be included under the definition of interoceptors are issues related to interoceptive function. What is the breadth of the interoceptive functional realm? For example, do interoceptors send information to the central nervous system at all times, or only when there is some deviation from the normal baseline, such as changes that produce pain or intermittent monitoring of homeostasis? What role might interoceptors play in disease processes (e.g., cardiac, gastrointestinal, and psychiatric and psychosomatic disorders)? Such issues will be addressed in subsequent chapters.

Interoception is closely related to awareness. Two broad areas of conditioning and learning exist in which interoceptive involvement in behavioral control occurs, one in which awareness seems not to be essential and the other in which it probably is—at least in some sense, such as intentionality. The first of these relates to classical, or Pavlovian, conditioning, and the other to variants of the operant, or Skinnerian, conditioning method. In later chapters the relationships of these two types of conditioning to interoception will be addressed, and the issue of awareness will be revisited at the end of the book. Suffice it to say here that, relevant to Pavlovian conditioning studies, the majority of which have been done in animals, the dependent variable is a change in behavior or physiological function, and the question of the conscious awareness of the occurrence of the conditioning by the subject is not considered directly relevant (but see Furedy et al., 2000).

Details of the important contributions of Pavlov and his students to research on interoception will be provided later. For present purposes a couple of important points need to be made. First, Pavlov hypothesized the existence of cortical "analyzers," regions or functions in the brain—Pavlov believed the cortex—that were the sites at which central nervous system representation of these interoceptive (and other) sensory events were represented. Second, these analyzers took part in the conditioning process, which was understood to be the formation of "temporary neural connections" between the conditioned and unconditioned stimuli. Pavlov and his students hypothesized that any unconditioned reflex could become conditioned. If that is so (an issue not fully resolved to this day), and if unconditioned visceral or interoceptive reflexes exist, then conditioned or learned visceral reflexes should exist as well.

It is widely believed that consciousness, or awareness, requires the function of the cerebral cortex, although this belief is at best an oversimplification. While it is beyond the scope of this book to attempt to resolve this issue generally, it should be noted that higher nervous activity, such as the ability to effect a conditioned Pavlovian response, does not require full cortical function. Despite Pavlov's assertion that the conditioned reflex occurred in the cortex, data from his own students in his own laboratory, including interoception studies, disproved this hypothesis. These findings demonstrate that even without conscious awareness, fundamental changes in these higher cortical functions and complex behaviors can and do occur. Most germane to the theme of this book, despite this fact that awareness, or consciousness, is not required, it is possible—even likely—that sensory information from the body feeds into the central nervous system, either intermittently or continuously, and influences brain function, behavior, and conscious experience, especially the so-called sense of self. (This issue has been discussed by Damasio, 1994, 1999.)

In the other major conditioning paradigm that has been used (operant, or Skinnerian), the organism is required to make a response when and if a visceral sensory event is detected, and either a different response or no response if it is not. This is often done in humans with verbal responses but can be done in animals as well, with the response signifying a detection being such things as bar presses or key pecks. Various psychophysical measurement methods have been used, for example, the signal detection technique (e.g., Clark, 1994; Adam, 1998). Although a strict operant analysis would discuss only the organism's behavior and would not include any appeal to a subject's awareness, the need for the organism to be aware in the usual sense of the word seems essential to the operant conditioning paradigm.

two

THE JAMES–LANGE THEORY OF EMOTION

Instinctive reactions and emotional expressions thus shade imperceptibly into each other. Every object that excites an instinct excites an emotion as well.

William James, *The Principles of Psychology*, Volume 2, 1950, p. 442

ANTECEDENTS

Claude Bernard, a French physiologist, is credited with developing the concept of the milieu intérieur in the mid-nineteenth century. It states that a major function of the body is to maintain constancy of the various internal conditions and functions. Implicit in this concept is the idea that the body must have many highly reliable body–brain mechanisms to sense and monitor the multiple physiological conditions that require control, as well as mechanisms to change these states and conditions whenever necessary. His theory highlighted the importance of the border formed by the skin and the lining of the gastrointestinal tract between the organism and the external world, the self and the non-self. It also served to emphasize the integrative nature of the control of the internal environment. For example, if the blood glucose is too high, not only must there be one or more means of detecting and decreasing it, but other physiological functions that tend to increase it must be damped—but not so much as to produce other imbalances. Consistent with the theme of this volume, almost 150 years ago Bernard was already focusing attention on the existence of sensory monitoring of the visceral environment by both nervous system and hormonal means.

About 75 years after Bernard, Walter Cannon addressed essentially the same issue with his concept of homeostasis. Cannon extended this issue by involving the behavior and emotion of the organism in the overall scheme of control of the internal environment. He studied in depth the fight-or-flight reaction, a normal reaction that primates, other mammals, and animals across the phylogenetic scale have to stressful circumstances that prepares the body for energy expenditure and for action. Neither Bernard nor Cannon, however, addressed the question of the milieu intérieur adequately because neither fully understood that even the normal resting function of the internal environment does not involve a completely static physiological state. Changes such as circadian rhythm fluctuations and even the fight-or-flight reaction itself demonstrated that normal function included predictable changes in internal conditions as demands on the organism varied. With regard to the present theme, Cannon is important for several reasons: *(1)* his research reinforced the emphasis on the importance of the body, *(2)* he brought behavior and emotion into the equation, *(3)* despite not addressing it himself, his research provided important evidence that the normal state of the milieu intérieur is not static but rather fluctuating in ways that are partly endogenous and partly due to environmental factors, and *(4)* he was the main antagonist to the James–Lange theory of emotion (see below).

A second important antecedent development was Darwin's theory of evolution by variation and natural selection. Charles Darwin was very interested in emotion and wrote a whole treatise on the subject (Darwin, 1965). While Darwin's main focus was on a functional analysis of emotion from an evolutionary point of view, discussion and speculations about possible underlying physiological mechanisms occur in many places.

The third and fourth antecedents were the developments of the neural sciences—neuroanatomy, neurophysiology, and clinical neurology—and the development of techniques by early psychologists for studying sensory and perceptive processes in humans and animals. Of particular note was the identification and study of the autonomic nervous system by John N. Langley early in the twentieth century. He described this as a motor system, a system purely of information outflow from the brain to the visceral organs. This description may be correct if one wishes to define it this way, but it has been misleading for many years. (Langley's writings demonstrate that he understood the importance of sensory autonomic function, but he did not study them.) Also of note are the methods developed to study sensory–perceptive processes. These include the various methods of psychophysics and the application of methods of careful behavioral control in animals obtained by the use of operant techniques.

These four threads—*(1)* understanding of the milieu intérieur and homeostasis, *(2)* evolutionary theory and its application to emotions and behavior, *(3)* advancements in the neural sciences, and *(4)* developments of methods to study sensory–perceptive processes—came together in two individuals whose theories and data provide the starting points for the study of interoception. The first is William James and the second is Ivan Pavlov. The James–Lange theory of emotion will be described here, followed by a summary of its critique and support. In the next chapter, Pavlov's theory of classical, or respondent, conditioning will be presented.

THE JAMES–LANGE THEORY OF EMOTION

William James is probably most famous for two ideas, his theory of emotions published in the journal *Mind* (1884) and later (James, 1894) and also presented in 1890 in the chapter on "The Emotions" in his *Principles of Psychology* (1950, Vol. 2), and for his concept of the stream of consciousness (1950, Vol. 1; see also Epstein, 2000). James saw emotions as very closely related to instincts and instinctual behavior, as illustrated by the quote at the beginning of this chapter.

James's theory of emotion was strongly based in physiology, focusing on bodily changes in the muscles and viscera as causing the feeling component of the overall emotional experience rather than being a result or concomitant of it. A similar theory, but based more narrowly on vasomotor changes, was proposed by the Danish physiologist Carl Lange. The theory has come be known as the James–Lange theory of emotions. The James–Lange theory is intimately related logically to the concept of interoception. Therefore, a description in some detail of the theory itself as James presented it will be provided here, followed by a presentation of critiques made by James himself and others, especially the critique and alternative theory of emotion of Walter Cannon (and Philip Bard). The chapter will close with a commentary on the critiques and a discussion of the apparent status of this theory at the beginning of the twenty-first century.

It seems best to start to describe William James's thoughts about emotion by quoting from him directly. These quotes are taken from James (1950, Vol. 2).

> Emotions, however, fall short of instincts, in that the emotional reaction usually terminates in the subject's own body... (p. 442)

> Our natural way of thinking about these courser emotions is that the mental perception of some fact excites the mental affection called the emotion, and that this latter state of mind give rise to the bodily expression. My theory, on the contrary, is that the bodily changes follow directly the perception of the exciting fact, and that our feeling of the same changes as they occur IS the emotion. Common-sense says, we lose our fortune, are sorry and weep; we meet a bear, are frightened and run; we are insulted by a rival, are angry and strike. The hypothesis here to be defended says that this order of sequence is incorrect, that the one mental state is not immediately induced by the other, that the bodily manifestations must first be interposed between, and that the more rational statement is that we feel sorry because we cry, angry because we strike, afraid because we tremble, and not that we cry, strike, or tremble, because we are sorry, angry, or fearful, as the case may be. Without the bodily states following on the perception, the latter would be purely cognitive in form, pale, colorless, destitute of emotional warmth. We might then see the bear, and judge it best to run, receive the insult and deem it right to strike, but we should not actually feel afraid or angry. (p. 449–50)

> I now proceed to urge the vital point of my whole theory, which is this: If we fancy some strong emotion, and then try to abstract from our consciousness of all the feelings of its bodily symptoms, we find we have nothing left behind, no 'mind-stuff'

> out of which the emotion can be constituted, and that a cold and neutral state of intellectual perception is all that remains. (p. 451)

James then goes on to address the difficulty of testing the theory experimentally and considers three objections.

> A positive proof of the theory would . . . be given if we could find a subject absolutely anaesthetic inside and out, but not paralytic, so that the emotion-inspiring objects might invoke the usual bodily expressions from him, but who, on being consulted should say that no subjective emotional affection was felt. (p. 455)
>
> First Objection. There is no real evidence . . . for the assumption that particular perceptions *do* produce wide-spread bodily effects by a sort of immediate physical influence, antecedent to the arousal of an emotion or emotional idea. (p. 456–7)
>
> Second Objection. If our theory be true, a necessary corollary of it ought to be this: that any voluntary and cold-blooded arousal of the so-called manifestations of a special emotion ought to give us the emotion itself. (p. 462)
>
> Third Objection. Manifesting an emotion, so far from increasing it, makes it cease. Rage evaporates after a good outburst; it is the pent-up emotions that "work like madness in the brain." (p. 466)

James replies to these issues as follows. He notes that "anaesthetic" "cases . . . are extremely hard to find" but that the few that have been reported support his theory. He then proceeds, over eleven pages, to answer the three objections by offering examples indicating that they are incorrect. His answers are too long to quote here. To this reviewer, James's answers do not seem to be completely convincing, but they do serve to emphasize that his early critics did not have fully convincing counter-arguments, either. James's theory was immediately influential but also controversial. His most well known critic was Walter Cannon, the same of homeostasis and fight-or-flight fame. Cannon's concerns will now be reviewed.

CRITIQUE OF JAMES'S THEORY

In his book *Bodily Changes in Pain, Hunger, Fear, and Rage,* Walter Cannon (1953) described results of his research studying the effects of emotional reactions on several bodily functions such as digestion, blood sugar level, blood coagulation, and red blood cell concentration. He also described results of studies assessing how the body responds in an adaptive way to emotional states and pain, such as increases in adrenal gland secretion (adrenaline—synonymous with epinephrine). Especially germane to the present topic, he described what was known at that time as The General Organization of the Visceral Nerves Concerned in Emotion, the data at that time concerning the peripheral and central nervous system sources of sensations related to hunger and thirst, and A Critical Examination of the James–Lange Theory of Emotions. After providing his critique of the James–Lange theory, he offered an alternative theory that hypothesized that the primary source of the emotions was in the central nervous system rather than in the periphery. He believed that emotional expression results from action of the subcortical centers, especially the thalamus.

In critiquing the James–Lange theory, Cannon offered five concerns. The first was that "[t]otal separation of the viscera from the central nervous system does not alter emotional behavior." Cannon quoted Sherrington, who cut the spinal cord and the vagus nerves of dogs, claiming that these two transections cut off all afferent connections with "all the structures in which formerly feelings were supposed to reside." Cannon and associates removed the sympathetic division of the autonomic nervous system in cats. This operation was performed in these cats because Lange had hypothesized that the vasomotor system was the system specifically associated with the genesis of emotion, and it prevented the reactions that Cannon had demonstrated were associated with emotion—secretion from the adrenal medulla, piloerection, diminution of gastrointestinal activity, and secretion of glucose from the liver. Cannon claimed that neither his cats nor Sherrington's dogs demonstrated any loss of emotional response.

In Cannon's second critique he stated that "the same visceral changes occur in very different emotional states and in non-emotional states." Cannon argued that preganglionic sympathetic outflow is diffuse, and therefore visceral activation is also generalized and diffuse instead of particular and specific. In other words, he argued that the sympathetic system is activated as a unit. Cannon noted that very similar visceral reactions occur in multiple circumstances—great excitement, fever, cold exposure, asphyxia, hypoglycemia. Cannon is referring here to situations in which sympathetic activation is quite robust. More recent data raise doubts about whether the sympathetic system really reacts as a nonspecific all-or-none unit.

The third critique states that "the viscera are relatively insensitive structures." Cannon observes, for example, that "we can feel the thumping of the heart [only] because it presses against the chest wall, we can also feel the throbbing of blood vessels [only] because they pass through tissues well supplied with sensory nerves. . . ." This critique anticipates an issue that will be raised in discussion of operant visceral conditioning and biofeedback—the question of mediation of apparent awareness of the functioning of the heart by sensory innervation of non-visceral organs, such as the chest wall.

The fourth critique states that "visceral changes are too slow to be a source of emotional feelings." Cannon argues that the latency period for emotional reactions by humans to pictures with affective content is less than a second, which is too short to be mediated by the slowly reacting visceral functions such as smooth muscle and various glands.

Finally, the fifth and last critique states that "artificial induction of the visceral changes typical of strong emotions does not produce them." The data to support this claim consist of studies involving the experimental administration of adrenaline, intended to mimic sympathetic nerve activation. Cannon summarized results observed, indicating that various expected physiological changes are commonly seen, associated with the expected subjective sensations. However, he stated that there is in most cases only an "indefinite affective state coldly appreciated, and without real emotion." Subjects described the feelings they experienced "as if" they were having various real emotions but not the full emotion. Cannon claimed that the subjects described their feelings as true emotions only when they had been primed ahead of time to be in some particular mood state by having discussions with affect-containing content prior to adrenaline administration.

After raising these five issues, Cannon concluded that acceptance of the James–Lange theory was not warranted and that afferent information from the viscera associated with emotion was minimal. Consistent with Cannon's overall research interests and program, this critique focuses primarily on sympathetic nervous function and adrenaline, especially critiques 1, 2, and 5.

The James–Lange theory of emotion and its critique by Cannon have been extraordinarily influential in emotion theory and research for more than a century. Before going forward with a discussion of more recent scholarship, it is essential to restate clearly what the James–Lange theory (especially James's version of it) really says. The James version, at least, is about feelings. It is about where the subjective feeling part of emotion comes from. It is not necessarily about where the emotional cognitions or behaviors come from. James readily admitted that someone might see a bear and run away even if visceral involvement were somehow removed, but the person would not feel afraid. If one trisects the emotional experience into behavior, physiology, and subjective awareness, James's theory is about the relationship between physiology and behavior with subjective awareness.

Fehr and Stern (1970) reviewed the status of research and theorizing about the James–Lange theory 30 years ago. They specifically addressed four of the five critiques originally made by Cannon. (They did not address the question of whether latency of visceral response was excessive.) More recently, Lang (1994) provided a more up-to-date discussion. Many others, such as Van Toller (1979), McNaughton (1989), Ellsworth (1994), and Damasio (1999), have reviewed and critiqued both the James–Lange theory itself and its critics. Their expositions are too long to summarize in detail here. Several points germane to the topic of this volume, however, are worthy of emphasis.

James and Lange viewed the visceral events in question as the cause of the emotional feelings, the emotional experience. It should be noted that James viewed his theory as more relevant for some emotions—the "courser" emotions—than for others. While it might be disputable that these visceral events are the cause, few who have written about this theory have argued that visceral events are not part of the experience, and an important part. Whatever the functional reason for their occurrence, whatever occurred evolutionarily to link these visceral sensory experiences to the cognitive and behavioral aspects of what is called emotion (for example, the preparation for action that is the purpose of fight-or-flight reaction, fast and hard heart beat, increased respirations, piloerection, sweating, etc.), they do occur in conjunction with (correlate with) these other aspects of emotion, and contribute in a major way to the overall experience. Furthermore, while Lange viewed the relevant events as limited to the cardiovascular system, James viewed them much more broadly, including not just changes in the functioning of various visceral organs but also feedback from the whole body, including the muscles and joints—that is, proprioception.

The correlations of physiology and behavior with subjective awareness are often, but not always, as robust as might be expected were the James–Lange theory correct. For some individuals there do appear to be meaningful associations, while for others there do not (see Lang et al., 1993, Lang, 1994; Fig. 2–1). Individual differences are prominent. The fact of these differences is important, but the interpretation is not resolved. Does this reflect a closer association between the cognitive–behavioral and the physiological changes during emotion in some individuals than in others (i.e., do some individuals have more interoceptive sensitivity than others)? If so, does

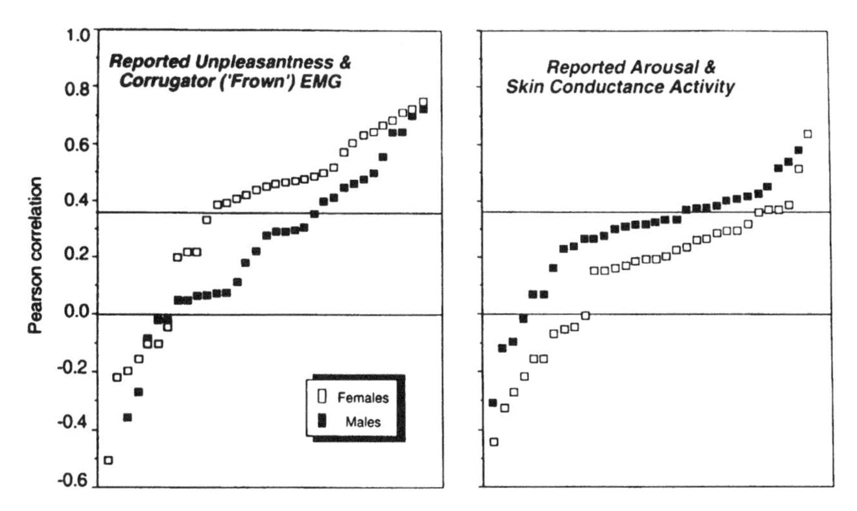

Figure 2–1. Correlations of subjective unpleasantness with EMG and subjective arousal with skin conductance. For both sets of correlations, the majority were either significantly positive or in the predicted direction. Gender differences were observed—correlations were stronger for women with unpleasantness, but for men with arousal. [From Lang, 1994, reprinted with permission of the holder of the copyright.]

this vary among emotions (for example, in one person there might be a closer link for anger than for fear, while in another the opposite might be true)? In persons who appear to have weak linkages, might they be stronger if other physiological variables were measured? In this context two factors do seem consistently to appear and account for a meaningful amount of variance in the autonomic and somatic changes associated with emotion. One is arousal, which represents a continuum of level of activation, and the other is valence, which represents an affective continuum from negative (unpleasant) to positive (pleasant). These seem to do better than attempts to account for the physiological data by identifying specific subjective emotions.

Concerning Cannon's specific critique that total separation of the viscera from the central nervous system does not alter emotional behavior, four major issues should be noted. First, the methods used by Sherrington and Cannon did not produce total separation; not only was some afferent nervous input from the body not eliminated by their procedures, but hormonal feedback was left largely untouched. Second, even if total separation were attained, it would not be a refutation because the theory is not mainly about behavior, it is about feelings. Thus, animal research is unlikely to be able to resolve this issue. This might also be a problem with humans because use of verbal reports perhaps incorrectly assumes that these feelings are reliably describable. Third, concerning feelings per se, studies in humans that have assessed effects of spinal transections, for example, have found that individuals sometimes do describe some diminution in their emotional experience after the transection (see Chapter 14), and this diminution correlates with the level of the transection. In other words, it does appear that interoceptive input does relate to emotion. Lastly, the experiments of Sherrington and Cannon have been criticized as artificial. For example, the operations performed modify not just visceral sensory input but also somatic input and motor outflow.

In the critique that artificial induction of the visceral changes typical of strong emotions does not produce them, Cannon and many subsequent investigators have assumed that because strong emotion under natural circumstances typically involves sympathetic activation, then experimental induction of sympathetic activation adequately recreates the natural situation. That is incorrect. Not only is there sympathetic feedback, there is also parasympathetic and hor-

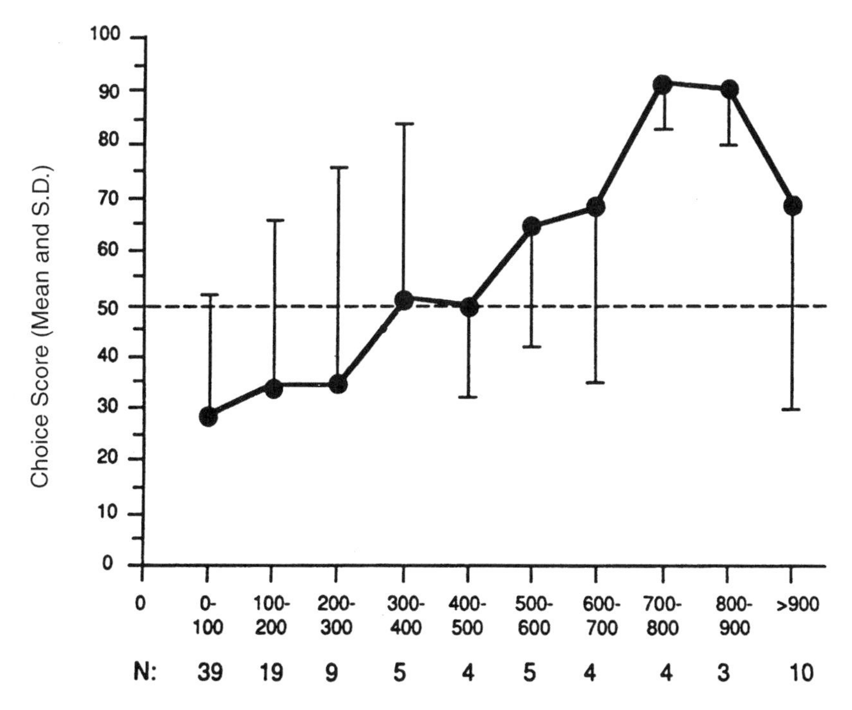

Figure 2–2. Subjective estimate (mean and standard deviation) in normal subjects of whether an infusion of epinephrine or placebo was given (*Y*-axis: 0 = sure placebo, 100 = sure epinephrine, 50 = guess), based on plasma level of epinephrine attained (*X*-axis) with different bolus infusion concentrations (*N*: number of observations at each plasma level). [From Cameron et al., 1990a, reprinted with permission of the holder of the copyright.]

monal, of which Cannon and subsequent investigators were (or should have been) aware. And, since then, new discoveries have substantially lengthened the list of possibilities—for example, circulating peptide substances of many kinds. Furthermore, it appears that the implicit assumption that the sympathetic–adrenergic symptoms that do occur in humans are due to increases in adrenaline is often incorrect. As has been shown (see Fig. 2–2), the level of epinephrine (adrenaline) needed to produce such symptoms exceeds that usually seen in circumstances of stress. If this is incorrect, then experiments that attempt to recreate these emotions by adrenaline injections are not recreating the normal mechanism.

Cannon, Bard, and many other influential researchers—including Paul MacLean and James Papez—have focused on the contributions of the central nervous system to the overall psychobiology of emotion. Others, such as Schachter and Singer (1962) have focused on the contributions of cognitive factors. Nothing here is intended to refute or even de-emphasize any of that. The point is that this is not an either–or question. Visceral input, interoceptive feedback, is also an essential part of the normal emotional experience.

three

CONDITIONING AND LEARNING, ATTENTION AND AROUSAL

For the things that we have to learn before we can do them, we learn by doing them.

Aristotle

Interoception can involve innate and/or learned visceral–somatic sensory processes. The ability of visceral afferent information to affect behavior and/or cognitive processes and to reach conscious awareness (not logically the same thing) might be hard-wired into the organism by genes and development or it might be learned. In fact, it is undoubtedly true that both types of mechanisms are essential. Further, as the quote above acknowledges, and as will be discussed later, sensory feedback from behavior affects learning. Later in the book the hard wiring will be addressed. As background to this the issue of learning will be discussed here. It is presented first because it arose first historically. The relationship of conditioning (a basic type of learning) to the study of interoceptive processes goes back to Pavlov and his colleagues in the early part of the twentieth century, while detailed study of the brain and body mechanisms that provide the hard wiring is more recent.

PAVLOVIAN (CLASSICAL, RESPONDENT, ASSOCIATIVE) CONDITIONING

The essential paradigm of Pavlovian conditioning is as follows. Any number of innate physiological responses to various specific stimuli exist that have been programmed into the organism by evolution. They have adaptive value. One such response is the increase in production of saliva that occurs when dry food is introduced into the mouth. As a physiologist of the gastrointestinal system, Pavlov was already studying the production of saliva as well as other gastrointestinal substances. The offering of food to the canine subjects or the direct injection of dry meat powder into the dogs' mouths, induced salivation. In the Pavlovian conditioning paradigm, the food or meat powder is the unconditioned stimuli (UCS or US—i.e., it produces a response without conditioning and is therefore innate), and the increased salivation is the unconditioned response (UCR or UR—also innate).

If another stimulus that is originally neutral with reference to the UCS (i.e., it does not produce an innate response similar to the one produced by the UCS) is paired with the UCS, it, too, comes to produce a response that is similar (usually) to the UCR, in this case, salivation. For example, if a bell is rung just before the meat powder is injected, after several pairings the bell will also produce an increase in salivation similar to the meat powder. The increase induced by the conditioned stimulus (CS) is referred to as the conditioned response (CR), also called the conditioned reflex. Pavlov's laboratory apparently started to study the conditioned reflex in 1899 and also studied habituation, called the "what-is-it" or, later, the orienting reflex. Pavlov's theory of the brain mechanism of the conditioned reflex is illustrated in simplified form in Figure 3–1.

Of relevance to the theme of interoception, the question arises as to what responses can be conditioned and with what stimuli? The list of such stimuli and responses is long (see Kimble, 1961, p. 51). The fundamental importance of Pavlovian conditioning to interoception is due to the observation that physiological functions can be conditioned and that stimuli applied to the internal organs can function as conditioned stimuli. Indeed, stimulation almost anywhere in the afferent (sensory) side of the US–UR reflex arc could serve effectively as a US, while stimulation on the efferent side could not (see Loucks, cited in Ruch, 1965, p. 518). Data documenting that

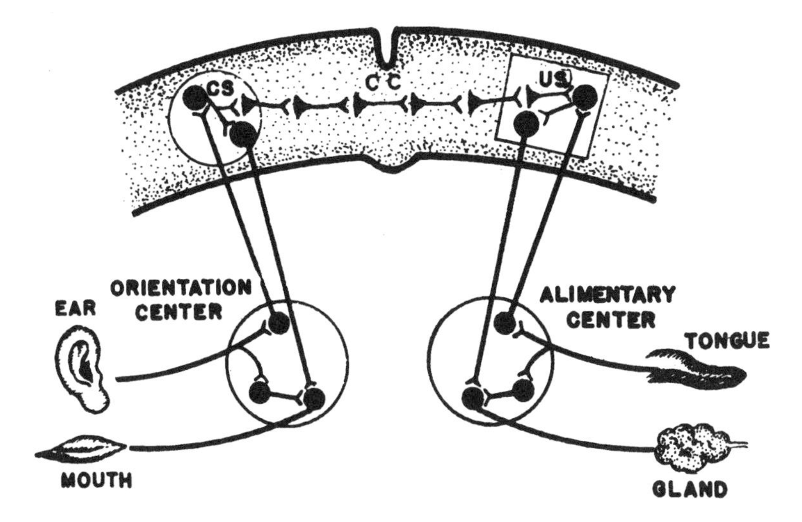

Figure 3–1. Schematic of Pavlov's theory of the mechanism of conditioned reflex (CS: conditioned stimulus; US: unconditioned stimulus; CC: cerebral cortex). [From Asratyan, 1961, reprinted with permission of the holder of the copyright.]

conditioned physiological effects can occur will be reviewed briefly here. Data relating to the internal organs as responsive to sensory stimuli will be analyzed in detail later.

A variety of physiological functions can be conditioned, that is, can become conditioned responses. One major category is conditioned drug effects. In Pavlov's laboratory it was demonstrated that morphine and apomorphine could both function as unconditioned stimuli and that pairings of drug administration with originally neutral stimuli conditioned the effects of the drugs to the neutral stimuli (Pavlov, 1960, p. 35). Conditioned morphine withdrawal has been demonstrated in rats (Trost, 1973; Baldwin and Koob, 1993), monkeys (Goldberg and Schuster, 1970), and humans (O'Brien et al., 1977). The opiate system also has been implicated in the mediation of placebo-induced analgesia (Levine et al., 1978), a phenomenon that probably involves conditioning. Drugs such as the stimulants amphetamine (Turner and Altshuler, 1976) and cocaine (Bridger et al., 1982), ethyl alcohol (Le et al., 1979), and haloperidol (Poulos and Hinson, 1982) all can produce conditioned effects. Cameron and Appel (1972, 1976) demonstrated that conditioning occurred more readily to some drug states than to others. Endogenous chemical substances including corticosteroids in rats (Ader, 1976; Coover et al., 1977), catecholamines in monkeys (Natelson and McCarty, 1980), luteinizing hormone and testosterone secretion in rats (Graham and Desjardins, 1980), gastric acid in rats (Weingarten and Powley, 1981), and insulin-induced blood glucose decreases in humans (Fehm-Wolfsdorf et al., 1993; Stockhorts et al. 1999) also show conditioning effects. These are just a few examples of the many demonstrations of various drug states and changes in endogenous substances that are conditionable by Pavlovian procedures.

In addition to conditioning effects associated with changes in various drugs and chemical substances found in the body, other physiological functions are conditionable as well. Three of the various possible examples are respiratory changes, immunosuppression, and the galvanic skin response (GSR—sweating). Ader and Cohen (1975, 1982) demonstrated that cyclophosphamide-induced immunosuppression could be conditioned to the taste of saccharin. Justesen et al. (1970) reported conditioned changes in breathing pattern and airway function in guinea pigs. Ohman (1988) reviewed classical conditioning studies of autonomic responses, including GSR, in humans. These data demonstrate that various physiological functions are modifiable by classical conditioning methods.

The temporal delay between onset of the CS and the US in successful Pavlovian conditioning is usually very brief, measured in seconds. However, one form of associative learning, taste aversion learning, has an optimal interval measured in minutes to hours. It is of interest here because the unconditioned stimulus is a visceral sensory experience involving the gastrointestinal tract that is usually (but not always) unpleasant, for example, nausea and emesis (Rozin and Kalat, 1971; Bernstein, 1999). It is an example of prepared learning, that is, there are some types of learned associations that the organism seems more ready to make than others and that it is prepared to make, presumably based on innate tendencies (Garcia et al., 1974). Aversion learning with a visceral UCS appears earlier in development than does aversive conditioning with a UCS in the periphery—that is, it is perhaps more basic or fundamental (Haroutonian and Campbell, 1979) and appears to be based on a different neural substrate (Jensen and Smith, 1985).

In taste aversion learning, a stimulus that produces a sensory experience in the gastrointestinal tract is paired with a novel or distinct flavor and/or smell. The organism comes to avoid that taste or smell, even though the gastrointestinal experience comes minutes to hours after the taste or smell. This type of learning is very robust, as are certain other forms (especially forms of aversive learning), often occurring after only one pairing. It has even been claimed that it can occur in an anesthetized or unconscious organism. In humans there is interest in the possibility that taste aversion learning contributes to cancer anorexia and the anticipatory nausea that cancer patients often develop when receiving chemotherapy (Morrow and Morrell, 1982; Bernstein, 1978, 1999). Anatomically, in the brain mixed evidence exists for involvement of the nucleus of the solitary tract (NTS) and the area postrema (which is involved in chemoreception) in taste aversion learning, while it appears that the pontine parabrachial nucleus is required (Bernstein, 1999). It also appears that some higher centers are involved, including the hypothalamus, the amygdala, and the gustatory part of the insular cortex (Gaston, 1978; Phillips and LePaine, 1978; Bernstein, 1999). These are brain regions that are part of the visceral afferent system and are implicated in interoception. Several sources have summarized this research (Bykov, 1957; Razran, 1961; Bykov and Kurtsin, 1966; Chernigovskiy, 1967; Adam, 1967, 1998).

INSTRUMENTAL (OPERANT) CONDITIONING AND ONE- VERSUS TWO-PROCESS LEARNING THEORY

Pavlovian conditioning is one of the two primary forms of associative learning. The other is instrumental, or operant, conditioning, also known as Skinnerian conditioning or the Law of Effect. In the instrumental paradigm, the organism receives a reward (some environmental event that decreases the intensity of some motivational state such as hunger—see Chapter 4) for making some response. This response is said to be reinforced, defined as any consequence of the response that raises its probability of occurrence (punishment is any consequence that decreases the probability).

In Pavlovian conditioning, responses are said to be elicited by the unconditioned and conditioned stimuli (UCSs and CSs). In instrumental conditioning the responses are said to be emitted by the organism. Pavlovian conditioning is considered to be stimulus-based because the unconditioned stimulus is not arbitrary. It is reflexive, "hard-wired" in the organism's biology. In contrast, instrumental conditioning is considered to be response-based. The response is considered to be flexible and arbitrary rather than reflexive and is controlled by its consequences instead of by some preexisting biological determination. This distinction, however, is a matter of degree. Even in instrumental conditioning, organisms have some predispositions. Some responses are more likely than are others for organisms of a particular species. Habit hierarchies and prepared learning (for example, taste aversion learning) do exist.

Pavlovian and operant conditioning paradigms are important to interoception primarily for two reasons essential to understanding when and how visceral organ function changes can be modified by learning and experience and the role visceral sensory processes play. First, stimulation of the internal organs can function as CSs or UCSs in conditioning (see Chapter 5). Second, a long-standing debate continues about whether these two types of conditioning procedures—classical and operant—actually represent the same or different types of learning, and basic to understanding interoception, are conditioned changes in functioning of visceral organs and systems effectable only by Pavlovian procedures, as is argued in two-process learning theory (see Chapter 6)?

A major issue related to the question of whether these two conditioning procedures actually represent different types of learning is the question of whether the results follow the same rules and laws. Do they demonstrate similar results? Kimble (1961, pp. 78–108) examined more than a dozen phenomena that are features of conditioning, such as patterns of response acquisition, extinction, and spontaneous recovery; stimulus and response generalization; and discrimination. He concluded that in these phenomena the two different procedures lead to very similar outcomes. However, some differences were apparent. For example, in the Pavlovian procedure, the conditioned and unconditioned responses were usually very similar, whereas in operant conditioning the learned response was arbitrary, based on what response was reinforced, and had only a partial relationship to innate responses. A second difference was that Pavlovian conditioning typically required continuous reinforcement (i.e., every presentation of the CS followed by the US) for optimal learning to occur, whereas after initial learning had occurred with the operant paradigm an intermittent reinforcement schedule actually produced stronger learning (i.e., greater resistance to extinction).

Most relevant to interoception, it was believed at that time that learned control over involuntary (e.g., visceral–autonomic) responses could only be effected by Pavlovian conditioning (i.e., two-process theory). It appears that one reason this dichotomy was considered was that the autonomic nervous system was believed to be separate from that part of the nervous system that controlled voluntary responses. Another factor, undoubtedly, was that Pavlov's research did not use what is conventionally thought of as behavior, but rather used a clear physiological response, salivation. An important confounding factor in trying to resolve this issue is the fact that creating pure examples of either Pavlovian or operant conditioning is, from a practical point of view, extremely difficult. These procedures in "pure culture" are possibly different, but in the "real world" they appear to have many similarities because, in fact, all instances of conditioning are instances of both types of learning at once. (Indeed, even a student of Pavlov demonstrated instrumental components to the classically conditioned salivation response—Zener, 1937.)

A great deal of attention has been paid to the role of two-process theory in providing an explanation for fear conditioning. This, of course, is important to the topic of interoception because fear is so often associated with visceral symptoms and signs of autonomic activation. The reason two-process theories focus specifically on fear and aversive conditioning is that in many situations the motive factor does not need to be learned. It is assumed to be innate. But in fear or aversive conditioning it apparently does need to be learned. For example, an animal must learn to be fearful of many situations in order to escape or avoid, whereas it does not need to learn to be thirsty in order to drink. (Clearly, however, innate fears exist as well, such as certain young birds' innate withdrawal from a shadow in the shape of a hawk. This distinction is a matter of degree.)

In a strong version of behaviorist theory, the question of what motivates the organism to escape or avoid does not call for an answer involving biological variables or inferred states. In many conditioning theories, however, an answer to that question does involve inferred emotional conditions, motivational variables, and drive states. It is in these types of theories that two-process questions most directly arise. Stated simply, it is hypothesized that fear (the inferred state, the intervening variable) is conditioned by classical conditioning, and the operant response is reinforced by the escape or avoidance response because the response lessens the fear. Thus, both classical and operant mechanisms are occurring, that is, two-process learning. Several books and articles discussing two-process theory are available, to which the interested reader is referred for more information (Mowrer, 1947, 1960; Kimble, 1961, pp. 266–277; Rescorla and Solomon, 1967; Herrnstein, 1969; Gray, 1975a; Rakover, 1979; Levis, 1989). The topics of emotion and motivation will be discussed in more detail in Chapter 4.

A tremendous amount of research has gone into attempting to determine if one-process or two-process learning theory is correct, but a final resolution remains elusive. One approach that was very influential in the 1960s and 1970s was to ask whether visceral conditioning can be produced by operant procedures? If the question could be answered in the affirmative, if visceral responses could be produced by operant techniques, that would be fairly strong evidence against one important aspect of two-process theory. The attempt to make such a demonstration, and the fate of that attempt, will be described below (Chapter 6).

The data on the relationships among fear, Pavlovian conditioning, and autonomic–visceral responses are far too extensive to review here, but a brief discussion can be provided. Many types of visceral changes have been reported in association with fear conditioning. These include skin resistance changes (GSR), changes in heart rate (both increases and decreases) and occasionally heart rhythm, stroke volume, systolic and diastolic blood pressure changes (usually increases), changes in respiration, changes in skin temperature (which reflect vascular changes), pilomotor erection, changes in gastrointestinal activity (especially increases in defecation), changes in urination, pupil dilation, changes in metabolic rate (typically increases), and changes in various biochemical–endocrine indices, such as blood glucose, catecholamine, and glucocorticoid changes. This list is not complete. The essential observation is that these autonomic–visceral changes are associated with fear. The essential questions are: Are they present because they were learned, and if so, were they learned strictly by Pavlovian conditioning? The alternative is: They are there by other means (i.e., they are innate), the association with fear having been evolutionarily selected for because of an advantage it provides (i.e., they are hard-wired). The answer to this essential question is: It is still not known.

What does all this have to do with visceral sensory processes, afferent information from the viscera to the brain? What is the importance of the question of one- versus two-process learning to interoception? First, if visceral processes can be conditioned, that would imply that visceral sensory information reaches high enough in the brain to participate in processes involved in learning. Even though very simple organisms can be conditioned, compelling evidence exists that in higher animals learning involves, and perhaps requires, the participation of higher centers, including the cerebellum and various other structures in and above the diencephalon. For orderly, predictable functional changes in visceral–autonomic systems to be learnable, the centers in which learning occurs must be able to monitor what events are actually occurring in these organs and systems. In other words, visceral sensory information must be reaching these centers, that is, must be feeding back the changes in these visceral organs and systems to the anatomical areas in the brain in which learning is occurring. There must be a closed loop. This need not reach consciousness, but by the broader definition, it clearly qualifies as interoception.

Second, a substantial effort was made in the 1960s and 1970s to test the question of one- versus two-process learning theory by determining if visceral functions could be conditioned with operant procedures. This is part of the larger question of whether any type of conditioning can affect visceral function. The implication is that any type of successful conditioning would necessarily involve afferent visceral sensory processes. This second question is a subset of the first. It asks specifically, Can operant conditioning affect visceral function?

The answer to the first, larger question is a tentative "yes." It is clear that Pavlovian procedures can affect visceral function, although it is not as clear if the visceral functions are modified directly by the conditioning procedure or if something else (e.g., brain functions) is being directly modified—learning—and the visceral functions are secondarily modified. In other words, they might be hard-wired to the brain processes and simply be "along for the ride" when the brain processes are changed by the Pavlovian procedure. It would seem that regardless of whether they are learned directly, even to be "along for the ride" there must be afferent visceral information (i.e., interoception) coming into the brain to make this "ride" operate in an orderly fashion.

The answer to the second question remains very unclear. A definite answer to this question is not required in order to examine the role of interoception in the linkage between visceral function and learning. Nonetheless, it has been very important historically and remains important today not only in the general area of interoceptive processes. For that reason this issue will be revisited later, in Chapter 6.

ATTENTION AND AROUSAL

Attention can be defined as the psychological action by which observation, mental processes, or consciousness are focused on a single object or thought, often selected from an array of two or more choices and usually with the purpose of improving clarity by virtue of the focusing. Arousal can be defined as the state of being excited or stirred to activity, provoked, or stimulated. In psychobiological usage, comparison to certain metaphors can clarify these meanings. In the case of a metaphor of a car, arousal is the motor and how fast the car is moving at a given time, while attention is the destination the car is heading toward. In the case of a radio, arousal is the volume control, while attention is the channel selector control.

How do these issues relate to interoception? Concerning attention, the question arises of what effect focusing of attention, and attentional processes in general, have on the ability to sense the functions of the internal organs? Concerning arousal, the concept includes potential changes in the function of these organs, not necessarily but possibly with awareness. For example, when one is said to be aroused, it is common to expect increased sweating, heart beat, respiration, muscle tension, and so on.

A finding essential to understanding the brain's control of attention was made in cats. It was found that after cutting the brain at the level of the colliculi, the EEG recorded from the cortical region above the cut resembled the sleeping cat (*cerveau isolé*), whereas cutting the brain at the level of the medulla (*encéphale isolé*) produced a cortical pattern similar to an alert cat (Fig. 3–2). This implied that the cortical alert pattern seen in the *encéphale isolé* was being generated by a portion of the brain region between the two cuts. This brain region has been called the ascending reticular activating system (ARAS). The ARAS region is strongly involved not

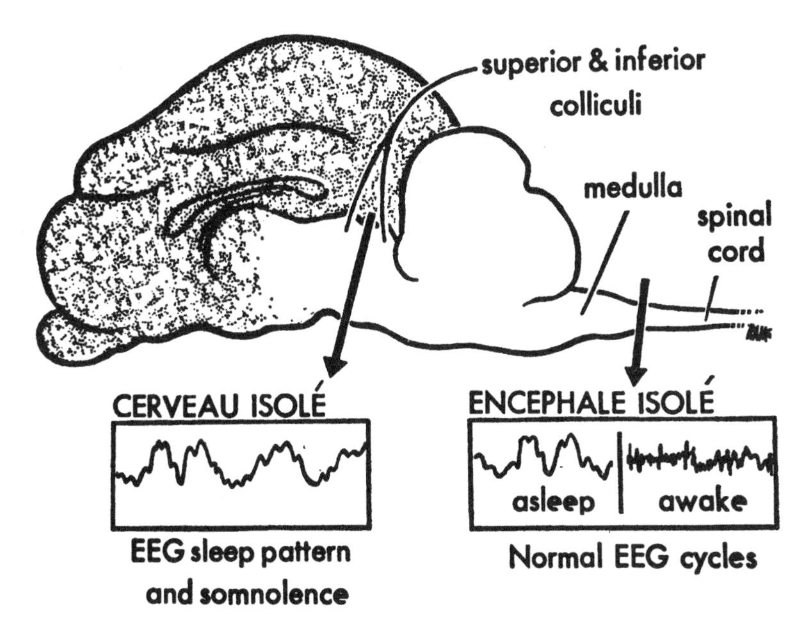

Figure 3–2. Schematic examples of effects on EEG of cerveau isole and encéphale isolé cat preparations. [From McCleary and Moore, 1965, reprinted with permission of the holder of the copyright.]

only in cats but in mammals generally in the control of alertness, wakefulness, and sleep, and therefore in attentional processes (Steriade, 1996). Studies involving electrical stimulation of the ARAS demonstrated that increasing activity in this brain region improved performance on tasks requiring close attention (see Hilgard and Bower, 1966, pp. 439–446). Activation of the ARAS and intralaminar thalamic nuclei is involved in attentional processes in humans as well (Kinomura et al., 1996). Other brain regions important to attention include several nuclei in the thalamus, regions of the so-called limbic system, the anterior cingulate gyrus, and the frontal, insular, and somatosensory regions of the cortex (Newman, 1995; Smythies, 1997; Coull, 1998), all regions specifically implicated in visceral sensation.

The ARAS region is the location of the cell bodies for the major monoamine neurotransmitter systems (norepinephrine: NA; dopamine: DA; and serotonin: 5-HT) as well as the source of ascending cholinergic pathways (Fig. 3–3). These pathways have all been implicated in attention and arousal (e.g., Mason, 1980; Clark et al., 1987; Harley, 1987; Robbins, 1997). Robbins reviewed literature on

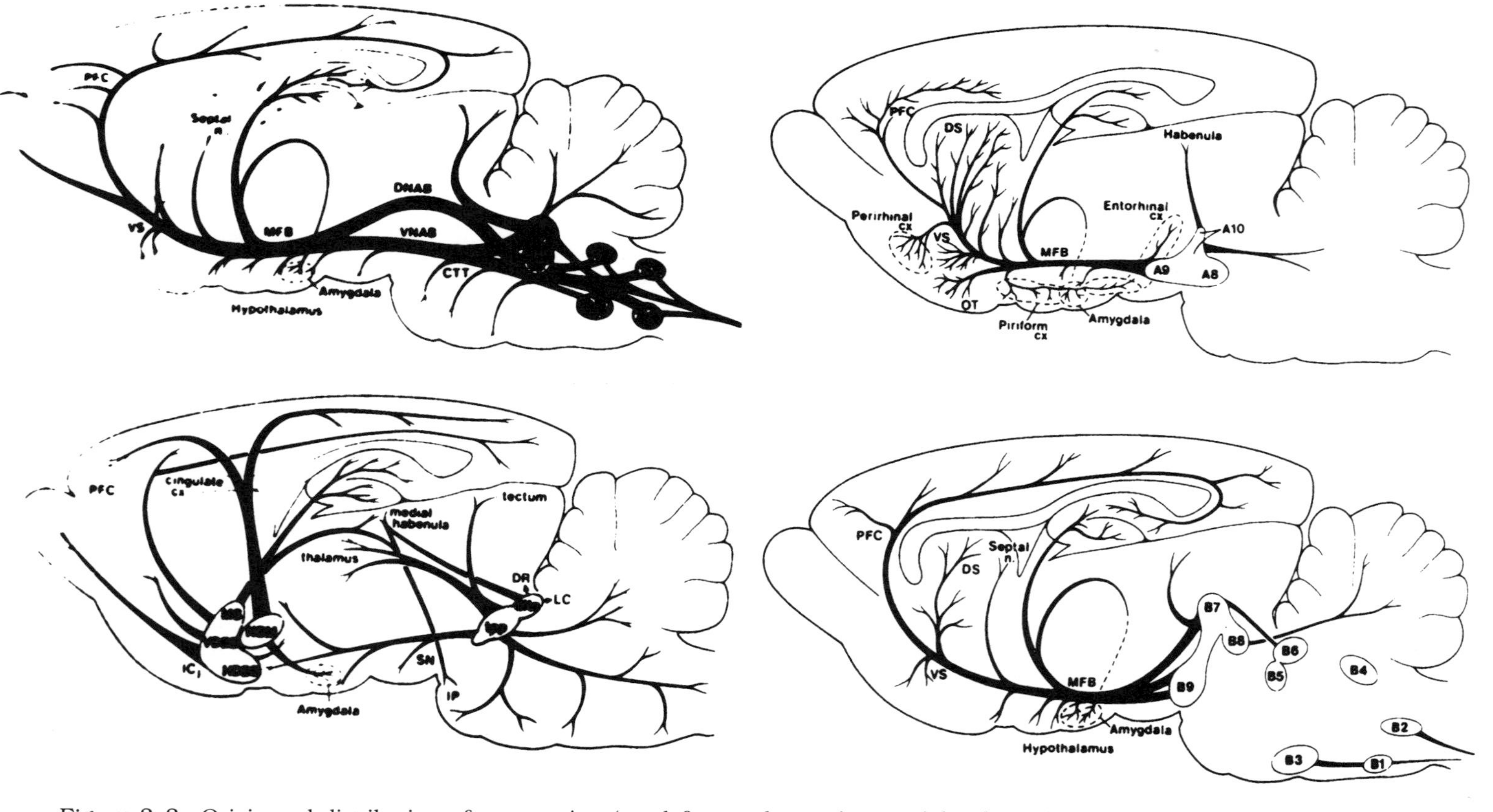

Figure 3–3. Origin and distribution of monoamine (top left: noradrenergic; top right: dopaminergic; bottom right: serotonergic) and cholinergic (bottom left) pathways in the rat brain. (See original publication for key to abbreviations.) [From Robbins, 1997, reprinted with permission of the holder of the copyright.]

the function of these various neurotransmitters and arrived at the following conclusions: "The . . . NA system seems to have a protective function of maintaining discriminability in stressful or arousing circumstances or maintaining 'alertness' to salient external stimuli; the . . . DA systems play a role in the activation of output, whether cognitive or motor in nature . . . the cholinergic systems appear to enhance stimulus processing at the cortical level . . . and the 5-HT systems may serve to dampen the actions of each of the others . . ." (Robbins, 1997, p. 67). Aston-Jones and colleagues (Valentino et al., 1994; Aston-Jones et al., 1999; also see Svensson, 1987) arrived at similar conclusions about the functioning of the NA system and the locus caeruleus. The NA system appears to enhance the signal-to-noise ratio of incoming stimuli being attended to and to screen out irrelevant stimuli. The highest locus caeruleus electrical spike activity was observed during the orienting reflex, then in descending order of activity were wakefulness, drowsiness, slow-wave sleep, and paradoxical (REM) sleep. Finally, an increase in locus caeruleus activity has been observed in response to a stimulus after that stimulus has been conditioned, that is, is a CS. Overall, these systems tend to serve a neuromodulatory role, by which their main function is to secondarily modulate the effect (e.g., the gain) of other neurotransmitter systems such as the inhibitory neurotransmitter (GABA).

Attention is considered to be specific. The organism attends to something in particular. Arousal, on the other hand, is considered to be a generalized activator or excitor towards behavior. Increased arousal, for example, can activate spinal reflexes, the startle reflex, or general exploratory behavior not directed to a particular object. Too much arousal can disrupt behavior. As discussed in the chapter on emotion and motivation (Chapter 4), however, that is not considered correct by all. For example, if an organism is in an aroused state produced by hunger or by the presence of an estrous female, these aroused states are not nonspecific. They are focused on eating or copulatory behavior, respectively.

The relevance of the arousal concept for interoception is due to the observation that arousal typically involves increases in visceral functions, such as rises in GSR and heart rate. The intercorrelations among the various visceral functions, both within and among individuals, when they are aroused can be quite variable, which has led some to question the concept of arousal as a legitimate organizing concept, but the tendency for visceral activation of one kind or another to occur is very well documented. With reference to interoception, it should be noted that attention and arousal do not require full awareness of the subject. For example, it has been shown in a dichotic listening situation (different auditory input to the two ears) that subjects instructed to attend carefully to the input to one ear typically cannot report the input to the other ear, but that input can nonetheless have an effect, such as producing a GSR response when emotional content is present in the auditory input to the ear opposite the one being attended to.

four

MOTIVATION AND EMOTION

> "You don't believe in me," observed the Ghost. "I don't," said Scrooge. "What evidence would you have of my reality, beyond that of your senses?" "I don't know," said Scrooge. "Why do you doubt your senses?" "Because," said Scrooge, "a little thing affects them. A slight disorder of the stomach makes them cheats. You may be an undigested bit of beef, a blot of mustard, a crumb of cheese, a fragment of underdone potato. There's more of gravy than of grave about you, whatever you are!"
>
> Charles Dickens, *A Christmas Carol*

Chapter 3 ended with the topic of arousal. Arousal was described as the concept that one important aspect of the control of behavior is a nonspecific activating factor, a motivator to behave, to act, to move. The question of whether there is a prompt to behavior that is truly nonspecific is not fully resolved. Little doubt exists, however, that behavior includes an aspect of prompting. In other words, for example, concerning learned or conditioned behaviors, these behaviors are not performed continuously. They are performed only when some prompting circumstance or situation is present. The topic of motivation addresses the question of what those circumstances or situations are.

Emotion is a topic closely related to motivation. Emotion refers to subjective feeling states of various kinds, such as fear, anger, disgust, hope, and contentment. The definitions of all the emotions, or even a list of them, is not fully agreed upon. They tend to shade into each other. Some investigators believe that there is a finite set of basic or fundamental emotions, while others do not. The relevant relationship between emotion and motivation is that emotional states are motivators, prompts to actions that are relatively specific to the particular emotions in question. For example, in Cannon's fight-or-flight formulation, fear is likely to produce flight if it is possible (if not, fight or freezing will often ensue), whereas anger will usually be associated with fight, or at least an urge to do so.

This chapter provides a brief, general, relatively simple discussion of motivation and emotion. Why discuss these topics as part of an exposition on interoception? Simply because motivational states including emotions almost always are associated with demonstrable physiological changes and bodily feelings. Early theories about such motivational states as hunger and thirst often invoked bodily sensations, for example, gastric motility changes and dry mouth, respectively, as the hypothesized major contributors or causes of the various motivational states. These early theories were typically found to be incorrect or, more commonly, not sufficient in themselves to explain the motivational states in question (although Davidson, 1993, for example, argued that conditions of food deprivation do produce interoceptive signals), and theories focusing on central nervous system (CNS) control are now more common. Nonetheless, these theories remain about the body. Damasio (1994, p. xvi) argued that "The mind had to be first about the body, or it could not have been." That is, the CNS is constantly and precisely monitoring body function as one of its highest priorities, and somatic markers of change in bodily activity exist that are fundamental not only to emotional expression but to cognitive and behavioral function as well.

Both motivation and emotion are hypothetical constructs, intervening variables invoked to explain behaviors and the various physiological changes and reported subjective feeling states associated with the various behaviors. It is predictable, although not invariably the case, that a person who is eating will describe a subjective feeling called hunger and that the person's stomach will be relatively empty and the blood glucose level in the normal to low range. Motivational states are usually considered to have a fundamental innate component. Organisms are hard-wired to feel the way they do when certain situations arise, but they also are hard-wired to be able to learn motivational/emotional states as well. Both the innate and the learnable/conditionable aspects of these states are adaptive to the organism.

MOTIVATION

Conditioning, learning, attention: These all put the "what" in behavior. They give behavior a direction, a specificity, a goal. Motivation, on the other hand, gives a "why," a prompt to act. Motivation is different from arousal, however, in that motivation is not really nonspecific. There might or might not be only one arousal state, but there are definitely multiple motivators, prompting the organism to act in specific ways—hunger, thirst, excess body heat or cold, sexual urges, and so on. In other words, there is both a "why" and a "what" to motivation. The questions of the extent to which, and the circumstances under which, deprivation of type A can motivate behavior usually motivated by deprivation B remain an issue.

Motivation, Homeostasis, and the Milieu Intérieur

Bodily feedback messages are motivating signals. They are the messages from the body indicating that one or another of the physiological variables have moved too far from the set points that define the correct homeostatic state of the milieu intérieur. Not all set points are completely fixed. The proper level of circulating cortisol in adult humans shows a circadian variation of 3–5-fold. The set point for body weight drifts upward as one ages. But set points exist, and the distance that a physiological state moves from the set point to a large extent creates and defines a motive and its intensity. Interoception is about how the brain does its work by knowing about the status of the body.

Maintenance of core body temperature, hunger, and thirst are examples of sensory interactions between brain and viscera (Kupfermann and Schwartz, 1995). The control of these internal states involves a combination of afferent information from the body and interpretation and processing of this information by the brain. For example, thirst involves feedback from the body that is largely based on vascular volume, tissue osmolality, and circulating hormonal factors such as angiotensin. Several brain regions do not have complete blood–brain barriers and are thus able to sense changes in bodily chemical variables. One, the area postrema, is in very close proximity to the nucleus tractus solitarius in the medullary brainstem, a major nucleus in the visceral afferent neuronal pathways (see Chapter 7). Blood pressure changes also affect fluid volume regulatory mecha-

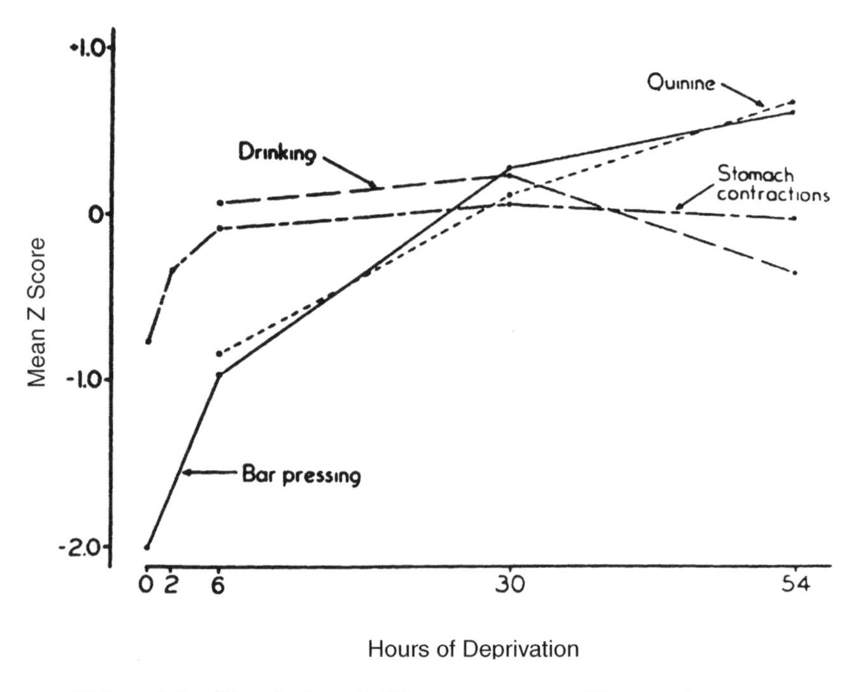

Figure 4–1. Dissociation of different measures of hunger in response to different durations of deprivation—Drinking: volume of enriched milk; Bar pressing: rate on a variable-interval schedule; Quinine: concentration in enriched milk needed to inhibit intake; Stomach contractions: measured by an implanted balloon. [From Miller, 1956, reprinted with permission of the holder of the copyright.]

nisms. Like other motivational variables, thirst and drinking are multidetermined, involving not just sensing of body chemistry directly by the brain but also responses to dry mouth, high body temperature, sweating, conditioned anticipation, and emotionally associated and adjunctive (i.e., behavioral schedule induced) drinking. Sometimes factors can be dissociated from one another, as illustrated in Figure 4–1 for hunger.

Centers in the Brain

Stimulation of certain areas of the brain is associated with autonomic changes (e.g., Brooks et al., 1979), especially the hypothalamus (see Morgane and Panksepp, 1979, 1980a, 1980b, 1981). Other implicated regions, some associated with characteristic behavioral changes (e.g., sham rage), include the septum, hippocampus, periaqueductal gray, anterior cingulate gyrus, and medial prefrontal and insular cortical regions. The association of these areas with autonomic changes implicate them in both motivational and interoceptive processes and indicate that discrete centers for control of these functions exist in the brain.

The hypothalamus, most frequently involved, is a small, very complex, medial brain structure that lies on both sides of the third ventricle below the anterior thalamus. Various anatomical subdivisions of the hypothalamus have been proposed (Card et al., 1999). The structure of the hypothalamus is illustrated in medial (sagittal) view (Fig. 4–2) and frontal view (Fig. 4–3). It has been called the head ganglion of the autonomic nervous system, involving both afferent input and efferent output. The hypothalamus is essential to integration among the three primary modes of bodily information transfer, the nervous system, the endocrine system, and the immune

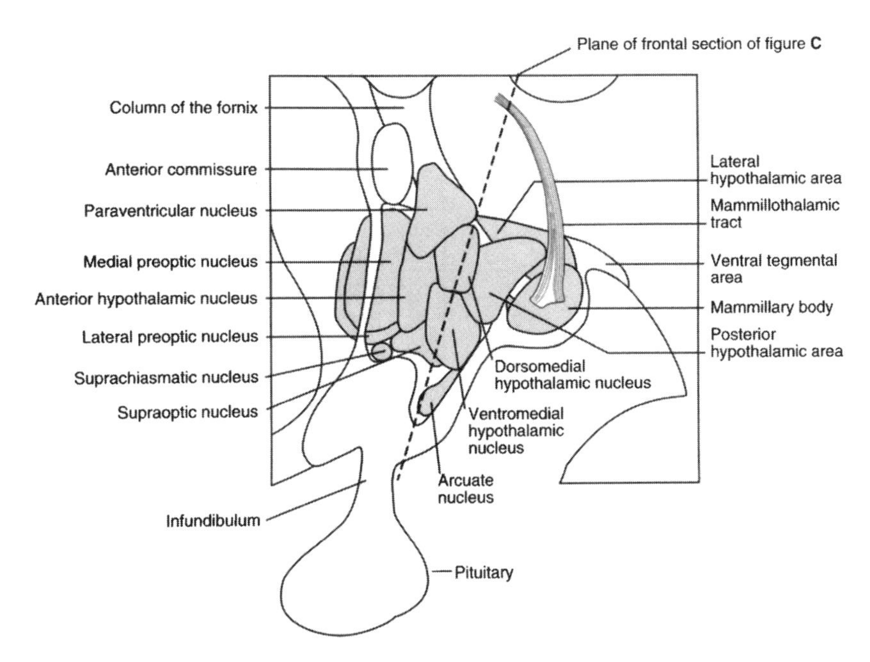

Figure 4–2. Internal structure of hypothalamus: medial sagittal view. [From Kandel and Kupfermann, 1995, reprinted with permission of the holder of the copyright.]

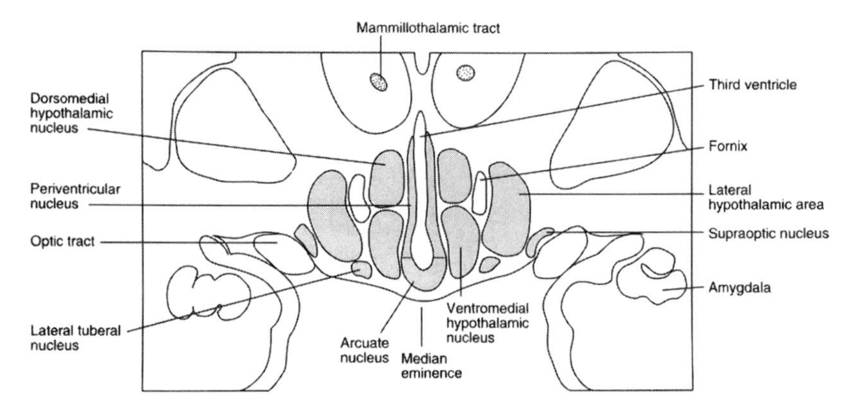

Figure 4–3. Internal structure of hypothalamus: frontal view, cut at plane shown on Figure 4.2. [From Kandel and Kupfermann, 1995, reprinted with permission of the holder of the copyright.]

system. Hypothalamic anatomy, function, and efferent outflow will be discussed here. Afferent pathways of relevance to interoception will be discussed both here and in Chapter 7.

The hypothalamus has neuronal connections with a number of brain regions that are relevant to interoception. Typically, reciprocal efferent and afferent connections exist. In addition to connections specifically with visceral organs, it has olfactory projections, connections to and from various limbic system structures, brainstem projections, and linkages to the circumventricular organs (which lack a full blood–brain barrier and function as chemoreceptors), all of which are relevant to motivation and thus, at least indirectly, to interoception.

The hypothalamus and the limbic system are intimately connected. For example, the fornix, which is considered part of the limbic system, passes through the hypothalamus, medial to the lateral hypothalamic area (Fig. 4–3). The medial forebrain bundle, a group of fibers arising in the brainstem and closely associated with the region called the ARAS (see Chapter 3, and Figs. 3–2, 4–2, and 4–3), passes through the lateral aspect of the hypothalamus. As already described, the medial forebrain bundle is partly a neuronal system that includes the locus caeruleus, involves monoamine neurotransmitters, and is implicated not only in motivational but also sensory–attentional functions. Olfactory function is closely related to the larger realm of overall gastrointestinal function, a system highly implicated in interoceptive processes. Pathways between the hypothalamus and the olfactory system are polysynaptic, including connections with the limbic system and brainstem structures such as the amygdala and medial forebrain bundle.

There are several ways in which the control of hormonal systems is of probable relevance to interoception. First, the tone of the body (e.g., metabolic effects of thyroid hormone on heart function) probably affects visceral sensory feedback. Second, hormonal changes are involved in most or all motivational effects. Third, hormonal changes, especially those from the adrenal cortex (cortisol in humans) and medulla (adrenaline), occur almost always in association with stress and various emotional changes (especially aversive reactions).

Hunger, thirst, and thermoregulation are just three examples of the many bodily functions that are controlled to a substantial extent by the hypothalamus. Immune function, such as responses to pyrogenic substances, is another example. The hypothalamus is also involved in control of motivationally controlled behaviors. The hypothalamus is likely to be involved in various emotional behaviors such as anger and aggression, as demonstrated by the fact that stimulation of limbic regions linked to the hypothalamus can produce sham rage, including autonomic and other bodily changes usually associated with aggressive behaviors.

Drug Abuse

Upon first consideration, it might not be obvious how drug abuse could be related to interoception. Several reasons exist to consider it so. First, the question should be asked, Is the common assumption correct that drugs are self administered because of subjective experiences they produce by entering the brain and having pharmacologic effects directly on drug receptors in the brain? Although the simplest answer to the question would appear to be "yes," drugs have many effects outside the brain as well. Consider the effects of stimulants such as caffeine and amphetamine on the heart, which are most likely produced by a combination of effects mediated by the brain and effects on cardiac receptors directly. Consider the effects of withdrawal from various types of drugs. These withdrawal reactions often appear to include effects which are mediated peripherally. Consider the long lists of side effects associated with many types of psychotherapeutic drugs, at least some of which are mediated peripherally.

Second, not only do psychotropic drugs, including drugs of abuse, have peripheral effects, but a further question should be considered: Are any of these peripheral effects reinforcing, that is, do they contribute to self administration? It is assumed, probably correctly (but possibly incorrectly), that drugs have self administration and abuse potential only if they produce discriminable states (i.e., that the drug state feels different than the non-drug state). The topic of drug-state discrimination will be reviewed in Chapter 11.

The third reason to consider the potential relevance of drug abuse to interoception is due to the close relation between drug abuse and motivational processes. In other words, to the extent that motivational processes, including sensory, motor, and reward factors, are relevant to interoception, so drug abuse is likely to be as well.

EMOTION

Emotions are motives. Both emotions and motives have behavioral, cognitive, subjective feeling (affective), and physiological components. As motives, they answer not only the "what" questions, but especially the "why" questions of behavior. Although the concepts of *emotion* and *motive* are almost synonymous, the words are often used differently. First, although motivation and emotion both denote all of the components listed above, when the word motivation is used it is most often meant to refer to a prompt to behavior, while emotion is used more often to refer to the other components, especially the affective. Second, reactions to aversive situations are more often considered under the rubric of emotion than motive. Fear is a motivator of escape and avoidance, but it is more often thought of as an emotion than as a motive. It is probably mainly for these two reasons that this distinction is made.

Like motives generally, emotions prompt behavior, but it is not completely resolved how they interact with learning. Emotional states that are more than mildly intense often disrupt learning, in part by disrupting attention. As motives, emotions tend to produce general arousal. Aversive emotional states are particularly potent in producing systemic autonomic activation.

Physiological and Cognitive Factors

In Cannon's critique of the James–Lange theory, two of the five specific points that Cannon made were *(1)* "Artificial induction of the visceral changes typical of strong emotions does not produce them" and *(2)* "The same visceral changes occur in very different emotional states and in non-emotional states." In making these two critiques, Cannon was relying on two sets of observations. Concerning the first critique, Maranon (1924), with later replications by Cantril and Hunt (1932) and Landis and Hunt (1932), found that systemic adrenergic activation produced by injections of adrenaline (probably also containing noradrenaline—synonymous with norepinephrine) failed to produce reports of emotional experiences in two-thirds of the subjects studied. The remaining one-third did report emotion-like reactions, but they were described as cold, or "as if," because the physiological component was present but not the full experience. Concerning the second critique, Cannon's own research indicated that fairly homogenous sympathetic reactions occurred during various great emotions (Cannon, 1953). In attempting to clarify this disagreement, investigators examined other possible contributors to the emotional experience. Major possible contributors were cognitive and interpersonal factors. One important and influential study designed to address this question was performed by Schachter and Singer (1962).

To view sympathetic activation induced by epinephrine injections as an adequate model for the physiological component of all emotions makes two significant errors. First, it assumes that epinephrine injections mimic the full sympathetic activation pattern that occurs during various emotional states. That assumption has not been confirmed and is undoubtedly incorrect. Second, it assumes that all emotions that are associated with any somatic–visceral activation pattern are completely or predominantly sympathetic. That assumption also has never been verified and also is almost certainly incorrect or at least highly oversimplified. A simple example of an error in the Schachter and Singer formulation is: If sympathetic activation is so important to the experience of emotion, blockade of adrenergic function with beta-adrenergic blocking drugs should substantially diminish the emotional experience, for example in people with anxiety disorders; that does not reliably occur (Tyrer, 1988; Munjack et al., 1989).

The importance of the Schachter and Singer experiment and the critiques that followed emphasize that an all-or-none judgment about the role of physiological changes in the genesis of emotion is undoubtedly an oversimplification. Not only are physiological and nonphysiological factors important, but viewing the physiology of emotion as strictly related to sympathetic activation is too narrow as well. Bodily changes of some kind or another are probably not sufficient to experience an emotion, but they probably are necessary. In that context, interoception becomes a fundamental aspect of emotion. Scrooge was afraid of Marley's Ghost, and he (and Dickens both) knew more than 150 years ago that fear was somehow linked to a "gut reaction."

Markers of Emotion and Interoception

Both sympathetic and parasympathetic autonomic nervous system functions, and probably functions of the enteric nervous system of the gastrointestinal tract as well (Gershon, 1998), are affected by emotional changes. Although various markers are associated with emotion, there are substantial variations in their occurrence. Individual differences exist among organisms (e.g., pattern differences for human men versus women—see Fig. 2–1), and differences exist among various emotions (e.g., appetitive versus aversive emotions).

Many hormone levels change in response to emotion and stress (e.g., Donovan, 1988), but the most extensively studied are the catecholamines epinephrine and norepinephrine and the hormones of the hypothalamic-pituitary-adrenocortical (HPA) axis—in humans, ACTH and cortisol. Despite the fact that the most prominent symptoms of stress and great emotion are sympathetic, the increases in systemic catecholamines are not sufficient to produce the sympathetic symptoms observed (Cameron et al., 1990a; see Fig. 2–2). Sympathetic symptoms seem largely due to the neuronally mediated collective activation of the organs that are sympathetically innervated, a fact of obvious direct relevance to interoception. Central nervous system catecholamines probably do contribute to the symptoms of emotion, stress, and anxiety (see Mason, 1984; Glavin, 1985;

Goldstein, 1994; Bremner et al., 1996a, 1996b; Tanaka, 1999; Sullivan et al., 1999). Also, they are involved in the systems in the brain that control peripheral sympathetic activity, but they do not make any significant direct contribution to the peripheral catecholamine levels themselves (nor do the peripheral contribute to the central) because of the blood–brain barrier. Other hormones, including the HPA-axis hormones, do not seem to make significant contributions to subjective symptomatology, but they could contribute to interoceptive processes even if they do not produce discriminative stimuli that reach awareness.

Autonomic Nervous System Anatomy

A neuroanatomical distinction is usually made between the skeletal nervous system and the autonomic nervous system. It is widely believed that these major parts of the peripheral nervous system are separate and that the skeletal part, controlling the skeletal muscles, mediates voluntary actions, while the autonomic part mediates involuntary functions. This distinction has already been described above. The examples of motoric changes associated with other bodily changes in emotion are just one type of evidence that the functions of these two parts of the peripheral nervous system are not completely independent and often act in concert. The voluntary–involuntary dichotomy is more appropriately considered as a continuum.

Visceral nervous system afferents are an integral part of the physiology of emotion. For example, for the brain to control increases in heart rate or epinephrine release, it must know the magnitude of the effects occurring in the periphery. The pathways involved must include all branches of the peripheral nervous system, the spinal cord, the brainstem, and multiple regions of the brain above the brainstem. Especially important regions are the hypothalamus, the limbic system, including especially the amygdala, and various cortical structures. In the remainder of this chapter, the limbic system, fear conditioning and the amygdala, and the role of certain cortical structures in emotion will be discussed. A brief review of functional imaging and emotion will be presented. A more detailed presentation of the neuroanatomy specifically relevant to interoception will be given in Chapter 7.

The Limbic System

Mega et al. (1997) have pointed out that the concept of a neuroanatomical limbic system ("cerebri limbus"—border of the cerebrum) dates back to the mid-seventeenth century. In the late nineteenth century P. Paul Broca defined the *"grand lobe limbique."* Then, in the 1930s and 1940s, Papez, P. I. Yakovlev, and MacLean modified and refined the concept both anatomically and functionally (Fig 4–4). They hypothesized that this set of structures was highly involved in visceral and emotional functions.

In the initial formulation by Papez, the structure of the limbic system was relatively simple, involving a hypothesized circuit that included the cingulate gyrus, the hippocampal formation, the mammillary bodies, and the anterior thalamic nuclei (see Fig. 4–4; Kandel and Kupfermann, 1995). Subsequently, other specific structures have been considered to be part of the limbic system, including the

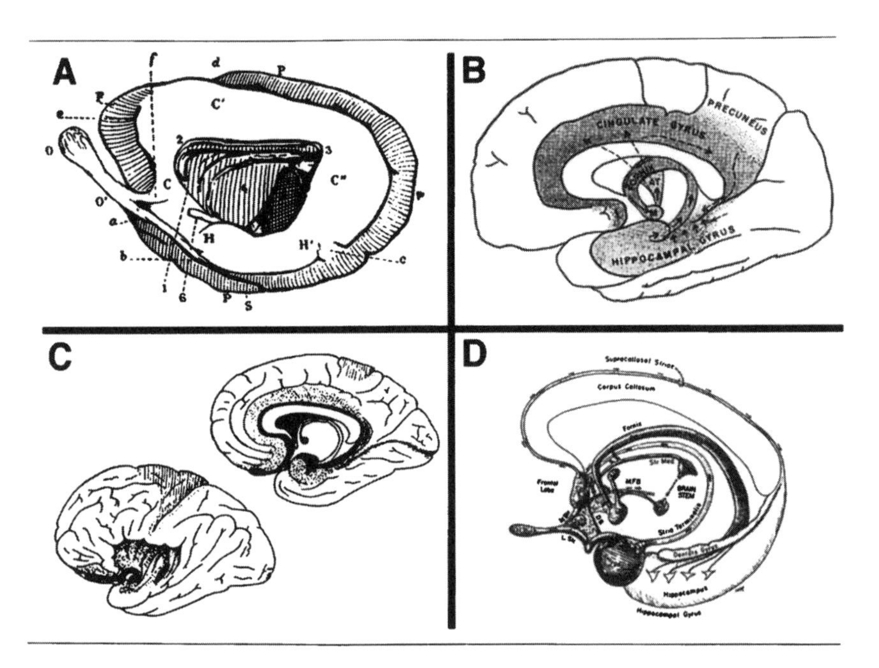

Figure 4–4. Four different conceptions of the limbic system (A: la grand lobe limbique of Broca, 1878; B: the Papez circuit, 1937; C: three functional zone proposal of Yakovlev, 1948; D: MacLean's, including update and modification of Papez, 1949). [From Mega et al., 1997, reprinted with permission of the holder of the copyright.]

Table 4–1. Two Model Examples of Anatomic Divisions of the Limbic System

Yakovlev's Three-Layered Model
Contains three hypothetical levels with different neurons, neuronal organization, evolution, anatomical structure, and functions. The inner layer subserves basic physiological functions, the intermediate layer is related to personality and emotion, and the outer layer is involved in detailed cognitive functions.
Paralimbic Divisions
Orbitofrontal Division: amygdala, anterior parahippocampus, insula, temporal pole, infracallosal cingulate—visceral, appetitive, affective, and social functions
Hippocampal Division: hippocampus, posterior parahippocampus, retrosplenium, posterior cingulate, supracallosal cingulate—cognitive, attentional, skeletomotor, and motivational functions

[Adapted with permission of the holder of the copyright from Mega et al., 1997.]

prefrontal and association cortices, the hypothalamus, and the amygdala. Fiber bundles connecting these structures, such as the fornix, stria terminalis, mammillothalamic tract, dorsal longitudinal fasciculus, and medial forebrain bundle are also considered part of the limbic system. The monoamines are prominently involved in limbic functions.

In addition to the specific structures that define the limbic system, various ways of defining subdivisions of the limbic system have been proposed. Mega et al. (1997) pointed out that Papez viewed different combinations of structures as being involved in the streams of movement, thought, and feeling, with feeling involving medial structures. Trimble et al. (1997) suggested that ". . . the medial limbic circuits mediate information more closely related to internal states, whereas the lateral limbic circuit is more involved with information concerning the body surface, external world, and social–personal interactions" (p. 124). The validity of these subdivisions remains to be determined. Table 4–1 contains two proposed ways of subdividing the limbic system suggested by different authors, adapted in modified form from Mega et al. (1997).

Fear and Conditioned Fear

The overall topic of fear is very broad. Many reviews are available (e.g., Sluckin, 1979; Davis, 1997; LeDoux, 1998; Gewirtz and Davis,

1998; Fendt and Fanselow, 1999). Those listed here generally emphasize the biological (unconditioned, innate, genetically based) aspects since that is more directly relevant to the physiology of fear, the autonomic nervous system, and thus to interoception. Fear is an especially important emotion to address in the context of interoception because of the prominence of visceral changes associated with it, and also because it is so basic to the need for, and theory of, two-process learning theory.

Fear can be learned as well as innate. In two-process learning theory it is learned fear that explains avoidance behavior (Masterson and Crawford, 1982). Numerous studies show that fear conditioning follows Pavlovian conditioning rules and procedures very similar to appetitive conditioning, including the observation that, within limits, more intense CSs and UCSs lead to stronger conditioning—another observation that links fear conditioning to general arousal.

Second-order conditioning occurs with fear. In this paradigm, an initial CS (CS1) is paired with the aversive UCS, then a second CS (CS2) is paired with CS1. Second-order conditioning is said to have occurred if the CS2 shows a qualitatively similar conditioned response as does the CS1. Higher-order fear conditioning has been reviewed recently by Gewirtz and Davis (1998), and they offered a hypothesized schema for the neuroanatomical circuit underlying first-order and higher-order fear conditioning (Fig. 4–5). Their schema involves many of the neural structures likely to be involved in interoception.

The relationship between fear and autonomic–visceral responses is closely related to an organism's ability to escape or avoid the aversive stimulus. For example, as the escape or avoidance behavior of a rodent or primate becomes more efficient, autonomic activation decreases. The conditioned emotional response (CER), on the other hand, shows high and persistent autonomic activation because the aversive stimulus is not fully avoidable or escapable. Further, studies with rats (Weiss, 1971a, 1971b) showed that control of escape or avoidance from shock had more to do with the development of gastric erosions than did the amount of shock received.

Amygdala. The renaissance of attention to the psychobiology of emotion during the 1980s and 1990s included a dramatic increase in interest and understanding of the anatomy and functions of the amygdala (Aggleton, 1993; Ono et al., 1993; Knuepfer et al., 1995;

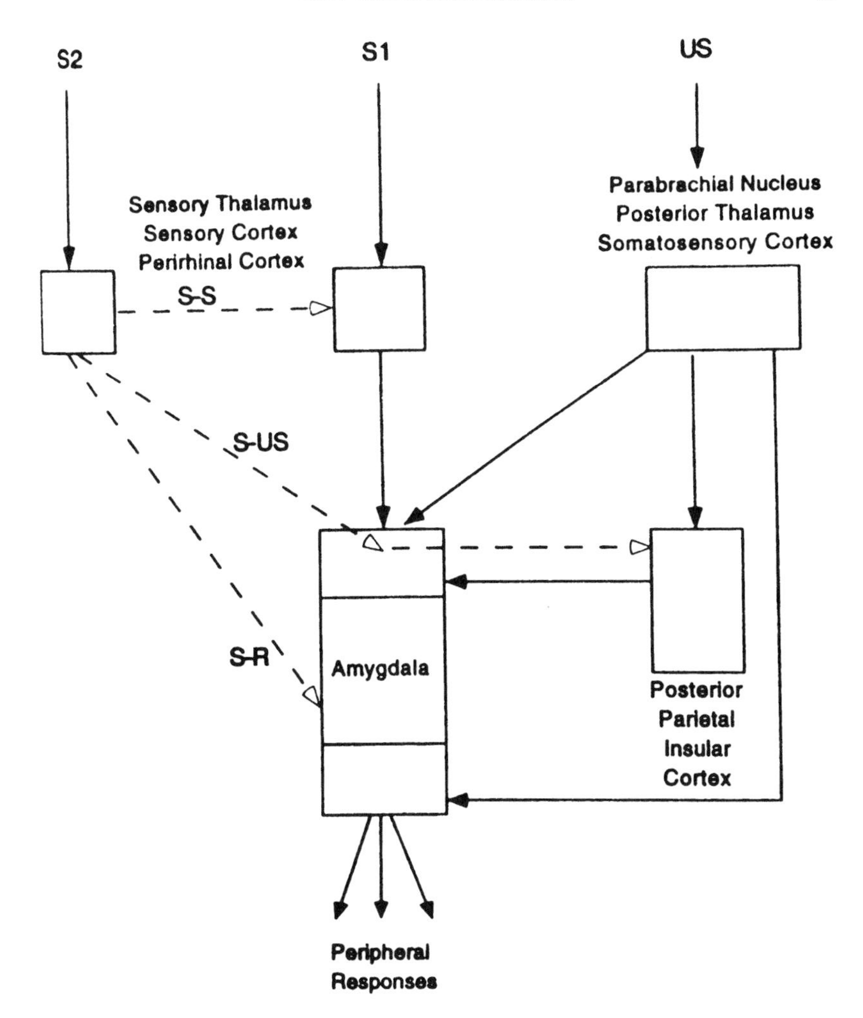

Figure 4–5. Neural circuit model of first-order and second-order Pavlovian fear conditioning. Note the involvement of several neuroanatomical structures implicated in interoceptive processes. [From Gewirtz and Davis, 1998, reprinted with permission of the holder of the copyright.]

Baklavadzhian et al., 1996; Armony et al., 1997; Benarroch, 1997; Davis, 1997; LeDoux, 1998; De Olmos and Heimer, 1999; Adolphs, 1999; Fendt and Fanselow, 1999). There has been increased understanding of the neuronal inputs and outputs to the amygdala as well as the internal structure of the amygdaloid nucleus itself. Table 4–2 and Figures 4–5 and 4–6 identify the relevant structures. The amygdala has been found to have a complex internal structure, with in-

Table 4–2. Outline of Amygdala Regional Inputs and Outputs

Inputs				
Brainstem	Basal Forebrain	Thalamus	Cortex Hippocampus	
Amygdala Nuclei Pathway				
Lateral	→	Basolateral and Basomedial	→	Central
Outputs				
Brainstem	Hypothalamus	Cortex		

[Adapted with permission of the holder of the copyright from Davis, 1997.]

puts primarily to the lateral nucleus, output from the central nucleus, and the basolateral region providing linkage between input and output. The concept of an extended amygdala has been developed, acknowledging that there are structures highly connected to, and functionally related to, the amygdala, especially the stria terminalis (De Olmos and Heimer, 1999). The inputs and outputs (see below) are largely to cortical and subcortical regions that have been implicated in emotional functioning and conditioning and in interoception.

The major function of the amygdala is its role in fear and aversive behavior, probably including normal and abnormal anxiety states (see references in the preceding paragraph). The amygdala is essential in both innate and conditioned fear (Gewirtz and Davis, 1998). Output affects the behavioral, autonomic, and endocrine aspects of aversive function. Not only is the amygdala involved in emotional functioning, but in the recognition of emotional display in others. In addition to the direct emotional manifestations of amygdaloid function, its overall function includes cognitive–memory, appetitive, and social behavior changes. For example, in Kluver–Bucy syndrome, monkeys show changes in sexual behavior and position in the dominance hierarchy. The cognitive–memory aspects (i.e., learning aspects) of amygdala function help account for the plasticity of function often seen in this system (Nader and LeDoux, 1997; Maren et al., 1998; Maren, 1999). Finally, evidence exists for a short loop fear circuit through the amygdala that mediates rapid reflex-like fear responses without the need for conscious awareness, and a long loop circuit through the amygdala that includes cortical structures and is involved with awareness. In other words, not only has it been known for a long time that fearful people are not always aware

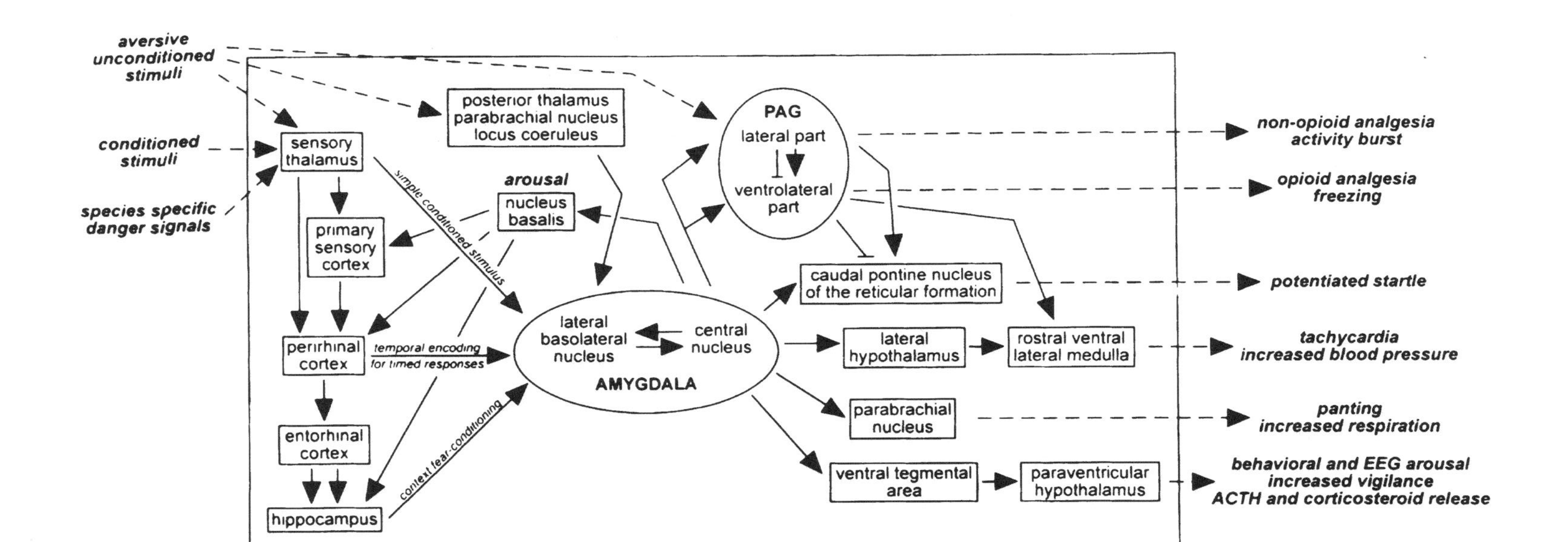

Figure 4–6. Theoretical model of central nervous system circuits underlying various aspects of conditioned fear. [From Fendt and Fanselow, 1999, reprinted with permission of the holder of the copyright.]

of what they are afraid of, but now evidence exists for an anatomical basis for this lack of awareness (LeDoux, 1996).

Other Anatomical Regions. In addition to the amygdala, a variety of other structures are involved with fear and fear conditioning, including the hippocampus, thalamus, hypothalamus, and brainstem structures including the locus caeruleus and the peri-aqueductal gray region. Fendt and Fanselow (1999) hypothesized that different neuronal circuits are involved in unconditioned (startle) versus conditioned fear (Fig. 4–6). Even though this review has focused on the emotional (affective, motivational) aspects of fear and the neuronal structures mediating fear, essential cognitive and memory functions are involved in these functions as well (Lovand et al., 1993). The resistance to extinction observed with conditioned fear is substantially a function of memory processes.

A number of regions exist in the cortex that are implicated in amygdaloid function. In conditioned fear, the cortical region related to the sensory modality of the CS (e.g., auditory and visual) is involved. More generally, cortical regions that have connections with the amygdala include the prefrontal cortex, especially the orbitofrontal region, the insular cortex, the anterior cingulate gyrus, and the perirhinal cortex. As described in Chapter 7, these are regions also implicated in higher visceral sensory function (i.e., interoception). Derryberry and Tucker (1992) reviewed the ascending and descending neural mechanisms associated with emotion, including interoceptive functions, noting the broad involvement of brainstem, limbic, paralimbic, and neocortical regions and peripheral endocrine, autonomic, and motor systems.

Right Hemisphere Predominance of Emotional Processing. An important aspect of cerebral function is hemispheric asymmetry. Some cortical and subcortical functions are localized on one side of the brain. The most obvious example in humans is the language function. Of relevance to interoception are findings that physiological and emotional processing show lateralization (see Davidson and Hugdahl, 1995; Ross, 1996). For example, amobarbital-induced inactivation of the left hemisphere in humans leads to heart rate increase, while right hemisphere inactivation leads to heart rate decrease (Zamrini et al., 1990). More specifically, stimulation of the left insular cortex led to heart rate decrease, while stimulation on the right led to increase (Oppenheimer et al., 1992).

Emotional processing also shows lateralization, mainly of associated physiological functioning. Hemispheric differences occur in autonomic responses and conditioning to emotional facial expressions (Spence et al., 1996; Johnson and Hugdahl, 1991, 1993), and reduced autonomic reactions to slides with emotional content after right hemisphere damage (Meadows and Kaplan, 1994). Canli (1999) reviewed studies indicating that positive emotions are lateralized toward the left and negative emotions toward the right.

Anxiety and Anxiety Disorders

Anxiety is usually considered to be closely related to fear, although not necessarily the same. From the point of view of anatomy and physiology, anxiety shows very similar autonomic–visceral changes to fear. Based in part on the observation that anxiety often is of longer duration and has less explicit cues than does normal fear, Davis (1998) suggested that abnormal anxiety might be mediated by the stria terminalis, which is separate from the amygdala but closely related to it. Neurotransmitters implicated in anxiety and anxiety disorders include GABA, the effects of which are enhanced by the anxiolytic benzodiazepine class of drugs (Tallman et al., 1999; Lambert, 2000) and norepinephrine (Southwick et al., 1999; Sullivan et al., 1999), the bulk of which comes from the locus caeruleus and is involved in attentional and interoceptive processes. Benzodiazepines affect escape and avoidance behavior (Dinsmoor et al., 1971; Essman, 1973; Feldman et al., 1973; Grilly et al., 1984) and decrease the somatic and visceral as well as the affective and cognitive symptoms of anxiety (e.g., Tyrer and Lader, 1974; Hoehn-Saric et al., 1989a, 1989b; Taylor et al., 1990; Berntson et al., 1997).

Imaging Studies of Emotion

Over the past 20 years functional imaging techniques have become very important in the study of emotion. Imaging studies will be reviewed in more detail in Chapters 8 and 9.

Function of Emotion vis a vis Interoception

What is the functions(s) of emotion, and how might this relate to interoception? Because emotion is a combination of behavior, cognition, affect, and physiology, any explanation must account for all these factors. There is no final explanation. One very unclear aspect of this question is the role consciousness plays in any of this. Regardless of the role of the other factors, why do emotional states produce awareness? This is part of the much larger question of the role awareness plays in general. No answer to this question exists at this time that is even close to satisfactory. The issue will be revisited in Chapter 13.

It would seem that an evolutionary answer to the question of the function of emotion would be most appropriate. Porges (1997) offered one possibility. It will be described here as an example of how this question might be answered. Whether Porges's theory is correct remains to be determined.

Porges calls the theory he proposes the "polyvagal theory of emotion." Briefly, the theory posits that a phylogenetic shift occurs in central nervous system control of the autonomic nervous system, from simple control of vegetative functions such as digestion and behavioral immobilization in response to threat, through a second stage involving development of the fight-or-flight reaction, to a third stage involving more flexible response to various types of affiliative and threatening circumstances. The theory hypothesizes concurrent evolution of control of cardiac and other visceral organs along with development of a broader and more flexible repertoire of emotional reactions. The importance of visceral sensory processes in this theory is obvious.

Not only is the function of emotion not yet fully clarified, but the neuroanatomy underlying emotion is still debated. Despite the fact that the concept of the limbic system has been discussed for more than 100 years, the autonomic nervous system for almost a century, and the ascending reticular activating system for almost 50 years, full agreement does not exist even as to the validity or usefulness of these concepts. Blessing (1997), for example, argued that none of these three systems "possess[es] functional and anatomical reality," and that "we must be careful not to substitute names for explanations." (p. 238). In his proposal for "frameworks for understanding bodily homeostasis" (p. 235), he argues that in order to understand visceral functions, the most appropriate approach is to identify afferent and efferent neuronal systems, and then, starting there, investigate the central circuitry that ties these two systems together. A full understanding of the peripheral and central afferent systems would provide a great advance in understanding interoception. Whether the autonomic nervous system, the ascending reticular activating system, and the limbic system are valid concepts is related to, but not essential to, understanding either emotion or interoception.

five

ROOTS IN RUSSIAN AND EASTERN EUROPEAN STUDIES

Interoceptive conditioning may best be defined as classical conditioning in which either the conditioned stimulus (CS) or the unconditioned stimulus (US) or both are delivered directly to the mucosa of some specific viscus.

Gregory Razran, *The Observable Unconscious and the Inferable Conscious in Current Soviet Psychophysiology*, 1961, p. 81

The primary topic of this chapter is the role of Pavlovian conditioning in visceral sensory physiology, that is, a review of the evidence that an organism can sense visceral afferent information, demonstrated by the fact that visceral sensation can play a role in Pavlovian conditioning. Much of this research was completed in the first half of the twentieth century (see Bykov, 1957, 1966; Razran, 1961; Adam, 1967; Chernigovskiy, 1967), but a monograph updating these reviews was recently published (Adam, 1998). This review depends mainly on these secondary resources because most of the primary sources are not generally available and have not been translated into English.

It is important to note the following: The ability to demonstrate Pavlovian conditioning in which the conditioned stimulus (CS) involves stimulation of a visceral sensory receptor is not necessary but is sufficient to demonstrate the existence of interoception in the sense that is meant in this volume, i.e., the ability of visceral afferent information to either reach awareness or affect behavior. The successful demonstration of interoception does not require awareness of any visceral sensation. Modification of behavior (e.g., the occurrence of a conditioned response) is sufficient.

Early investigators of interoceptive functions consistently observed that sensory impulses from the visceral organs produce either no awareness or only vague or faint sensations. From one point of view—that is, that there must be sensations to which to respond in order for visceral function to affect behavior—this observation would seem to call into question any possibility of a psychology of visceral sensation. Two reasons exist, however, why this logic is incorrect. First, not only do such faint sensations appear to occur, but under some circumstances they are much more than vague. Second, in the context of Pavlovian conditioning, the only overt requirement for conditioning to occur is a specific type of functional, temporal (i.e., signaling) contiguity between the conditioned stimulus and the unconditioned stimulus (UCS). Even though the function of the CS is considered or hypothesized to be a signal, the relationship between this signal function and awareness is not necessarily one in which the organism is aware of the signal in the sense in which awareness is usually meant (e.g., in humans, that it could be identified as occurring in consciousness and be described as such verbally).

The concept of a reflex in the physiological, or unconditioned, sense goes back at least to the late 1700s. The first study of conditioned reflexes that demonstrated interoceptive conditioning was reported by Bykov and Alekseev-Berkman in 1926. In this study with dogs, the unconditioned response was a diuretic reaction and the unconditioned stimulus was injection of water into the rectum. The experimental setup itself was the conditioned stimulus, and the conditioned response was also diuresis. Before describing subsequent studies, a review of some aspects of the physiology of interoceptors will be presented.

PHYSIOLOGY

Perhaps the most extensive review of interoceptors, at least through the first 60 years of the twentieth century, was a volume by Chernigovskiy (1967) aptly titled *Interoceptors*. It contains more than 2000 references, of which roughly ⅔ refer to Russian literature. Although the book addresses issues related to conditioning, it focuses more on the physiology of the interoceptive process than on the psychobiology. It provides an appropriate point of departure in reviewing some of the literature on Pavlovian conditioning and interoception, starting with a review of the more physiological aspects of the topic. In addition to the extensive literature review provided by Chernigovskiy, the two volumes by Adam (1967, 1998) provide shorter reviews. Additional reviews of visceral sensory physiology can be found in Newman (1974), Haines, (1997), Zigmond et al. (1999) and Kandel et al. (2000). This review will focus on the material referred to in Adam (1967) and Chernigovskiy (1967) because this is the information that was available as background to the investigators who performed the conditioning studies described below.

Adam (1967) has argued that there has been increasing phylogenetic differentiation of interoceptors. Along with this increasing differentiation has been the development of structures in the central nervous system that receive this afferent sensory information. He states that ". . . visceral afferentation is present already in the lower vertebrates . . ." (p. 8) and that there is clear advancement of this function in mammals.

Chernigovskiy subdivided the interoceptors into four major types based on the kind of stimulus they respond to—mechanoreceptors, chemoreceptors, thermoreceptors, and osmoreceptors. Chernigovskiy acknowledged that proprioception, including vestibular function, was an important aspect of interoception but did not review it separately. He also noted the importance of nociception. Chernigovskiy described the concept of an interoceptive zone or field, by which he meant the distribution of interoceptors that affect a single system. He wrote ". . . the whole cardiovascular system is provided profusely with receptors which, taken together, form a single interoceptive field that influences reflexly the whole vascular system." (p. 20). Finally, Chernigovskiy pointed out that the difference between interoceptors and exteroceptors was sometimes not distinct. For example, where in the gastrointestinal tract are sensory nerve endings exteroceptors, and where does interoception start? (One imaging study—Aziz et al. [2000]—found that proximal and distal esophageal distention activated different parts of the somatosensory and cingulate cortices, suggesting that such a location can be identified.)

Adam included volume receptors separately in his list and commented on the possibility of multimodal receptors (e.g., responsive to both mechanical and chemical stimulation). The focus of the discussion was on receptors as they are found outside the central nervous system, but chemoreceptors, including receptors that respond to osmotic changes, and some thermoreceptors also reside within the central nervous system. Before discussing the functional aspects of the various interoceptor types, a few comments about morphology are appropriate. Additional comments on the anatomic and functional aspects of specific systems will be made when appropriate, for example, in Chapter 10 about pain.

Nerve endings can be found throughout the viscera, including a substantial innervation of the vasculature, and in such unexpected places as the liver (Fuller et al., 1981), spleen (Ackerman et al., 1989), and immune system (Felton et al., 1987). Morphologically, unencapsulated nerve endings are found in broad distribution. Various kinds of encapsulated endings, for example the Pacinian corpuscles, are also observed. The histology of these nerve endings is often non-specific. They are often diffuse, do not differentiate one organ from another, and do not even differentiate afferent from efferent endings.

Mechanoreceptors are widely distributed and essential to the normal function of the cardiovascular, respiratory, alimentary, and urogenital systems. Mechanoreceptors are mainly responsive to stretch. Functionally, in the lung, for example, these receptors can be differentiated into different groups based on adaptation to stimulation, at least into slowly adapting and rapidly adapting groups. A subdivision into three groups, based on adaptation rate, has also been proposed. These receptors probably play some role in motivational processes, such as the contribution to satiety made by stretching of the stomach.

It has been observed that stimulation of these receptors sometimes leads to responses not just in the organ from which the nerves arise but from other organs as well. For example, stimulation of the alimentary tract can lead to changes in cardiovascular function. The mechanisms by which this occurs are often not clear. Are there reflex connections at the level of the brainstem? Is this cross-organ response mediated at higher central nervous system levels? Is the response specific or nonspecific (e.g., a response to pain or pain-like experiences)?

Some chemoreceptors are typically considered exteroceptors

(taste and smell receptors in the mouth and nasal cavity, respectively), some are in the central nervous system in the brainstem and hypothalamic regions (including osmoreceptors), and some are classical interoceptors in the viscera. Chemoreceptors are sensitive to oxygen, carbon dioxide (osmoreception), glucose, amino acids, and other substances. In the gastrointestinal tract, for example, glucoreceptors may play a role in hunger and satiation. While it is common to consider the sensation of hunger and satiation, for example, as occurring primarily in the central nervous system (especially in the hypothalamus—see Chapter 4), it contradicts data as well as parsimony to assume that afferent visceral sensory impulses play no role in such motivational states.

Thermoreceptors are present in the gastrointestinal tract, as is obvious to anyone who has swallowed liquid that is too hot or too cold and can feel the sensation in the esophagus and perhaps also in the stomach. There may also be thermoreceptors in the vascular system. Thermoreceptors are present in the central nervous system as well. The ability of a warm-blooded organism to control core body temperature demonstrates that it is sensing it, and very accurately.

Osmoreceptors can be considered either as a type of chemoreceptor or as a separate type of interoceptor. Osmoreceptors are found in the hypothalamus. To what extent specific osmoreceptors exist outside the central nervous system is still not completely clear. They may exist in the liver and elsewhere in the alimentary tract. In addition to the ability to monitor osmotic content, evidence exists that body water content (i.e., volume) can be monitored and adjusted as well.

Evidence of bi-modal and even poly-modal interoceptors in the alimentary tract has been found. These might be nonspecific, responding to pain. Multi-modal sensors, if they do exist, might contribute to the vagueness often associated with visceral sensation. Some specificity could be contributed from the central nervous system pathways, but there is substantial lack of specificity there as well, including overall visceral versus somatic sensation.

EXAMPLE EXPERIMENTS

In the early 1960s Razran (1961) was aware that a substantial body of important research existed that had been performed in Russia and Eastern Europe for three decades but was almost unknown outside those laboratories. It related to classical conditioning. He identified especially three areas of potential interest for review—interoceptive conditioning, semantic conditioning (i.e., conditioning involving words and their meanings), and the orienting reflex. For the purposes of this book, of course, the focus will be on the interoception research. In interoceptive conditioning, at least one stimulus is visceral (as defined in the quote at the beginning of this chapter). Before providing the specific reviews, he gave a more general background to classical conditioning methods.

Much of this work involved the gastrointestinal tract, and stimuli have included such things as inflating balloons to produce distention and stimulate mechanoreceptors, injections of water at different temperatures to stimulate thermoreceptors, and injections of drugs to stimulate chemoreceptors. In some studies, multiple stimuli were presented at different points in the gastrointestinal tract. Studies of this kind were performed in humans who had preexisting fistulas in place for medical reasons. Other systems can be studied in similar fashion, for example, by cannulation of the bladder. Exteroceptive stimuli such as skin scratching have also been used.

A wide variety of responses were found in these studies. Many of the studies included exteroceptive (i.e., motor) responses such as a response of the leg. Interoceptive (i.e., visceral) responses that were measured included changes in all of the systems studied—for example, changes in heart rate, respiration, blood pressure, urine production, and alimentary motility. Associated changes such as the electroencephalogram (EEG), galvanic skin response (GSR—sweating), and changes in pupil size also were measured during some of these studies.

Razran provided a definition of interoceptive conditioning that was quoted at the beginning of this chapter. Multiple combinations of stimuli and responses that could be, and have been, studied. In the case of stimuli, both the CS and UCS could be interoceptive, or one could be interoceptive and the other exteroceptive. Razran pointed out that the major interest occurred in interoceptive research when the CS was interoceptive (i.e., interoceptive signaling). For responses, there were three different possibilities—visceral, skeletal (i.e., exteroceptive), and in the case of humans, verbal (semantic). Obviously, for interoception research a visceral response must be included. Razran reviewed experiments that had been reported using many of these possible combinations and demonstrating the existence and parameters of Pavlovian interoceptive conditioning. Some of these results will be described here. Razran divided his discussion into experiments with animals first and then with humans, followed by a summary statement.

The first two experiments with animals demonstrated interoexteroceptive conditioning, that is, conditioning in which the CS is interoceptive while the UCS is exteroceptive. In one of these experiments the CS was scratching of the uterine horn and the UCS was presentation of food. In this experiment the interoceptive scratching stimulus came to elicit salivation. In the second experiment an air jet to the uterine horn was the CS and a shock to the paw was the UCS. The interoceptive air jet stimulus was successfully conditioned to produce the paw withdrawal originally produced by the shock.

The next experiment was said to be an example of both interointeroceptive and exterointeroceptive conditioning. In this experiment the UCS was hypercapnia induced by forcing the subject dogs to breath 10% carbon dioxide (interoceptive—respiratory). Two different CSs were used. The interoceptive CS was experimentally induced rhythmic distentions of the intestinal loops of the animals, while an exteroceptive CS (auditory tones) was also used. Conditioned hyperventilation initially produced by the 10% carbon dioxide was observed in association with both the intestinal distentions and the tones. Both types of conditioning were found to be resistant to extinction, and it was found that the dogs could even learn to discriminate among different rates of rhythmic motions.

Higher-order conditioning was demonstrable in these studies. In one study it was shown that with an exteroceptive US (shock-induced paw withdrawal), an interoceptive CS (distention of intestine) could be second-order conditioned to an exteroceptive CS (buzzer sound). In another study, an exteroceptive CS (metronome sound) could be second-order conditioned to an interoceptive CS (water irrigation of intestinal loops). Sensory preconditioning with interoceptive stimuli was demonstrable, the rate of initial conditioning seemed to be slower with interoceptive stimuli, interoceptive conditioned responses once formed seemed to be more resistant to extinction, and interoceptive conditioning could interact with and influence the performance of a complex reinforced (operant) behavioral chain.

Lastly, Razran highlighted a study performed by Cook et al. (1960) in which various drugs (and also jejunal pressure) were used successfully as CSs to condition paw withdrawal. Not only is this study important, as Razran says, because it was the first demonstration of interoceptive conditioning in the United States, but it demonstrated the potential importance of drug-induced internal states as interoceptive stimuli (which will be discussed in Chapter 11). Of note, the drugs used in this study included epinephrine and norepinephrine, which do not cross the blood–brain barrier. Thus, the drug-induced stimuli to which these animals responded were in the periphery. They were not primary, direct drug-induced changes within the central nervous system.

A number of studies using humans as subjects also were described. Verbalization was used as stimulus and/or response. In the first experiment described, inflation of the bladder with air or fluid was used as the UCS and both urine production and reports of urgency were used as the UCRs. Associated respiratory, vascular, and GSR responses were monitored. Sham physiological measurement dials were the CSs. Conditioning was clearly demonstrated. In another study, an inflow of cold air functioned well as the CS. Similar conditioning effects were demonstrated in the respiratory system. Furthermore, in the respiratory study, associated conditioned effects in another system, vascular changes, were also demonstrated, indicative of what was called a viscerovisceral vasomotor reflex. Other associated changes, such as GSR, were also observed, and it was noted that associated responses in different systems occurred at different intensities of the CS.

In a human study with balloon inflation in the alimentary tract as the UCS, subjects' reports of their subjective sensations followed a progression from no sensation to non-painful awareness to pain. Thus, a non-painful awareness definitely occurred. Also, using electrical stimulation at two different points of the gastrointestinal mucosa 8 cm apart, it was reported that shorter duration intervals between stimulations at the two points was not resolvable, but longer ones were, reminiscent of two-point discrimination data on the skin. Both temporal and spatial discriminations occurred at a longer interval for the spatial.

Using thermal stimuli, interoceptive-exteroceptive conditioning was demonstrable. This study demonstrated interactions between conditioning of interoceptive and exteroceptive responses, and it appeared that when opposing responses were called for, the interoceptive response was typically dominant. Finally, in several of these studies, natural conditioning was observed, that is, conditioning of naturally occurring physiological changes during the experiment to the defined stimuli.

Razran concluded this section of his review by offering six principles: *(1)* Interoceptive stimulation leads to largely unconscious reactions. *(2)* Interoceptive conditioning is readily obtainable. *(3)* Interoceptive conditioning is a built-in function, constantly generated and regenerated. *(4)* Interoceptive conditioning is slower to form but more resistant to extinction. *(5)* Interoceptive conditioned reactions are dominant over exteroceptive. *(6)* Exteroceptive and interoceptive stimuli with the same conditioned reaction tend to decrease the intensity of the conditioned effect. He viewed the last three of these as provisional, but the first three as clearly demonstrated. Uno (1970), in a study of skin conductance conditioning (GSR), reported results generally in support of principle nos. 1 and 4.

Razran's review specifically attempted to identify different kinds of phenomena that had been demonstrated, such as situations in which the CS and UCS were both interoceptive, situations in which the CS was interoceptive but the UCS was not, and situations in which verbal behavior was involved. In all of these types of situations, Razran concluded that interoceptive conditioning had definitely been demonstrated. But these are certainly not the only demonstrations or the only data. Further information will now be reviewed in three general categories—interoceptive conditioned reflexes themselves, the relationship of the reflexes to the central nervous system, and the relationship of these reflexes to behavior.

CONDITIONED REFLEXES AND VISCERAL FUNCTION

As described above, the first study that demonstrated interoceptive conditioning was a study in which both the UCS, kidney function and diuresis, and the CS, infusion of fluid into the gastrointestinal tract, were interoceptors. Initially, the water infusion induced a hyposmotic state, which prompted an unconditioned diuretic reaction. Eventually (after approximately 20–25 conditioning trials), introduction of water that was then rapidly removed via a fistula tube to avoid absorption produced a diuresis as well. These two visceral systems—gastrointestinal and renal—have been involved in a substantial amount of interoception research. It was demonstrated that interoceptive conditioning in these systems followed the laws of Pavlovian conditioning that had previously been shown to exist for exteroceptive conditioning. These included forms of inhibition such as extinction and differentiation (discrimination), for example. Much of this work has been summarized by Bykov (1957) and Adam (1967).

The kidney appears to contain baroreceptors and chemoreceptors. Further, it has been known since the middle of the nineteenth century that damage to the floor of the fourth ventricle in the brain can affect renal function, and many subsequent findings have documented various ways in which renal function is under central nervous system control. In conditioning experiments it was possible to demonstrate that salivation produced by food as a UCS could be conditioned to increased pressure in the renal pelvis as the CS, and that the subjects could learn to differentiate pressure in the renal pelvis from pressure in the ureter.

Over the course of several experiments involving kidney function, the following findings were reported. Extinction of the response occurred, similar to that in exteroceptive conditioning. It was shown to be possible to condition decreases as well as increases in urine production, and second-order conditioning was demonstrated. Different portions of the ureters seemed to vary in ability to demonstrate differentiation. Generalization and differentiation appeared to occur under appropriate training conditions between the two kidneys in the same subject. Functional or anatomic removal of one kidney after conditioning training led to a greater response in the other kidney. A hormonal substance that inhibited urine production could function as an effective UCS. Individual differences in animals' tendencies to become conditioned as well as differences in their general temperament (as expressed, for example, by differences in the extent to which novel stimuli would affect urine output) were observed.

The function of visceral organs other than the kidney can also be conditioned. For example, bile secretion can be brought under control of a CS. Again, differentiation between stimuli was demonstrated, as was conditioned inhibition of bile production. It was even claimed that two UCSs that elicited different bile concentrations or compositions would lead to similar differences in bile that was secreted to associated CSs.

Although data related to interoceptive conditioning and the central nervous system will be reviewed in more depth below, a few comments specific to the brain and conditioning of renal function will be provided here. These data were collected from electrodes chronically implanted to monitor cortical and reticular formation EEG activity. Initial distention of the renal pelvis was detectable by the dog, as indicated by the fact that EEG and behavioral arousal reactions were observed. After conditioning, presentation of the CS produced EEG evidence of arousal, which also could be extinguished and then reconditioned.

This is an example of interoceptive-interoceptive conditioning. Conditioning of another response in the alimentary tract, bile secretion, also was demonstrated. Central nervous system changes associated with this conditioning were described. A third example, which also included central nervous system monitoring, was conditioning involving the alimentary tract and the cardiovascular system. Changes in function of the vascular system, and especially the heart, are of particular interest because of their linkage to emotion.

Bykov (1957) took interest in the Pavlovian conditioning of the action of the heart using drug effects on the heart as UCSs. Large doses of morphine produced changes in the electrical activity in the heart, which was conditionable to sham injections. Changes produced by nitroglycerin were conditionable, as well, to an auditory stimulus. Consistent with Razran's observations, it took many trials to effect these conditioned reactions, but once present they extinguished very slowly. Other drugs that were reported to produce conditioning were strophanthin, acetylcholine, atropine, and epinephrine. Not only heart rate changes were reported to be conditionable, but even the pattern on the electrocardiogram. Unlike the slow conditioning that occurred with drugs, increased heart rate and work of the heart produced by exercise was conditionable in only a few trials.

In addition to conditioned effects on the heart, Bykov studied conditionability of blood vessels using thermal stimuli to control vasodilation and vasoconstriction. Again, conditioning was demonstrated, and it was possible to condition both vasoconstriction and vasodilation to different stimuli in the same animal at the same time. Interestingly, exhausting physical exercise temporarily damped the conditioned reactions. Lastly, the contractive effect of epinephrine on the spleen was shown to be conditionable.

Adam (1967, p. 42) pointed out that early demonstrations of interoceptive Pavlovian conditioning tended to focus on receptors in the walls of "hollow viscera," but that other visceral structures such as "vascular or tissue receptors" were potentially important as well. The carotid sinus was one such potentially important structure.

In an experiment involving the carotid sinus, mechanoreceptors in the sinus were stimulated by dilation of a balloon inserted in the sinus. The dilation was the CS. An exteroceptive auditory CS was also used. The UCS was food, and the response was an operant response, food retrieval from a dish (i.e., a motor reflex response related to the alimentary system). Cortical EEG recordings were done during the study. Consistent with the principles articulated by Razran above, both types of stimuli (auditory and carotid interoceptive) produced conditioned effects, and the interoceptive response took longer (i.e., more conditioning trials) to control behavior.

For the EEG-dependent variable, the response studied was blocking of resting cortical electrical activity (i.e., EEG desynchronization, indicative of arousal, attention, the orienting response). The EEG was recorded from across the cortex: fronto-temporal, temporo-parietal, and parieto-occipital leads. The effects described below were seen in all three leads.

First, habituation was required to the EEG response to carotid stimulation. In other words, initial dilations of the carotid sinus not only sent afferent impulses that reached the central nervous system, but they reached the cortex (from where the EEG signal is obtained). In other words, the dogs could sense the carotid dilation without conditioning, and this unconditioned sensory information affected brain function up to the level of the cortex. Habituation occurred to this EEG desynchronization, again showing a response consistent with the behavior adjustments seen in response to exteroceptive stimuli. Conditioned EEG desynchronization was demonstrated, and this conditioned response demonstrated extinction and disinhibition, as would be expected with a Pavlovian conditioned response.

Based on the results of the conditioning studies involving both the renal pelvis and the carotid sinus, Adam offered five conclusions (p. 49, paraphrased): *(1)* Impulses from these two visceral regions reach "higher nervous centers and influence higher nervous function." *(2)* Properties of the two conditioned responses were the same, indicating that responses to stimulation of hollow viscera versus other types of visceral receptors followed the same laws. *(3)* The characteristics of Pavlovian interoceptive conditioning were similar to exteroceptive. *(4)* Interoceptive stimulation produces activating (arousal) effects in the central nervous system, and these effects can be conditioned. *(5)* Operant behaviors can be influenced by Pavlovian interoceptive conditioning.

One issue not mentioned by Adam but that will arise later in the discussion of biofeedback and the potential to condition visceral responses with operant techniques is the question, What is being sensed? For example, did the procedure used to stimulate the carotid sinus produce any vibrations that the subject could sense by afferent nerves not part of the viscera and respond to? Not all aspects of interoceptive conditioning would be negated if this were true, but it is a possible confounding of substantial relevance and importance.

Conditioning in the digestive tract has already been discussed, but additional data are available. Pavlov demonstrated that after diverting the esophagus in experimental animals so that food that was eaten did not reach the stomach, or in humans with gastric fistulas or after sham eating, gastric juice was nonetheless secreted by the empty stomach. Indeed, salivation itself (from the initial portion of the alimentary tract), the main unconditioned response that Pavlov initially studied, might be considered a visceral response. Proprioceptive afferent impulses (passive leg flexion) could serve as CSs for conditioned salivation. Infusion of water into the stomach at two different temperatures could be differentiated. Inhibition of salivation could be conditioned. Merely the sight of food by the subject produced increased saliva, as the reader has undoubtedly experienced (and with smells as well). Conditioned pancreatic secretion was reported as well as gastric and biliary. Presentation of conditioned stimuli in association with monitoring of gastrointestinal motion at different points in the alimentary system demonstrated differential effects, for example, more activity in the small intestine than in the large in one study.

It has been shown that humans increase respiration in anticipation of physical exertion, and that increases in respiration produced by inhalation of carbon dioxide can be conditioned to an auditory stimulus in both humans and animals. Abnormal respiratory patterns induced by hypoxia in dogs could be conditioned as well. Unlike many of the other visceral systems, visceral conditioning of the respiratory system is complicated by the fact that breathing has both an involuntary and a voluntary component.

Many of the conditioned responses described in this chapter would now be considered as operant rather than Pavlovian, for example, a dog's leg withdrawal from a shock or retrieval of food from a dish. The nature of the response does not change the fact that a visceral stimulus was or was not sensed and did or did not come to control either a physiological change or a behavioral change (including verbal behavior in humans). One such example is metabolic conditioning in humans. It was shown that subjects could be conditioned to respond with an anticipatory increase in respiration in response to an auditory signal that work (exercise) was going to be demanded. Given that respiration can be either voluntary or involuntary, it is unclear if this should more properly be considered an example of Pavlovian or operant conditioning. What is important to clarify in such a situation is whether the behavior (in this case, increased respiration) was controlled solely by the auditory stimulus or if a visceral stimulus (e.g., a conditioned rise in carbon dioxide and/or decrease in oxygen in the blood) also contributed to the increase in respiration. In this particular experiment the investigators apparently were not sure. The experiment is described here to highlight the issue that there are sometimes multiple possible stimuli controlling behavior, and the active ones—including or excluding visceral ones—must be empirically determined and not assumed.

An intriguing finding of this line of research in dogs was that repeated exercise led to evidence that the dogs had a modification in their metabolism in response to the exercise, but that this fitness development was site-specific. The dogs appeared more fit in the situations in which exercise occurred, as measured by respiratory gas exchange, than in situations in which exercise did not occur. A similar phenomenon was seen in humans. In other words, this appeared to be conditioned fitness, a combined condition of the body and the brain. In another experiment related to metabolic changes, it was claimed that dogs could show conditioned metabolic changes, as measured by gas exchange, to repeated injections of thyroid hormone, even though the metabolic effects of the hormone injection took two days to peak. This delay is reminiscent of the delay that is hours long in taste aversion conditioning (see Chapter 3), but this study calls for careful replication before uncritical acceptance.

Another example of conditioning of a metabolic process is thermoregulation, which in the case of dogs is highly related to breathing and salivation, which also serves to dissipate heat. Results were similar to conditioned respiration, and the mechanistic issues are the same. Both increases and decreases could be conditioned. Despite Pavlov's theory that conditioning occurred at the level of the cortex, knowledge of the control of homeostatic functions such as thermoregulation clearly implicate subcortical structures as very important and highly involved in these conditioned reactions. Unlike virtually all the other research reported in this chapter, which was done in dogs and humans, thermoregulatory conditioning was also demonstrated in pigeons and mice.

As a last example of visceral conditioning, in an attempt to begin to understand the mechanism of conditioned changes in physiological functions, investigators assessed whether changes in saliva composition (permeability to iodine) could be conditioned with food as the UCS. It was reported that this could be done, but the investigators were unable to take this line of research much further, although it was also reported that urine sugar content could be modified with conditioning.

In summary, a variety of physiological and behavioral responses were shown to be conditionable in which visceral or interoceptive stimuli were involved. These studies documented that visceral afferent impulses can reach those regions of the brain involved in these complex psychobiological processes and can participate in these processes.

INVOLVEMENT OF SPECIFIC BRAIN REGIONS

Even though Pavlov's theory that conditioning occurred specifically at the level of the cerebral cortex has turned out to be not completely correct, it is certainly true that the central nervous system (above the level of the spinal cord and spinal reflex arcs) is essential to the conditioning process. For example, more recent research addressing the role of the amygdala and pathways involving the amygdala in fear conditioning has already been described. A number of other studies in the Pavlovian tradition have endeavored to elucidate the central nervous system mechanisms underlying classical conditioning. Some of this research will be reviewed here. It has focused mainly on two anatomical regions, the midbrain reticular region and the cortex. Consistent with a great deal of other electrophysiological research, some of this research was done in the cat.

It has already been noted that stimulation to the renal pelvis or to the carotid sinus produced EEG desynchronization and orienting behavior (until habituated), indicative of arousal and demonstrative of the fact that sensory afferent impulses originating in these visceral structures reach levels of the central nervous system associated with these changes. Although the EEG is measured from the cortical surface, which is indicative of activity at that level of the brain, it has been assumed that the origin of diffuse cortical desynchronization in the EEG is the midbrain ARAS (ascending reticular activating system). In other words, desynchronization is evidence of effects occurring at both these levels, but primarily at the midbrain.

Of methodological note, in order to demonstrate conditioning, habituation of any unconditioned response to the stimulus to be used as the CS must be done first, but once it has been done, continued physiological and/or behavioral response to the CS is strong evidence of the occurrence and conditioning (i.e., effects occurring at and above the midbrain). Furthermore, habituation to one stimulus and lack of habituation to another is evidence that the two stimuli can be discriminated (see Fig. 5–1).

While carotid sinus stimulation usually produced EEG desynchronization, sometimes synchronization was observed, leading to the conclusion that both excitatory and inhibitory sensory impulses were produced. In another system, the alimentary tract, it was shown that dogs could discriminate separate points in the intestine where distention occurred as close together as 7 cm (Fig. 5–1).

There is additional evidence for the importance of the reticular formation in visceral sensory processes. Electrical stimulation of the reticular formation appears to facilitate the conduction of exteroceptive and proprioceptive information to the cortex, but to inhibit interoceptive (which might relate to the relative lack of interoceptive information reaching awareness). Further, the arousal produced by direct electrical stimulation of the reticular formation could be conditioned so that mechanical stimulation to either the ureter or the intestine also produced conditioned arousal. This visceral conditioning paradigm produced very rapid conditioning (unlike the multiple trials usually required).

Concerning the question of the central nervous system level at

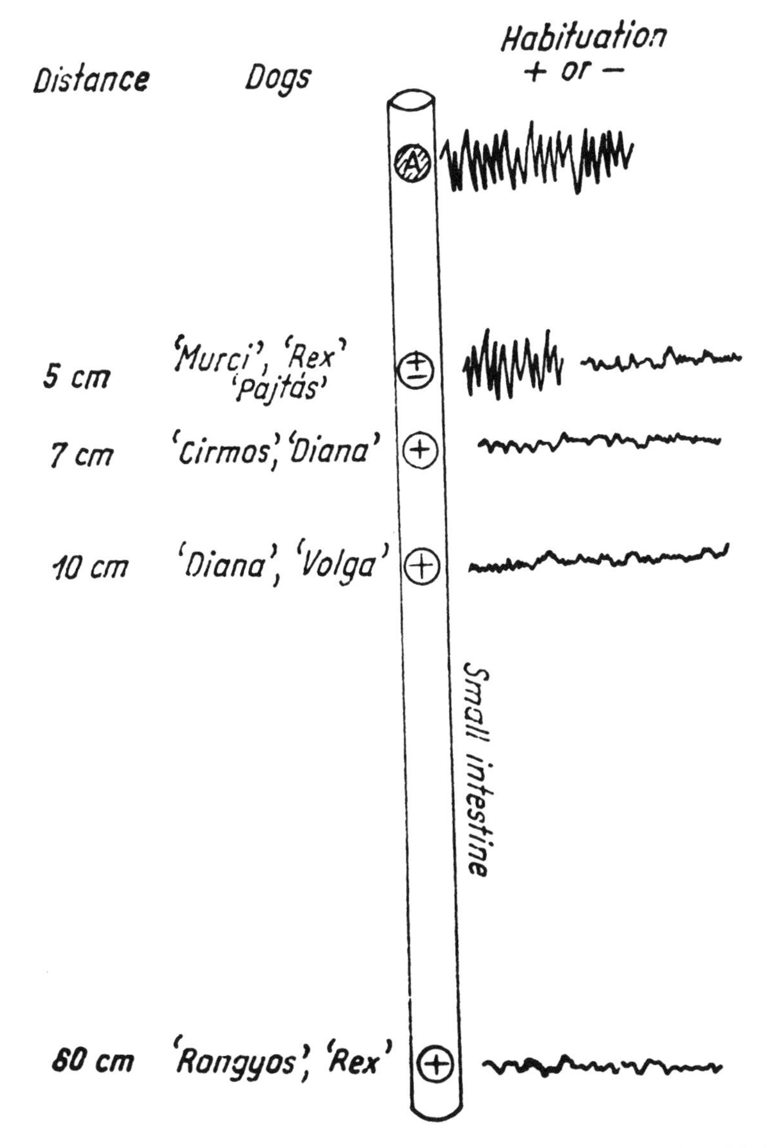

Figure 5–1. Central nervous system activation demonstrated by EEG desynchronization (low voltage EEG) after habituation (high voltage) due to balloon distention in the canine gastrointestinal tract. After habituation at point A, desynchronization due to distention at another point demonstrated dogs' ability to interoceptively discriminate distentions 7 cm, and possibly 5 cm, apart. [From Adam, 1967, reprinted with permission of the holder of the copyright.]

which conditioning occurs, Adam (1967, p. 79) reported that Zeleny, a student of Pavlov's working in Pavlov's laboratory, demonstrated conditioning in a decorticate dog as early as 1911. Subsequent reports of conditioning in decorticate dogs and cats were reported in the 1930s. Some even claimed that conditioning was demonstrated in spinal animals, but the question of whether or not this qualified as learning (versus just sensitization) was raised even at that time. Adam reported success in conditioning a respiratory response using nerve stimulation as the CS and UCS in mesencephalic cats (brain transection at the level of the superior colliculi), although it required 100 conditioning trials to obtain a conditioned response and more than 200 to obtain a stable response. Extinction of the response supported the fact that it was a conditioned response. In animals that had the conditioning done first, then had the transection operation, evidence of conditioning was present immediately after recovery, indicating that even in the non-transected animals retention of the conditioned learning occurred at least in part subcortically, below the level of the transection. Additionally, it was reported that stimulation of the reticular formation below the level of the transection facilitated learning. These data do not argue that the cortex is not involved in Pavlovian conditioning in the intact animal, but that the cortex is not necessary for Pavlovian conditioning to occur.

Even though the cortex, or a type of brain that normally has a cortex, is not necessary for Pavlovian conditioning, the cortex is of great interest in Pavlovian conditioning. Visceral afferent information does reach the cortex and associated structures under normal conditions.

Using evoked potentials, a crude cortical localization of the activation sites for the conditioned and the unconditioned stimuli could be studied. It appeared that the topographical pattern of evoked responses changed as conditioning occurred. A similar phenomenon occurred with subcortical recordings in the thalamus. It was even reported that changes in the firing pattern of single cortical neurons could be conditioned.

Surgical removal of the right sensorimotor and prefrontal cortex of the dog after establishment of interoceptive conditioning produced either a partial decrement in conditioning or did not substantially disrupt the conditioned reflex. In contrast, unilateral removal of the middle and dorsal portions of the cingulate cortex produced a decrease in intensity of both interoceptive and exteroceptive conditioned salivatory reflexes and a loss of differentiation specifically in the interoceptive conditioned reflexes (Figure 5–2).

The surgery described above was unilateral. Studies with pigeons in which both hemispheres were removed led to loss of the ability to demonstrate conditioning. Adam (1998) addressed the question of hemispheric laterality of visceral sensory processing, although not with specific reference to Pavlovian conditioning. Heart beat detection was the task. The data indicated that the right (usually non-dominant in humans) hemisphere was more involved than the left in this task, consistent with prior results, and that this effect was influenced by gender.

Another factor of potential importance in visceral sensory processes is the normal rhythmic fluctuations observed in most physiological processes. Given that a major function of visceral afferent signals is to allow the organism to maintain a homeostatic milieu intérieur, and given that the normal state of the milieu intérieur is not static (i.e., homeostasis is a state that fluctuates with rhythms of multiple amplitudes and cycle lengths), the central nervous system must have not simply a series of benchmarks to define normality for all the ongoing physiological processes but rather benchmarks that change across the day, the month, and so on. Thus, both the central nervous system control mechanisms and the visceral sensory afferent information must be of very high fidelity. So far, it does not appear that the interaction of visceral sensation with circadian rhythms (or endogenous rhythms of other cycle lengths) has been systematically investigated.

One more point should be made about the role of the central nervous system in visceral sensation and Pavlovian conditioning. The efferent system is often thought of as being a motor system, that is, a system that has a physiological or behavioral outcome as its result. Another potential efferent effect is modulating feedback onto the afferent system. With reference to Pavlovian responses, Bykov (1957, p. 131) called this "the correcting influence of the cerebral cortex" (believing that conditioning occurred in the cortex) and provided some evidence that feedback can prime a damped conditioned reflex or damp a primed one. As would be expected in such a system, it functions as a closed loop. Afferent impulses feed forward to the brain and the efferent system, and the efferent system feeds back onto the afferent. The idea that, somehow, the autonomic nervous

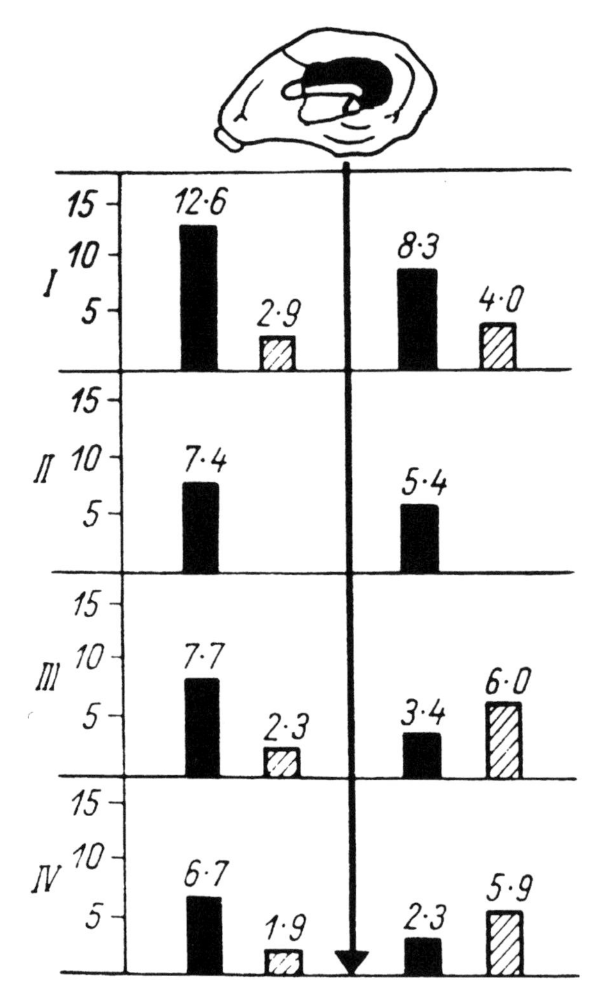

Figure 5–2. Effects of ablation of the middle and dorsal portions of the canine right cingulate gyrus on salivation conditioned to I: auditory CS (conditioned stimulus); II: visual CS; III: renal pelvis mechanical stimulus CS; or IV: small intestine mechanical stimulus CS. Black bars are reinforced stimuli, hatched bars are unreinforced. Left is before ablation, right is after. Larger disruption of conditioned effects were seen with interoceptive conditioned stimuli (III, IV) than exteroceptive (I, II). [From Adam, 1967, reprinted with permission of the holder of the copyright.]

system could have functioned properly as a stand-alone efferent system without information coming back to the central nervous system about what is happening in the periphery should have been recognized as highly improbable from the start. Langley apparently understood this 100 years ago, but it was then seemingly forgotten until fairly recently.

BEHAVIOR

So far in this chapter, the physiology of interoceptors, their role in Pavlovian conditioning, and some data about brain mechanisms related to them have been discussed. Now, more discussion of behavior in this context will be provided.

One of the main problems in understanding the relationships among visceral afferent impulses, effects on the higher centers of the central nervous system, and behavior is the fact that these impulses often do not produce any obvious subjective experiences. The data reviewed above provide strong evidence for an association between one kind of behavior, Pavlovian conditioning (and also effects on the midbrain and cerebral cortex as demonstrated by EEG and evoked potential data) and visceral sensory information. Despite this, comparison to the exteroceptive senses seems to make visceral sensation suspect because the exteroceptive senses usually seem to produce subjective experiences, and visceral sensations do not. Quantification of the stimulus is often difficult and can be contaminated with effects on somatic, non-visceral receptors, but usually it can be done. Quantification of the experience, however, creates very difficult problems and appears to be almost meaningless if no experience is produced.

Adam reviewed four methods of addressing the question of measurement of the subjective experience. One is straightforward introspection, that is, directing one's attention to the viscera or to a specific organ, such as the heart, and reporting what is felt. A second method is to bypass the experiential aspect and use such physiological measures as electrical changes in the brain. The other two methods also involve introspection but involve additional aspects. One of them is signal detection, a method devised to separate sensory detection from criterion (i.e., motivational) variables. The other is a threshold titration method. Other methods are available, and a further discussion of methodology with special reference to detection of heart action will be provided in Chapter 8. Also, a discussion of the issue of psychological factors affecting perception and reporting of internal sensation and symptoms will be provided in Chapter 8. Suffice it to say here that reports of subjective sensations or symptoms are influenced by beliefs, expectations, attentional processes, etc., and not just by the stimulus.

As reviewed above, interoceptors are not morphologically different from other sensory receptors and generally seem to follow the same laws of Pavlovian conditioning as exteroceptors, but there are differences. As already noted, interoceptive conditioning generally takes longer to establish, but once established there is more resistance to extinction and a greater tendency toward disinhibition. Evidence also exists for interactions between interoceptors and exteroceptors. The presence of an exteroceptive conditioned motor response seems to facilitate the development of interoceptive motor conditioning, while the existence of the conditioned interoceptive motor reflex sometimes tends to inhibit the exteroceptive response. If both stimuli were presented together and the conditioned responses to both stimuli were the same, either summation or inhibition of the conditioned response could occur.

The question of the relationship between visceral sensation and one kind of learning, Pavlovian conditioning, has already been addressed. The question of the role of visceral conditioning in operant conditioning, other forms of learning, and cognition needs consideration as well. The connection of visceral sensation to operant conditioning will be addressed in the chapter on visceral operant conditioning and biofeedback, below, and the issue of these kinds of relationships generally will be at least in the background in all the remaining chapters. Only a few points related to this question will be made here.

Cognition often connotes the idea of conscious processing of mental data, but it is clear that much (perhaps most) of what is called cognition goes on unconsciously. For example, one is able to remember some fact but is unable to describe how it was recalled from memory. Visceral perception would appear to be a prime example of unconscious mental processing. As is clear from the involvement of interoception in Pavlovian conditioning, visceral afferent impulses can participate in learning and affect central nervous system processes at levels usually implicated in cognitive function.

Adam (1998) pointed out how visceral sensation, including visceral pain, has a vague protopathic, as opposed to the more precise epicritic characteristic, of exteroceptive sensations. The observation that the non-verbal, non-dominant hemisphere might be dominant in the processing of visceral perception is consistent with the vague, or non-conscious, characteristics of this type of sensation. This representation of visceral perception in the non-dominant hemisphere, if true, might also relate to the difficulty of putting these feelings into words.

It has been suggested, and it would not be surprising, if learning affected visceral perception. For example, children presumably must learn to be aware of rectal and urinary pressure at a young age in order to develop bowel and bladder control. These abilities appear to have a critical period for learning involving maturational processes in the nervous system. It is not clear to what extent learning versus maturation is necessary. Non-human animals, at least mammals, develop control as well, and it is not clear how great a role learning plays in this.

One intriguing example of visceral learning in humans was described by Adam (1998, p. 96). Using either stimulation of mechanoreceptors in the gastrointestinal tract or electrical stimulation of the uterus, it was found that EEG desynchronization occurred at stimulation levels that initially were not subjectively detectable. However, with verbal reinforcement, the threshold for subjective awareness became more acute and approached the level at which desynchronization occurred (Fig. 5–3). It took 50–70 trials for the final subjective threshold to be reached. A second experiment with colostomy patients supported this finding. Thus, it appears that non-conscious visceral afferent impulses can be made conscious with verbal reinforcement or feedback. Visceral sensation was described as at the borderline of consciousness, but learning can improve awareness. The role of stimulation of non-visceral, somatic receptors in this process is unclear.

Many of the experiments described above involved overt, sometimes voluntary, behaviors as the conditioned responses. Some studies have attempted to directly address ways in which interoceptive conditioning affects motor behaviors. Two specific examples will be mentioned. Before describing them it is important to note that, separate from the involvement of muscles in motor responses, the muscles have sensory receptors as well, usually considered part of the

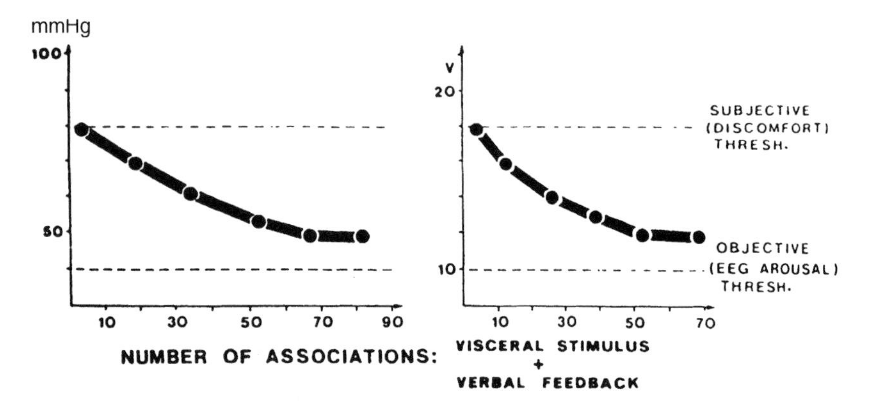

Figure 5–3. Demonstrations of human verbal learning effects in visceral sensation, using pneumatic duodenal (left) or electrical uterine (right) stimuli. Subjective thresholds are greater than EEG arousal thresholds at the start of experiments but approach the EEG thresholds over 50–70 trials with verbal feedback. [From Adam, 1998, reprinted with permission of the holder of the copyright.]

proprioceptive sensory system. Proprioception will be discussed below in Chapter 10. Additionally, Bykov (1957) observed that efferent effects (and afferent as well) could be due to humoral instead of, or in combination with, neural impulses.

As the first example, the effects on an exteroceptive conditioned response of removing one form of visceral sensory afferent input to the central nervous system was assessed. Deafferentation of the carotid sinus facilitated the tendency of rats to perform a motor response to an auditory or visual signal.

Second, the ability of gastrointestinal mechanoreceptor stimulation to function as a discriminative stimulus for an operant response, bar pressing for food, was tested in monkeys. It was shown that after habituation as measured with EEG, subjects could learn to respond when the interoceptive stimulus was on and food was available, and refrain from responding when the stimulus was off and responding was not reinforced. Subsequently, however, it became apparent that there were potential problems with this result (see Adam, 1998, p. 95). First, it was difficult to replicate the study in rats. Second, replication became possible only when intensity of stimulation was increased. Third, while it was claimed that the stimuli needed to be aversive rather than pain producing to show behavioral control in the rat study, it is difficult to see how the investigators could have known that pain was not occurring. Finally, it was also difficult to determine if other non-visceral, somatic stimuli were or were not present and influencing behavior. It is clear that motor behavior can be involved in visceral conditioning and learning, but it is also clear that the details are far from being well understood.

CONCLUSION

Of course, investigations of the phenomena and mechanisms of the physiology of interoception, especially Pavlovian conditioning and interoception, did not stop at the end of the 1950s. Before the 1960s, this conditioning research was done in relative isolation in Russia and Eastern Europe, almost all of it was published only in those languages, and little was known of it in the West. After that time, it became more integrated into general psychobiology.

An essential concept held by Pavlov and his students concerning the mechanism by which conditioned reflexes were formed was the idea of temporary connections. It was believed that a stimulus became a CS by virtue of a reversible linkage formed with the UCS in the cortex of the brain, based on repetitive associations. It was also believed that these connections were not permanent. Extinction, that is, subsequent lack of association after a conditioned reflex was formed, would lead to a breakage of the connection, an apparent loss of conditioning (although there are reasons, such as the existence of disinhibition, to believe that these linkages might not be lost completely). For the purposes of this discussion, the important issue in this theory was the belief that "any organ can form temporary connections under [appropriate] experimental conditions" (Bykov, 1957, p. 232), including visceral organs. In fact, it might or might not be true that any organ can participate in conditioning, but it is almost certainly correct that sensory afferent impulses from many organs, including many visceral organs, can functions as CSs—as signals that come to influence and control behavior.

The question of how conditioning or learning might be temporary (if it ever is), that is, if and how neuronal connections might be broken once formed, has apparently not been studied or even thoughtfully considered. The question of how visceral sensory processes might be dysfunctional, on either an innate or a conditioned basis, and how that might contribute to medical (and, especially, psychosomatic) disorders is just beginning to be assessed.

Last, as already stated, the relationship of visceral sensation and Pavlovian conditioning to awareness is very important. It is clear that the occurrence of Pavlovian (or operant) conditioning carries no assumption that the conditioning process requires any awareness on the part of the organisms in which learning occurs. In other words, visceral sensory processes may participate in important ways in higher mental processes regardless of the organism's ability to be aware, or learn to be aware, of these sensations. As both Razran (1961) and Adam (1967) discussed, the apparent (at least relative) silence of visceral sensations in one's consciousness does not imply silence in affecting thought processes and behavior. The conclusion that visceral sensory receptors participate not only in physiological reflexes involving the central nervous system, but also in higher nervous functions, including conditioning and behavioral control, is strongly supported.

six

OPERANT CONDITIONING OF VISCERAL FUNCTION (BIOFEEDBACK)

> I'd rather leave all my automatic functions with as much autonomy as they please, and hope for the best. Imagine having to worry about running leukocytes, keeping track, herding them here and there, listening for signals. After the first flush of pride in ownership, it would be exhausting and debilitating, and there would be no time for anything else.
>
> Lewis Thomas, Autonomy. *N Engl J Med* 287; 90–92, 1972

The initial positing of two-process learning theory did not rely on any substantial body of evidence. It seems to have been more an inference garnered from observations that operant methods could and do affect skeletal responses and Pavlovian conditioning does affect visceral–autonomic responses, combined with the subjective experience in humans that the skeletal responses seem to be voluntary while the visceral seem not to be. Logically, however, for the two hypothesized learning processes to be different, it should also be true that *(1)* Pavlovian conditioning could not directly affect skeletal responses and *(2)* operant conditioning could not directly affect visceral responses. It was already clear from Pavlov's studies in which motor responses such as leg withdrawal were used as the unconditioned response that *(1)* above was probably false. What was not known as of the 1960s was, Is *(2)* true or false? Several investigators, probably most influentially Neal Miller, set out to resolve this question. The results of that attempt and how this issue relates to interoception are the topics of this chapter. Commenting on the belief that individuals could learn to control their automatic functions with biofeedback techniques, Thomas observed that even if he could learn this, he would not want to (see quote above).

Aversive conditioning, including learned fear, has been a dominant theme in this area of research for two main reasons. First, although innate fears certainly exist, learning plays a very important role in many kinds of fear. Second, autonomic–visceral functions are often strongly involved with fear and aversive situations in general. Of course, visceral changes are probably importantly involved in appetitive situations as well (i.e., hunger and thirst), but visceral changes such as cardiovascular responses have more strongly gained the attention of investigators of aversive processes, perhaps because of the importance attached to sympathetic activation in fight-or-flight responses by Cannon and subsequent investigators. It is not necessarily the case, however, that experiments designed to test the ability to condition visceral responses with operant techniques must use aversive experimental paradigms. Many did not, as will be described below.

The parenthetical title of this chapter, Biofeedback, comes from the design of the original studies devised to determine if operant techniques could be used to successfully modify visceral responses. In operant conditioning of skeletal responses, an essential part of the conditioning paradigm involves, indeed requires, that the response to be conditioned be rewarded. In other words, the organism obtains feedback of the consequences of its behavior by virtue of the fact that the occurrence of reward (e.g., water to a thirsty rat) feeds back these consequences to the organism.

The term biofeedback has taken on a much broader, more clinical meaning than it originally had. The term is now used in many quarters as almost a synonym for a wide variety of behavioral techniques used in the treatment of many clinical disorders. In this chapter the focus will be on studies that used valid feedback methods. Clinical uses will be briefly described at the end of the chapter but are usually only of indirect relevance to the issue of interoception.

Preparation of this chapter presented a dilemma. Biofeedback research, much of it addressing the issue of two-process theory, was very active for approximately a decade starting in the mid-1960s. After that, interest waned substantially, except in the arena of potential clinical use. The main reason for this was the inability to replicate many of the more important early findings. For example, Miller and Dworkin (1974) published a review in which they presented data indicating that over a five-year period from 1966 to 1970, the magnitude of heart rate change that could be conditioned with operant methods in curarized rats fell from more than 20% to less than 5%. In a subsequent paper, published a decade later, the same investigators (Dworkin and Miller, 1986) report on a very large-scale attempt to replicate these original findings. They state bluntly, but with a caveat, in their abstract that "After more than 2,500 rats were studies, it is concluded that the original visceral learning experiments are not replicable and that the existence of visceral learning remains unproven; however, neither the original experiments nor the replication attempt included the necessary controls to support a general negative conclusion about visceral learning."

Various possible reasons for this inability to replicate may exist. The methods are technically very complicated, especially the use of the muscle relaxant curare with the attendant required procedures to maintain the animal while controlling multiple physiological factors. In any case, much of this research has been left in doubt regarding reliability and validity. It would be an overstatement to claim that the initial findings were incorrect, but in many cases they have not been and could not be replicated.

Given the lack of replicability of these studies, the dilemma is, What should be reported in this chapter? The issues of biofeedback and two-process theory are of both important historical and theoretical interest to interoception. Two-process theory is still an area of interest within the larger realm of learning theory, and interest in fear conditioning, including its autonomic–visceral aspects, is now very active. Therefore, relevant studies on this topic will be described, with comments as appropriate. The reader should bear in mind the tentative, doubtful, nature of some of the results. However, the reader should also consider the theoretical and actual importance of visceral sensory processes as underscored by this and the prior chapter. It must also be remembered that the inability to replicate was related largely to curarized subjects specifically, subjects treated so as to remove the possibility of mediating somatic behaviors, because these potential mediating behaviors would confound the attempt to determine if visceral responses per se could be operantly conditioned. These failures to replicate do not mean that biofeedback does not work. They mean that the possibility of mediating somatic responses cannot be excluded as the possible underlying mechanism.

The reader should also remember that lack of replication of these visceral conditioning studies is in no way a negation of the existence or importance of interoceptive processes themselves. Indeed, most of the research in biofeedback has not addressed the role of interoception in visceral conditioning and biofeedback more generally. In one sense this might be expected. The logic behind the use of biofeedback methods is to externalize visceral functioning so that it can be fed back to the organism through the external senses because internal sensation was thought to be unreliable (or even nonexistent). On the other hand, this whole logic rests on an assumption: that visceral sensation is unreliable or nonexistent. Investigators in this field eventually did address this question, and that research will be reviewed in subsequent chapters. However, few of the early studies acknowledged this issue. And even later, the question of how visceral sensation might interact with visceral conditioning did not seem to be adequately understood or investigated. It was not a completely overlooked problem, however. For example, Gannon (1977) discussed the role of interoception in learned visceral control. That review pointed out that two competing theories existed about the role external feedback plays in visceral conditioning. The first is that it does what is mentioned above, that is, externalize internal function. The other hypothesizes that feedback in not an externalization, but rather an enhancement of internal cues, that is, of interoception. In other words, one of these theories implies that interoception is an essential part of the physiology of visceral conditioning.

In order to orient the reader to this topic, a review of early studies will be provided. Many of these are precisely those studies that turned out to be unreplicable, which the reader must bear in mind. They will be described, nonetheless, because they constitute the history of this field. Studies in which visceral sensory processes were at least indirectly considered will also be discussed. Only a small number of such studies exist, and not all of them will be discussed, but those chosen for presentation will provide a sampling. After the studies themselves are discussed, a more general discussion of this issue will be provided.

HISTORY

The term vegetative nervous system dates back 200 years. Several reviews contain early histories of the sources of the idea that visceral functions cannot be conditioned with operant methods (e.g., Miller, 1969; Kimmel, 1974). Although Skinner is sometimes credited with this idea, it dates back at least as far as S. Miller and Konorski in 1928. Although Neal Miller was the most influential in questioning this idea, there was a report of conditioned vasodilation and constriction of a finger from a Russian investigator as early as 1960, described in Razran (1961) and Kimmel (1974). Studies by Mandler, Harwood, Shearn, and Kimmel (see Kimmel, 1974) were early attempts to produce conditioning of visceral responses—galvanic skin response (GSR) and heart rate. They produced a mixture of positive and negative results. Subsequent studies, done in the early 1960s, including GSR, heart rate, and vasoconstriction, produced generally positive and encouraging results.

Miller published his first studies of conditioning of visceral responses with operant methods in the mid-1960s (see Miller, 1969). He referred to his first clear-cut results as occurring in a study involving salivation and dogs. Subjects were reinforced for either increasing or decreasing salivation. Muscle paralysis with curare was not used. Although learning was slow (as is true with visceral conditioning generally), conditioning was unequivocally demonstrated, and the fact that both increases and decreases could be conditioned eliminated the possibility that this was simply Pavlovian conditioned response (Fig. 6–1).

This study was published in 1968, the same year this investigative group published studies of instrumental conditioning of heart rate in which curare was used. These animals were reinforced with direct brain stimulation. It was reported that both increases and decreases in heart rate could be conditioned, the response could be brought under discriminative stimulus control, and the learning was recalled for three months without additional training. In addition to heart rate changes to obtain reward by brain stimulation, it was reported that curarized rats could learn to change heart rate to escape or avoid painful shocks. Finally, it was observed that heart rate conditioning transferred from the curarized to the non-curarized state.

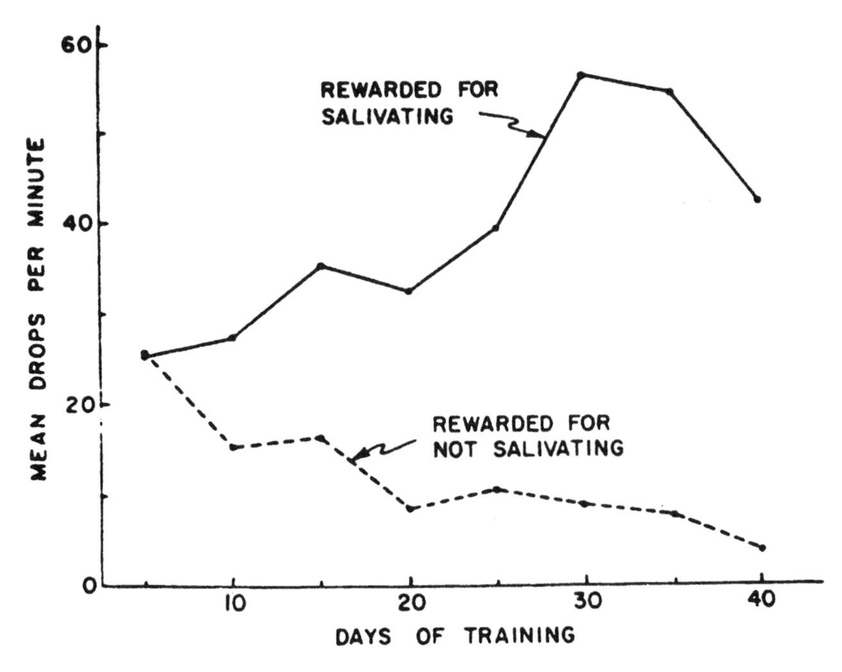

Figure 6–1. Results of an early biofeedback study without use of curare that showed that operant conditioning of salivation, although slow, could clearly be demonstrated. [From Miller and Carmona, 1967, reprinted with permission of the holder of the copyright.]

Miller and his colleagues investigated conditioning of other types of responses. Rats could learn to increase or decrease intestinal contractions. Furthermore, rats that were reinforced for intestinal contractions but not heart rate changes showed the appropriate increases or decreases in contractions but no change in heart rate. Different rats reinforced for the opposite responses did the opposite—increased or decreased heart rate but did not change contractions. Thus, the conditioned responses were specific to the system that was reinforced. In addition to the types of responses so far described, it was reported that kidney function, blood flow in the stomach, and peripheral vasomotor responses could be conditioned. It was claimed that blood pressure changes could be conditioned independent of heart rate. Lastly, using brain stimulation as a reward in cats, it was reported that EEG brain wave changes could be conditioned.

Because the studies using curare have not be able to be replicated, what is their importance? First, not all the studies that showed successful conditioning used curare (e.g., see Fig. 6–1). Second, as already stated, the lack of replicability does not disprove the possibility that visceral responses can be brought under control. What it appears to demonstrate is that mediating responses are necessary, although perhaps not sufficient in all cases (visceral plus somatic changes might be needed), to demonstrate these changes in visceral functioning. And, as already stated, neither these studies nor the failed replications fully clarify or even adequately address the issue of what role visceral sensory processes might play in this type of learning. Some studies that do begin to address the question of this relationship will now be reviewed.

STUDIES INVOLVING VISCERAL SENSORY PROCESSES

Ray (1974) replicated two earlier results that showed that people with internal locus of control (focus on the self) were better able to increase their heart rates, with or without external feedback. Subjects with external locus of control were better at decreasing heart rate. The strategies they used differed. In Bergman and Johnson (1972) subjects who received instructions to raise heart rate and subjects who received instructions plus the sound of their own heart beats did better than subjects not given instructions related to the heart. The interoceptive aspects of the study seemed more important than the exteroceptive.

The Autonomic Perception Questionnaire (APQ) is a scale of self-reported awareness of putative autonomic symptoms or sensations. Bergman and Johnson (1971) observed that subjects were able to both increase and decrease heart rates without feedback and that these changes were not due to respiratory or GSR changes. Subjects with middle APQ scores (possibly the most accurate) were best able to both increase and decrease heart rates. In McFarland (1975) all subjects were able to decrease their heart rates, and half were able to increase them. Decreases did not correlate significantly with either heart beat perception or APQ scores but were affected by respiration. A significant correlation was found between heart rate increase and heart beat perception. The APQ did not correlate with the increases, nor did respiration. McFarland favored the theory that external feedback augments already-existing internal (i.e., interoceptive) sensory information.

Can subjects reliably recognize the functional status of the heart, and does feedback assist that discrimination? Epstein and Stein (1974) found that asking subjects to indicate when heart rate was up or down did not lead to correct responses. However, when feedback was given subjects were able to make the discrimination. Thus, subjects were able to perform an interoceptive task if external information about the correctness was provided. These results are consistent with the hypothesis that subjects already had visceral sensory information about heart rate, but the feedback provided a link of particular sensory input to actual heart rate.

Many neuroanatomical structures, from the peripheral autonomic nervous system to the cortex, are involved in visceral sensory function. Several studies have assessed the effects of lesioning or stimulating these structures. Thornton and Van Toller (1973b) assessed the effects of pharmacologic sympathectomy on animals' ability to learn to change heart rate. They were able to demonstrate operantly conditioned changes in the control group but not in the sympathectomized group. It was not clear, however, how much of this lack of effect was due to a lack of conditioning and how much was due to an inability of the rats to produce even unconditioned heart rate increases.

Many of the studies designed to demonstrate visceral conditioning while controlling for mediating somatic responses used intracranial stimulation as the reinforcer for the visceral operant response (e.g., Miller and DiCara, 1967; Trowill, 1967; Hothersall and Brener, 1969; DiCara and Stone, 1970). While these studies turned out to be very difficult to replicate in general, it is interesting to note that in the review article by Miller and Dworkin (1975), their data (see their Fig. 16.1, p. 83) indicated that studies using intracranial stimulation had more consistent positive effects than did studies using other reinforcers. Whatever factors contributed to both the initial positive results and the later negative outcomes, intracranial stimulation tended to have a more potent effect than did other reinforcers. Additionally, intracranial stimulation modified brainstem norepinephrine content (DiCara and Stone, 1970). Rats trained to increase heart rate showed an increase of more than 50% in comparison to rats that had brain stimulation but were not involved in conditioning, while rats trained to decrease heart rate showed about a 20% decrease. To the extent that interoceptive processes are important in visceral conditioning (with or without mediating responses), might there be a relationship among interoception, intracranial stimulation, and conditioning?

Intracranial stimulation can function as a reinforcer, indicating that it is most likely closely related to motivational functioning in general, possibly including emotional processes (e.g., Deutsch and Deutsch, 1966; Grossman, 1967; Milner, 1991). The anatomical regions that support intracranial self-stimulation (as well as those for which it serves as a punisher) are the medial forebrain bundle, associated regions in or near the hypothalamus, and, in general, subcortical regions that are part of the midbrain and limbic structures. These structures are strongly implicated in visceral afferent pathways.

Although admittedly quite speculative and seemingly untested, this hypothesis presents itself: In addition to motivational effects—and perhaps even related to them—might intracranial stimulation in these anatomical areas produce physiological effects that mimic visceral sensory impulses by stimulating some of the same pathways? Whether or not they might contribute to the reinforcing effects of intracranial stimulation, does this type of stimulation produce visceral sensation-like experiences?

CARDIOVASCULAR–RESPIRATORY FUNCTION

Heart rate is the visceral response most extensively studied in visceral operant conditioning studies. Many other types of responses have been evaluated, including other cardiovascular responses (heart rhythm and vascular function), respiration, GSR, salivation, urination, gastric motility, EEG, and EMG. The linkage of these studies with interoception is sometimes explicit but many times is only implicit. Some of these studies will be reviewed here.

Fields (1970) reported successful operant conditioning of different intervals in the ECG of the curarized rat. The results indicated that *(1)* changes in these intervals could be conditioned, and independently of each other, *(2)* tail shock was more effective than brain stimulation, and *(3)* timing of the temporal association of the specific ECG parameter and the reinforcers must be very brief for effective conditioning. Some animals showed gradual conditioned effects and others showed all-or-none effects. The role of possible interoceptive cues was not assessed.

Weiss and Engel (1971) demonstrated in eight patients with premature ventricular contractions (PVCs) that all eight could learn some heart rate control, and five reduced PVC frequency. Pharmacologic analysis demonstrated that PVC generation and control can occur by sympathetic and/or vagal influences in different people. Debriefing of subjects did not reveal any consistent imagery or behavior by which subjects could effect control. Interestingly, in one subject, better PVC control was associated with awareness of its occurrence, and in another the subject was aware of the decrease in PVCs with conditioning and had to learn to feel comfortable with the decrease. Both this study and a subsequent one by Bleecker and Engel (1973), which involved learned control of ventricular rhythm, demonstrated that conditioning of cardiac function cannot be reliably done with a significantly damaged heart or conduction system. The results of the study by Bleecker and Engel indicated that learned control of ventricular function was under cholinergic control.

Concerning interoceptive processes, two related points relevant to the above studies are important. First, autonomic visceral function, highly involved in interoception, is also highly involved with conditioning of cardiac action, suggesting that these functions are likely to interact. Second, an overall neuronal loop must be intact to allow conditioning to occur. As described by Weiss and Engel, this loop must contain a combination of afferent and efferent elements, including ". . . (a) peripheral receptors which are stimulated . . . (b) afferents which carry the information to the CNS; (c) CNS processes to enable the patient to recognize the PVC . . ." This is an apt description of interoception.

It is clear that there are interoceptors associated with the vascular system, and one of the most extensively studied and well understood interoceptor is a vascular interoceptor—i.e., the baroreceptor. Blood pressure changes have been successfully conditioned with operant techniques in both animals and humans. For example, Shapiro et al. (1969) demonstrated that systolic blood pressure could be increased or decreased in humans. A few studies involving methods potentially relevant to interoceptive processes have been reported. For example, Brady et al. (1974) reported that human subjects could learn to decrease blood pressure with relaxation techniques. While relaxation is typically considered to relate to skeletal muscle, the possibility exists that it involves use of interoceptive cues to produce relaxation as well. In another study, Shapiro et al. (1972) found that subjects trained to change diastolic blood pressure with external cues could maintain the blood pressure decrease or even voluntarily decrease it further without continued feedback, suggesting that it might be done with internal cues. These studies have not explicitly addressed the role of interoception in blood pressure control, but they do raise the possibility that it plays a significant role.

Biofeedback studies of respiratory changes have been few. This might be due in part to the fact that control of respiration is a complex combination of voluntary and involuntary functions, making any attempt to separate them very difficult. One study by Luparello et al. (1968) assessed the role of suggestion on airway resistance in people with asthma and other lung disorders. Subjects given an aerosolized substance that was inactive but was described as an irritant known to produce bronchospasm caused increases in airway resistance in approximately half the subjects studied. While interoceptive processes are not discussed by the authors, it seems likely that the sensation of the aerosolized substance in the lung played a role in this response.

GASTROINTESTINAL FUNCTION

Studies were done to determine if curarized rats could learn to control intestinal contractions. Miller and Banuazizi (1968) demonstrated that stimulation of the medial forebrain bundle would reinforce either increases or decreases in contractions of the large intestine. The conditioned responses extinguished when reinforcement was discontinued. Shock per se did not cause such contractions. It was demonstrated that this learned response of the intestine was independent of heart rate changes. In a subsequent study (Banuazizi, 1972) it was demonstrated that similar effects could be demonstrated with tail shock as the aversive reinforcer.

Other changes in the gastrointestinal tract in animals related to biofeedback have been reported. For example, Weiss (1971a, 1971b) found that rats that could avoid or escape aversive tail shock developed less gastric ulceration than rats that could not escape. He also found that warning signals and feedback had a substantial effect on the severity of ulceration. Rats that received a warning that shock was imminent had fewer ulcers than rats with no warning. Furthermore, animals that received a feedback signal with each successful avoidance or escape response developed very little ulceration.

Human studies in both normal people and people with various gastrointestinal dysfunctions have also assessed the role of biofeedback in gastrointestinal activity. Whitehead et al. (1975) demonstrated that with a combination of external feedback and reinforcement normal subjects could learn to increase gastric acid secretion. Welgan (1974) similarly showed that gastric acid secretion could be modified with feedback in peptic ulcer patients. Muscle sphincter control has also been demonstrated (Engel et al., 1974; Furman, 1973). Hubel (1974) reviewed data showing feedback effects throughout the gastrointestinal system, and Whitehead (1992) reviewed the broad use of biofeedback techniques in many dysfunctions of the gastrointestinal system, especially the lower tract.

The question of the likely relationship of interoception to visceral feedback and operant conditioning was addressed early (Stunkard and Koch, 1964; Griggs and Stunkard, 1964). Using a signal detection model, obese and non-obese men and women were studied. Gastric motility was measured, and people were asked if they experienced hunger, both when motility was present and when absent. Non-obese women reported hunger during motility and no hunger during quiescence. Non-obese men showed a similar pattern but with a weaker correlation. In contrast, both obese men and women showed a very different pattern, apparently substantially affected by bias in reporting more than by actual motility. Both obese and non-obese people could perceive contractions, although there was more accuracy with the non-obese, and feedback influenced results. The results of these studies highlight several issues related to interoception, such as *(1)* possible effects of bias on reporting visceral (as well as other) sensations, *(2)* the potential differences that might be found in different populations—animals versus humans, men versus women, those without versus those with some dysfunction; and *(3)* the role of feedback in affecting perception as well as function (i.e., interoception as well as visceral conditioning). Subsequent studies addressing gastrointestinal sensitivity and awareness will be discussed in Chapter 9.

Following Pavlov's extensive interest in the salivatory response in dogs, salivation was one of the first responses shown to be conditionable with operant methods (Miller and Carmona, 1967; see Fig. 6–1). The investigators were not able to identify any associated motor behaviors, but the animals were not curarized. Subsequently, several groups (Brown and Katz, 1967; Frezza and Holland, 1971; Wells et al., 1973) demonstrated changes in salivation with operant methods in humans. The results of the study by Wells et al. is especially interesting because change in salivation was maintained by the subjects after external auditory feedback was discontinued. This result suggests that subjects can learn to change salivation and learn to use other than external cues to do so.

OTHER FUNCTIONS

One of the earliest attempts to demonstrate voluntary control of visceral function involved control of urination. Not only did Miller and his group study this system early, but an even earlier demonstration was reported. Lapides et al. (1957) curarized humans and showed that despite muscle blockade, these individuals were able to voluntarily start and stop urinating. Miller and DiCara (1968) demonstrated that curarized rats could be conditioned operantly using brain stimulation to either increase or decrease urine production. Glomerular filtration rate and renal blood flow changed in the appropriate directions, while multiple other physiological variables that might have mediated the urine production did not change. In a related study (Miller et al., 1968) it was shown that rats could learn to choose either injections of ADH (anti-diuretic hormone) or an inactive substance based on body tonicity. The results of this study strongly suggested that the rats could learn to sense their chemical tonicity state, in other words, chemical interoception (be it thirst or some other sensation).

A number of studies have demonstrated operant conditioning of electrodermal activity (GSR), most of which are not directly applicable to interoception. Several relevant findings do bear brief mention. It was found that actors were not able to control GSR responses by means of acted, imagined emotions (Stern and Lewis, 1968). A curarized human subject was able to be operantly conditioned to produce the GSR response (Birk et al., 1966).

Ultimately, interoceptive processes usually lead to an output, a motor response. Biofeedback techniques have been applied to attempts to operantly condition muscle activity. Fine control of very small motor units has been demonstrated (Cox et al., 1975; Hefferline et al., 1959; Basmajian, 1963; Leibrecht et al., 1973; Smith et al., 1974), including conditioning while curarized (Black, 1967; Koslovskaya et al., 1973). Biofeedback training, relaxation training, and placebo medication treatment all produced shifts towards internality on the locus-of-control variable (Cox et al., 1975).

INVOLVEMENT OF THE BRAIN AND EEG BIOFEEDBACK

The brainstem is involved in interoceptive processes. Another involved cerebral structure is the cortex. As already described, it has been demonstrated that Pavlovian conditioning does not require the cortex. The question of potential involvement of the cortex in operant conditioning of heart rate has been investigated.

DiCara, Braun, and Pappas (1970) performed experiments in curarized rats using both Pavlovian and operant conditioning methods to condition changes in heart rate and intestinal contractions. Experiments were performed with intact rats and in rats that had had their cortical tissue removed. The results of the study of intestinal contractions will be described below.

In the rats that were cortically intact, small increases in heart rate and large decreases in heart rate were demonstrated in response to operant conditioning with shock avoidance as the reinforcer. In the decorticate rats, no operant conditioning effect was observed. Consistent with earlier research, and thus demonstrating that the decorticate rats could learn, both intact and decorticate rats showed conditioned heart rate decreases with Pavlovian procedures. Histologic data showed that at least 90% of the neocortex, along with some thalamic and anterior cingulate tissue, were lost in the decorticate rats.

Thornton and Van Toller (1973a) used a method of reversible decortication, spreading cortical depression, to determine if they could replicate the results of DiCara, Braun, and Pappas (1970). Rats in the experimental group demonstrated that spreading cortical depression blocked both conditioned increases and conditioned decreases in heart rate. Subjects in a sham-operation control experiment did show conditioned effects. These results support the finding that a functioning neocortex must be present to allow successful operant conditioning of visceral responses. Concerning interoception, under the admittedly unresolved question of the validity of these studies with curarized rats, a tentative hypothesis could thus be offered: Visceral afferent information involved in conditioning must reach the cerebral cortex to participate in those physiological processes necessary for instrumental conditioning, while for Pavlovian conditioning visceral afferent information only needs to reach the brainstem.

Above, the study by DiCara, Braun, and Pappas (1970) was described, in which rats had conditioned changes in heart rate and intestinal contractions. Some of the rats had had their cortical tissue removed. It was shown that those rats were not able to learn the operantly conditioned intestinal contractions. Using a different learning paradigm, taste aversion learning, it was subsequently shown by Lasiter (1983) that removal of the insular cortex alone could disrupt learning. This result suggests that a much more precise lesion than the whole cortex would disrupt visceral learning in the gustatory–gastrointestinal system.

While interoception is a phenomenon that occurs initially in the periphery, the central nervous system is, of course, intimately involved as well. Changes in electrical activity in the brain produced by biofeedback techniques have been investigated in a large number of studies. Many of these studies have focused on the ability of individuals to control so-called EEG background, such as the alpha rhythm, often associated with quiet wakefulness and closed eyes.

Most of these EEG studies have involved humans, but a few have studied animals. Black et al. (1970) trained dogs with drug-induced paralysis to emit either more or less electrical theta waves from the hippocampus to avoid electric shocks. Presentation of the discriminative stimulus used during training after the paralysis had worn off produced not only more theta in the group trained to produce theta but also more skeletal activity, suggesting a possible relationship between conditioning of hippocampal activity and implicit mediating motor behavior.

Engel (1974) assessed the changes in EEG and other physiological variables in monkeys during operant heart rate conditioning. High-voltage EEG changes showed a negative correlation with heart rate, although it was not robust (and blood pressure showed a positive correlation). For both EEG and blood pressure, correlations became larger as learning progressed.

Rosenfeld and Hetzler (1973) demonstrated that rats could recognize and respond with bar pressing to large-component evoked potentials. The results of these three studies demonstrate that *(1)* EEG activity can be recognized by animals, *(2)* brain wave activity can be operantly conditioned, and *(3)* brain EEG changes can be associated with operant conditioning of peripheral visceral responses such as heart rate. Adam et al. (1966) reported that even the activity of individual cortical neurons in cats could be conditioned.

Humans have also been conditioned with operant methods to change the evoked potential (Rosenfeld et al., 1969). The humans were able to do this with and without visual feedback. Additionally, humans are able to learn to recognize different stages of sleep, such as REM from non-REM (Antrobus and Antrobus, 1967).

A number of human studies involving EEG and biofeedback included at least comments or basic assessments of the relevance of subjective states or experiences, aspects relevant to interoception. Some of these touched on the issues of mediation and the role of arousal as well. For example, Peper (1972) reported that subjects could gain some control over EEG alpha localized to one or the other hemisphere, and Beatty and Kornfeld (1973) demonstrated that subjects could gain control of EEG alpha independent of possible mediation by cardiac or respiratory changes (which differs from the result reported by Engel, above, with monkeys and heart rate).

The majority of studies of EEG alpha and feedback have reported that alpha occurs more frequently in a relaxed state (e.g., Brown, 1970; Nowlis and Kamiya, 1970), although not all the results have agreed with this. For example, anticipation of electric shock did not suppress alpha (Orne and Paskewitz, 1974). In another study Travis et al. (1975) reported that approximately 50% of the subjects found the experience unpleasant. One investigator reported that the alpha experience required not only the presence of alpha in the EEG but also the proper instructional set (Walsh, 1974).

One study (Plotkin and Cohen, 1976) addressed issues of EEG alpha directly related to body awareness and therefore also of relevance to interoception. It assessed the degree to which alpha

Table 6–1. Examples of Clinical Disorders Responsive to Biofeedback Treatment Techniques

Angina	Muscle Tension
Anxiety	Pain
Asthma	Raynaud's phenomenon (vascular)
Epilepsy	Substance Abuse
Headache	Vomiting

strength was associated with five dimensions of subject behavior—degree of oculomotor processing, degree of sensory awareness, degree of body awareness, deliberateness of thought, and pleasantness of emotional state. Results indicated that the strength of association occurred in the order in which the five factors are listed here, with degree of oculomotor processing strongest. The association of alpha with body awareness was marginally significant (in a small sample of only eight subjects). Heightening, or focusing on, body awareness suppressed alpha. Subjects were approximately 84% as likely to be in alpha if focusing on their bodies as if not focusing. Body awareness did not relate to the viscera per se but could probably be considered to include it.

Biofeedback has become a clinical treatment method that has been shown to provide benefit for many disorders. Research in the clinical arena has not explicitly focused on issues of specific relevance to interoception. However, many of the disorders that have been shown to benefit from biofeedback techniques are likely to involve interoceptive processes. Table 6–1 contains a brief list of these disorders. All of them, with the possible exceptions of headache, muscle tension, and epilepsy, would appear to be likely to involve visceral sensory changes.

METHODOLOGICAL ISSUES

A number of methodological issues are involved in biofeedback research. Some relate directly to the intersection of biofeedback and interoception. The types of subjects used is important. In studies attempting to determine if visceral–autonomic functions can be conditioned with operant methods, verbal reports theoretically might be helpful in understanding the mechanisms by which the conditioning occurs, but, in fact, they have often been only of minor benefit. However, in studies in which the involvement of interoception in the conditioning process is of interest, verbal reports of visceral–autonomic sensations would potentially be of great usefulness (despite the critique of verbal report in Chapter 8).

A very important issue is that the research with curare in rats turned out not to be replicable. The experiments with muscle paralysis were designed to address the question of mediation and two-process learning theory, as described below. It is important to note that not all studies failed to be replicated. Those using muscle paralysis did. Biofeedback modification of visceral-autonomic responses has not been disproved, only the question of the underlying mechanisms has been put in doubt.

Crider et al. (1969) and Black (1971, 1972) reviewed the issue of somatic mediation of conditioned visceral–autonomic responses. Two types of possible mediation have been considered, mediation of visceral–autonomic (involuntary) responses by skeletal (motor, voluntary) responses, and mediation of peripheral responses by central nervous system responses, such as cognitive changes. Much of these discussions relied on the assumption that the results of the curare studies were valid and therefore are not completely relevant now. The issue of mediation must be considered to be unresolved, leaving the question of the possibility of operant conditioning of visceral–autonomic responses also unresolved. Therefore the issue of the need for a two-process (versus one-process) learning theory to explain conditioning of fear and visceral–autonomic responses is unresolved as well—not proved or disproved, but unresolved.

What has or can biofeedback research tell about interoception? Under the basic assumption of biofeedback, visceral sensory processes do not play a role in visceral–autonomic conditioning. It was precisely because it was assumed that visceral sensory processes were not available for operant conditioning that biofeedback methods—connection of visceral activity to external, non-visceral sensory processes—were developed. But it was always an assumption, not a demonstrated fact, that visceral sensory processes could not play such a role. It was clear that they could play a role in Pavlovian conditioning, and the central question of the original biofeedback studies was whether operant and Pavlovian conditioning were essentially the same or different in their underlying neural and psychological mechanisms. That one-process versus two-process question remains unresolved. Thus, it would appear that the role visceral–autonomic sensory processes can play in operant conditioning remains unresolved as well. In either case, however, these visceral–sensory interoceptive processes are real and important.

PART TWO

The Essential Recent Science

seven

THE NEURAL BASIS OF VISCERAL PERCEPTION

"Where my heart lies, let my brain lie also."

Robert Browning

The history of the study of the visceral–autonomic nervous system involves a substantial overlap of anatomy and pharmacology. Galen, who lived in the second century, C.E. allegedly recognized a part of the nervous system as separate that is now known to be part of the sympathetic trunk. He also apparently believed that the body parts and organs functioned in "sympathy" and that these nerves coordinated this activity. In the seventeenth century it was recognized that nerve input to the heart contributed to the control of heart rate.

By the nineteenth century differences in function of the visceral from the somatic and the vegetative from the sympathetic aspects of nervous system function were beginning to be recognized. In the latter half of the nineteenth century anatomical details of this part of the nervous system were beginning to be defined, such as specific nerves and ganglia. Also at this time investigators were beginning to study the physiological and pharmacologic aspects of autonomic function in earnest. Bernard did his work in the middle of the nineteenth century, and James was writing at the end of the century. Connections between autonomic function and cerebral function were noted as early as the middle of the nineteenth century, when visceral changes associated with seizure activity were described by J. Hughlings Jackson, and recognized in poetic language well (see quote above).

Early in the twentieth century Otto Loewi demonstrated chemical transmission of nerve impulses in the heart, and Cannon also did his research then. Acetylcholine, adrenaline, and norepinephrine were identified as neurotransmitters in the sympathetic nervous system, and receptors for these neurotransmitters were identified. Research in the first half of the twentieth century demonstrated that lesions or stimulation of various parts of the central nervous system produced various types of changes in visceral–autonomic function.

From an anatomical point of view, probably the most influential investigator was Langley. Langley coined the term autonomic to apply to the nervous system functions he was studying. He recognized that there were two separate branches of the efferent, or motor aspect, of autonomic function, and that they had different neurotransmitters and different (often opposing) effects. He also recognized the existence of sensory nerves but was not able to study them. (The existence of afferent fibers running in autonomic nerves was verified in the 1920s and 1930s.) His focus on the efferent aspect had the de facto effect of defining the autonomic nervous system as a motor system, a functional definition that has been widely represented in textbooks of neuroanatomy and pharmacology up to the present day. The opposing names *sympathetic* and *parasympathetic* were given around this time. It was also recognized by this time that the gastrointestinal tract has an endogenous, relatively autonomously functioning enteric nervous system that is often considered a third branch of the autonomic system.

SOME GENERAL PRINCIPLES

It is often said that there are five senses: vision, hearing, touch, smell, and taste. Some add a sixth, proprioception. It is also often said that the function of the senses is to allow the organism to interact with the external world. It is clear, however, that even considering only these senses, that statement is oversimplified. Proprioception is about the position and movement of the body—in external space, but about the body. Taste is about things in the external world, but about to enter the body, as is smell in many cases. (Note: Substances in the gastrointestinal tract are not in the body. They enter the body only when absorbed.) In other words, the senses are about things that are external to the body, but also about the body in its interaction with the external world. Seeing or hearing something does not, per se, interact with or change it, but eating it does. This book is about the fact that other, visceral senses exist that are even more about the body, senses that are not contained in the list of five (or six), but are just as real. Pain is the most prominent but not the only visceral sensation. This chapter is about what is known of the neuronal pathways that carry this visceral sensory information.

An important distinction is usually made between sensation and perception. Sensation at its simplest represents activation of sensory receptors and the primary sensory pathways associated with these receptors. Sensory receptors vary in their structure and the type of information they respond to, and sensory information is coded not only by the receptor type stimulated but also by the temporal and spatial characteristics of the passage of information in the sensory pathways, presumably including interoceptive pathways.

Perception, on the other hand, describes the experience resulting from the sensory activation after central nervous system processing, affected, for example, by prior learning. The sensory–perceptual pathways make multiple interconnections but also follow relatively predictable routes, such that somatotopic organization is present for many nuclei and pathways. As described below, similar to other sensory–perceptual modalities, there are specific structures throughout the neuraxis up to and including the cerebral cortex that carry visceral afferent information. The relationship between sensation and perception with consciousness is complex for all sensory–perceptual modalities, including interoceptive awareness.

THE PERIPHERY

Descriptions of the efferent components of the autonomic nervous system, the sympathetic and parasympathetic branches, are widely available and are not the direct focus of this exposition. The enteric nervous system, although certainly connected in important ways to central nervous system afferents and efferents, normally functions in an autonomous fashion, controlling the movements and absorptive functions and characteristics of the gastrointestinal tract and associated structures such as the gall bladder and pancreas (Gershon, 1998).

Information about peripheral aspects of visceral afferents is available from several sources: Pick (1970), "Anonymous" (1986), Appenzeller (1990), Loewy (1990), Cervero and Foreman (1990), Freire-Maia and Azevedo (1990), Amann and Constantinescu (1990), Broman (1994), Janig (1996), Burstein (1996), Benarroch (1997), Hardy and Naftel (1997). Visceral sensory receptors are divided into two groups, pain receptors (nociceptors), which are free nerve endings, and physiological receptors, which monitor ongoing function of visceral organs and mediate visceral reflexes (e.g., baroreception). The physiological receptors are further subdivided into rapidly adapting receptors that monitor changes in status or function, and slowly adapting receptors that monitor ongoing status such as stretch. Additionally, there are other specialized visceral receptors such as chemoreceptors, osmoreceptors, and thermoreceptors. These three receptor types exist in the hypothalamus and elsewhere in the body as well. These figure very importantly in interoceptive processes. Pain will be addressed further as a separate subject in Chapter 10.

The afferent visceral–autonomic sensory fibers, which are mostly unmyelinated or thinly myelinated, generally run anatomically with the efferent fibers in both sympathetic and parasympathetic nerves. Their cell bodies are in the spinal and cranial ganglia. Fibers from physiological receptors are more common in parasympathetic nerves. The sympathetic nerves more commonly contain pain fibers. Afferent fibers run in the sympathetic system in the thoracic and upper lumbar dorsal root ganglia. Afferent fibers run with the parasympathetic fibers through the oculomotor nerve (III), the trigeminal nerve (V), the geniculate ganglion of the facial nerve (VII), the petrosal ganglion of the glossopharyngeal nerve (IX), the nodose ganglion of the vagus nerve (X), and the sacral dorsal root ganglia S2-S4. Overall, visceral afferent fibers are more abundant in parasympathetic nerves. For example, it has been estimated that more than 80% of the vagal fibers are visceral afferent, while less than 20% of the fibers in the greater splanchnic nerve (sympathetic) are visceral afferent. It has been estimated that, overall, the ratio of parasympathetic to sympathetic visceral afferent neurons is approximately 3:1. This difference appears to lend support to the idea that parasympathetic function has more specificity than does sympathetic (i.e., more precise differential control of one organ versus another), but it remains doubtful if this concept is correct. The relatively small number of visceral pain fibers, combined with the facts that *(1)* somatic and visceral sensory nerves converge in the dorsal horn and that *(2)* developmental patterns sometimes produce substantial anatomic separation between the origins of the somatic and visceral nerves entering the same spinal segments accounts for the typical poor localizability of visceral pain and the fact that pain of visceral origin is often experienced in somatic regions sometimes far distant from the source (e.g., cardiac pain—angina—referred to the left arm, or pain from inflammation of the diaphragm referred to the shoulder).

Evidence exists that excitatory amino acids, especially glutamate, mediate peripheral visceral afferent neurotransmission. Other excitatory amino acids and peptide neurotransmitters are also probably involved (Broman, 1994).

In addition to afferent information carried in neuronal pathways, some non-neuronal cells such as endocrine cells also respond to changes in internal states in order to restore homeostasis. Examples include control of extracellular osmolality, ion balance, glucose levels, and circadian rhythm activity (Wenning, 1999). Circulating substances affect homeostatic control through chemical receptors in the central nervous system as well (Benarroch, 1997; see also Chapter 4). In other words, visceral sensory control involves both neuronal and non-neuronal mechanisms.

Cervero and Foreman (1990, pp. 107–110) discussed the question of the encoding mechanism for visceral sensory receptors, an issue of fundamental importance to interoception. They pointed out that the complexity and apparent experiential and anatomical vagueness of both painful and non-painful visceral sensations, along with the fact that much visceral afferent information does not normally reach awareness, make this an unusually difficult problem to study. Stimulation of the receptors that control visceral reflexes evokes no sensations under normal circumstances. Whether individuals can learn to be aware of the function of these reflexes, and how and to what extent they influence behavior with or without consciousness, is discussed in several places throughout this book. Concerning receptors that do involve awareness, most appear to mediate both nonpainful and painful experiences, probably due to a combination of encoding of receptor specificity (e.g., in a particular organ), intensity of receptor stimulation, and/or summation of signals by the central nervous system. To summarize, there are three logically distinct effects of visceral sensation on awareness: no awareness, awareness without pain, and awareness with pain. The extent to which these categories are qualitatively separate from the point of view of biology and psychology is a central issue of interoception.

The overall organization of autonomic pathways into the brain includes reflexes and central integration (Loewy, 1990, their Fig. 6–1). For example, the afferents of the baroreceptor reflex seem to go as far as the nucleus of the solitary tract (nucleus tractus solitarius—NTS), then connect to brainstem descending efferents via interneurons. The second set of pathways leads higher in the neuraxis, producing functional integration by affecting hypothalamic, limbic, thalamic, and cortical centers that influence hormonal output and emotional–motivational, behavioral, and cognitive functions.

SPINAL CORD

Connections of visceral afferents in the spinal cord have been reviewed by Broman (1994), Cervero and Foreman (1990), Craig (1996), and Hardy and Naftel (1997). The afferents in the parasympathetic cranial nerves synapse in the brainstem, including cranial nerves VII, IX, and X, on the nucleus of the solitary tract. All other afferents enter the central nervous system and synapse at the level of the spinal cord. The tracts that enter the spinal cord then send ascending tracts to the brainstem, including connections to the solitary tract nucleus (see below).

Sympathetic visceral afferents in the cardiac and splanchnic nerves enter the spinal cord through the dorsal root via Lissauer's tract and synapse mainly in laminae I and V, which then project contralaterally in the anterolateral system, but they also synapse in laminae VII and VIII, which then project bilaterally in the spinoreticular fibers. The afferent parasympathetic fibers enter the spinal cord through the pelvic nerves and synapse in the dorsal horn. Lamina I also receives afferents from somatic regions—including skin and muscle, thus providing one region of convergence of visceral with other afferent information. Craig (1996) argued that the ascending projections in lamina I serve a broader function than just transmitting pain impulses. Rather, they constitute "an ascending general homeostatic afferent pathway . . ." (p. 225). The major ascending pathways within the spinal cord that carry visceral nociceptive, and probably other visceral afferent, information are the spinothalamic, spinoreticular, spinomesencephalic, and spinosolitary tracts (Cervero and Foreman, 1990). Spinal cord visceral afferent input is illustrated in Figure 7–1.

A number of neurotransmitter substances that are likely to be involved in visceral afferent processes are present in the spinal cord (Broman, 1994). The inhibitory neurotransmitters GABA and glycine are present in many neurons in the dorsal horn. The monoamines serotonin, noradrenaline, dopamine, and histamine are present. Taurine, D-aspartate, and glutamate are also implicated. The opioid peptides are strongly implicated in pain transmission, as is substance P.

BRAINSTEM

Anatomy

Long Tracts. As already noted, ascending visceral sympathetic afferents are mainly, although probably not completely, nociceptive pathways. Ascending visceral parasympathetic pathways carry both pain and more general visceral sensory information. Some pathways in the brainstem are known to carry visceral sensory information, and others (e.g., spinocervical, spinohypothalamic) are less well studied.

Several distinct brainstem tracts that carry visceral sensory information have been identified. Each is named for its destination in the brainstem or above. The spinothalamic tract, which passes through the brainstem, has two divisions, the lateral and the medial. The lateral terminates in the ventral and ventral posterior lateral areas of the thalamus and carries pain and possibly other sensation from relatively discrete visceral (and somatic) areas. The medial pathway projects to the medial and intralaminar thalamic nuclei and is thought to carry less discrete information, possibly associated with homeostatic, autonomic, and emotional processing of nociception.

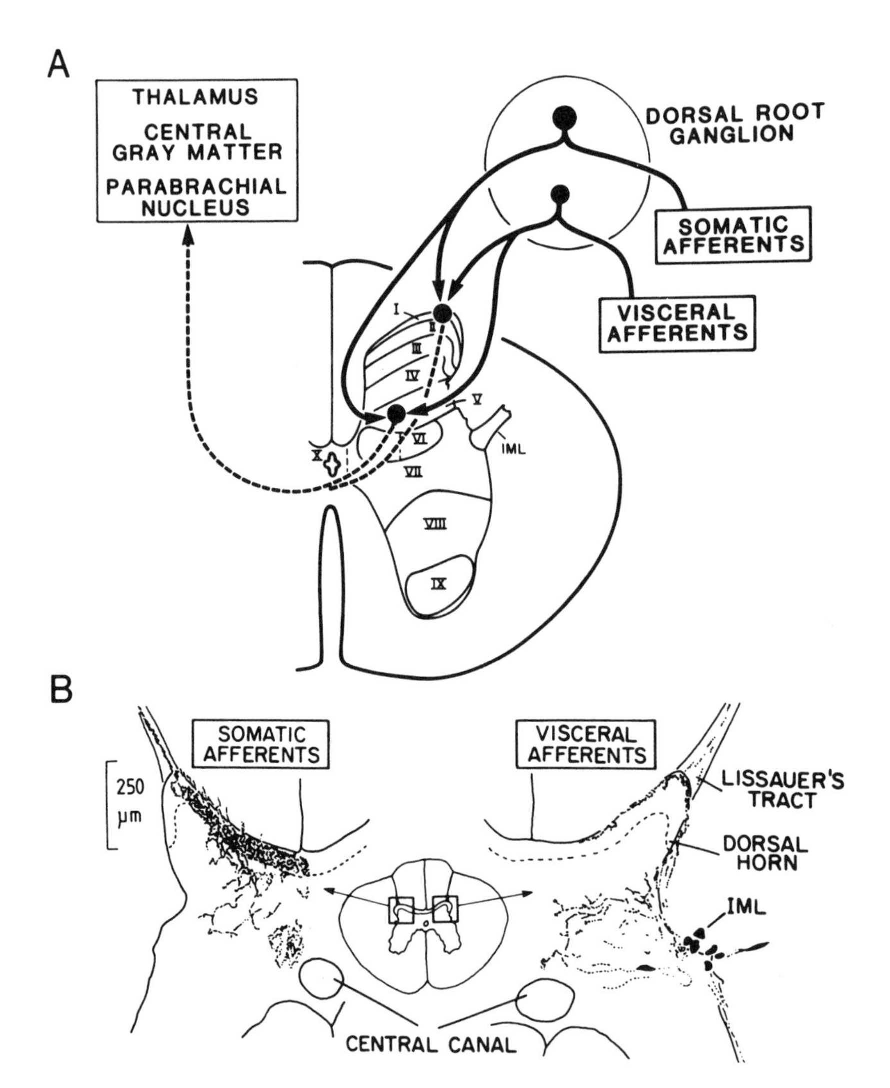

Figure 7–1. Entry of visceral (and somatic) afferents into the spinal cord via the dorsal horn, (A) onto laminae I and V, and (B) reconstruction of horseradish peroxidase tracing from splanchnic nerve at spinal cord level T9. [From Cervero and Foreman, 1990, after Cervero and Connell, J. Comp. Neurol. 230; 88–98, 1984, and Loewy, in Stober et al., *Central Nervous System Control of the Heart*, Martinus Nijhoff Publishers, distributed by Kluwer Academic Publishers, Norwell MA, 1986, reprinted with permission of the holder of the copyright.]

This tract is highly involved with visceral pain from the heart (angina) and the gastrointestinal tract.

The spinoreticular pathway receives input from both painful and non-painful stimuli from both visceral and somatic sensory afferents. It projects to the reticular formation and the caudal raphe nuclei, which in turn project mainly to the thalamic intralaminar nuclei. The spinomesencephalic tract projects to the periaqueductal gray and the parabrachial nucleus and is also mainly involved in nociception.

The Nucleus of the Solitary Tract. The nucleus of the solitary tract (NTS) is a very important structure in the visceral afferent pathways. The caudal portion of the NTS receives a spinal projection as the spinosolitary tract, which is thought to be involved in visceral–somatic sensory integration, including pain. The other major visceral afferent input to the NTS is from the solitary tract, which is formed by the cranial nerves (VII, IX, and X) that carry the parasympathetic visceral sensory information as they enter the brainstem.

The NTS lies at the level of the obex in the brainstem near the fourth ventricle, the dorsal vagal nucleus, and the area postrema. There are incoming fibers that are organ-specific to subnuclei and also multi-organ inputs to the commissural NTS. The organ-specific nuclei are mainly (or solely) involved in the autonomic reflexes, while the commissural NTS forms the ascending pathway to the higher forebrain integrative centers.

The functional organization of the input to the NTS is usually separated into gustatory (taste) pathways and general visceral pathways. All three of the visceral afferent cranial nerves carry taste information to the rostral and intermediate regions of the NTS. The ascending taste fibers from the NTS involved in central integration include a major projection to the parabrachial nucleus. Projections then go to multiple higher centers, including such regions as the area of the hypothalamus involved in hunger and eating behaviors (Ruggiero et al., 1998).

The general visceral pathways are divided into four tracts. These four tracts end in the NTS in distinct subnuclei—cardiovascular to the medial and dorsolateral, pulmonary to the ventral and ventrolateral, respiratory to the interstitial, and gastrointestinal to the parvocellular. These specific nuclei are involved in the general visceral afferent reflexes. These four inputs also converge in the commissural NTS and pass to higher centers, including the central nucleus of the amygdala, paraventricular nucleus and lateral area of the hypothalamus, and the cortex. The commissural NTS also receives input from the area postrema. The NTS also appears to receive olfactory input (Garcia-Diaz et al., 1988). Finally, as is true of most of the structures being reviewed, there is not only afferent input from lower structures, but the higher structures often send feedback down to the lower structures.

The Parabrachial Nucleus. The NTS has prominent projections to the parabrachial nucleus and the A5 and A6 noradrenergic nuclei in the pons, as well as the nucleus ambiguus at the same brainstem level as the NTS. (The A6 level is also known as the locus coeruleus.) The taste region of the NTS projects to the medial region of the parabrachial nucleus, while the general visceral tracts project to the lateral region including the Kölliker-Fuse nucleus. The parabrachial nucleus sends ascending pathways to the hypothalamus, several of the thalamic nuclei, the amygdala, the cortex, including the insular cortical region and several other areas, including effects on systemic catecholamine release (Saleh and Connell, 1998) and on cerebral blood flow and metabolism (Mraovitch et al., 1985). Like the NTS, the parabrachial nucleus also receives somatic afferent input.

The Central Gray. The midbrain periaqueductal gray is organized in columns that carry visceral and somatic afferent (and efferent) information. Different columns are implicated in different functions, such as cardiovascular and nociceptive functions. These columns have connections with the other brainstem structures.

Visceral–Somatic Connections. As already noted, there is a substantial amount of convergence of visceral and somatic afferents in the brainstem as well as in the spinal cord (Sato, 1997; Yates and Stocker, 1998). Sato (1997) demonstrated that somatic stimulation can produce activation of a variety of visceral organs, including the gastrointestinal tract, bladder, heart, and adrenal medulla. Cerebral blood flow and immune function can also be affected. Yates and Stocker (1998) reviewed studies that demonstrate that physical movement (somatic motor system) and cardiorespiratory activity often are coordinated, most likely by interactions involving brainstem mecha-

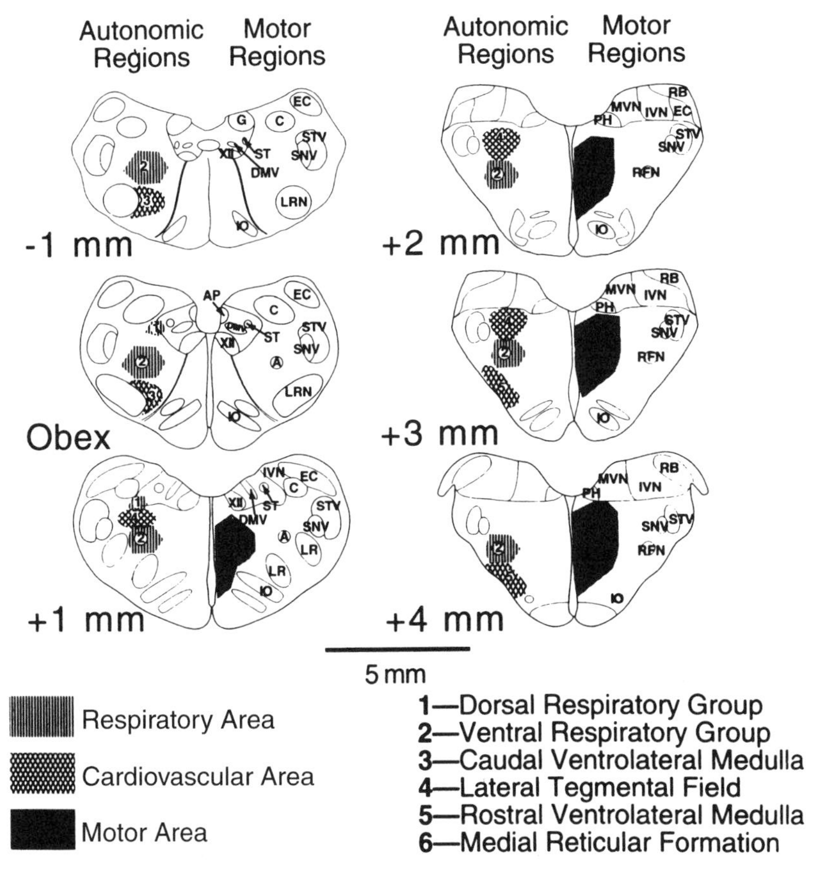

Figure 7–2. Traditional model of medullary control of respiratory and cardiovascular (visceral) functions shows non-overlap with areas of somatic musculature, but more recent studies indicate overlap and convergence of functions in this region. (See original for abbreviations of structures.) [From Yates and Stocker, 1998, reprinted with permission of the holder of the copyright.]

nisms and structures such as the NTS, the lateral reticular formation, and vestibular nuclei. It is likely that this convergence is a major source of such visceral sensory phenomena as referred pain. Figure 7–2 illustrates the apparent lack of overlap of the brainstem regions traditionally considered to mediate autonomic function with the regions mediating motor function. Visceral–somatic convergence demonstrates that these regions either do overlap (i.e., are not fully indicated by the traditional regions) or are functionally and anatomically linked. Holzl et al. (1999), for example, presented data from a study using a masking procedure and interoceptive stimuli from the colon and the abdomen indicating that, at least in this specific experimental situation, convergence occurred above the level of the spinal cord.

Pharmacology

The role of the monoamines in ascending functions was discussed above (Chapter 3). Noradrenergic pathways are particularly important in visceral afferent function, especially the pathways originating in the A5 and A6 nuclei (e.g., Svensson, 1987; Barnes and Pompeiano, 1991; Valentino et al., 1994). The A5 region is involved in cardiovascular, nociceptive, and thermoreceptive functions. The A6 region (locus coeruleus) has circumscribed input from the nucleus propositus hypoglossi, the nucleus paragigantocellularis, and the rostral ventrolateral medulla. It also receives polysynaptic input from multiple pathways that carry visceral afferent information through splanchnic, vagal, and pelvic nerves. The output of the locus coeruleus is exceedingly broad, including all layers of the neocortex, the hippocampus, the paraventricular nucleus of the hypothalamus, the median eminence, and other brainstem nuclei. Functionally, the locus coeruleus is involved in arousal, orienting, attention, and vigilance and is activated by various visceral sensory events such as bladder distention (see Svensson, 1987). Early theories indicated that norepinephrine had mainly inhibitory functions, but more recent data indicates that it functions in various ways at different synapses, often functioning as a neuromodulator and increasing the synaptic signal-to-noise ratio (Waterhouse et al., 1991). Other brainstem noradrenergic nuclei exist, but their function is less well understood in relation to visceral sensation. The A6 locus coeruleus provides more than 50% of brain norepinephrine and noradrenergic neurons.

The primary neurotransmitters in the brainstem (e.g., the dorsal column, the spinocervicothalamic, and the spinothalamic tracts) are GABA and glutamate. Other substances that are present include enkephalin and dynorphin, substance P, calcitonin gene-related peptide, cholecystokinin, galanin, bombesin, as well as the monoamines (Broman, 1994). Most of the spinothalamic neurons implicated in pain transmission respond to bradykinin.

BETWEEN BRAINSTEM AND CORTEX

Structures important in visceral–autonomic sensory processing and motivational–emotional function that are above the brainstem but below the cortex have been discussed (Chapter 4). These include such structures as the thalamus, hypothalamus, cerebellum, amygdala, and other limbic structures.

From the viscera to the brainstem, afferent pathways related to visceral–autonomic sensory processes are relatively straightforward to identify and track. Once one reaches regions of the central nervous system above the brainstem, tracking specific pathways related to visceral afferent function becomes more difficult and the results less clear due to the complexity of the tracts involved. While feedback loops certainly exist even at the level of the brainstem and below, interconnections among higher structures are much more complicated, to the point that identifying a particular tract as afferent versus efferent becomes very difficult. This is not true in all cases (e.g., clearly defined afferent pathways to the thalamus). Discussion will focus on regions and structures known to be involved with visceral sensory function.

Thalamus

Much of the ascending visceral and somatic sensory information, including most that reaches the cortex, goes first to the thalamus. Nociceptive information goes prominently to the thalamic ventral posterolateral nucleus. As already stated, visceral pain stimuli are not well localized in the body, and thus it can be inferred that there is probably less precise somatotopic localization for visceral pain sensory stimuli than for somatic in the ventral posterolateral nucleus. Also previously noted, in addition to the lateral branch of the spinothalamic tract sending fibers to the ventral posterolateral nucleus, a medial branch also sends fibers to the medial and intralaminar thalamic nuclei, which is thought to carry information more relevant to the physiological state of the organism (i.e., motivational–emotional information). The thalamus is not always listed as an important part of central autonomic pathways (Benarroch, 1997). However, if one includes pain and visceral–somatic convergence, the thalamus is undoubtedly involved to at least some extent (e.g., Bruggemann, 1998). The major neurotransmitters involved in visceral afferent input to the thalamus seem to be GABA and glutamate, with cholinergic, noradrenergic, and histaminergic input as well (Westlund et al., 1991; McCormick et al., 1991; Broman, 1994).

Cechetto (1987) reviewed data indicating that there is some general somatotopic organization to visceral afferent pathways in the ventral posterior lateral nucleus. Gustatory information from the parabrachial nucleus goes to a parvocellular region. Lateral to that region, cardiopulmonary information is received from the parabrachial nucleus. Finally, lateral and dorsal to these regions is a region receiving information from the dorsal column nuclei and the spinal cord. From these three areas information is relayed to distinct areas of the insular cortex.

Hypothalamus

The hypothalamus (see also Chapter 4) receives both monosynaptic and polysynaptic input from the periphery. It is likely that the monosynaptic input is involved in rapid response systems such as pain and, possibly, fear. The polysynaptic input is more likely to serve slower response systems. Thus, both slow and fast components of motivational–emotional response involve hypothalamic centers.

The hypothalamus appears to be an integrator of these visceral–autonomic–homeostatic (i.e., interoceptive) inputs that come from many sources within the central nervous system as well as from the periphery (in contrast to the thalamus, which seems to serve more as an integrator of input of sensory information from the external environment—exteroceptive—although this is only a simplified generalization). The sources of exteroceptive and interoceptive inputs in the brain and periphery to the hypothalamus include the olfactory, visceral sensory, visual, limbic, and circumventricular (chemoreceptors, osmoreceptors) systems, and the functions involved include hormonal, reproductive, behavioral, immune, thermoregulatory, gustatory (hunger and thirst), and biorhythmic, in addition to autonomic. It has been suggested that, as a simplifying generalization, the posterolateral portion of the hypothalamus serves various activating (arousal?) functions while the anteromedial portion serves the opposite ("vegetative," or "contentment") functions.

Tracts from the spinal cord and brainstem go to the hypothalamus (see Burstein, 1996). It has been demonstrated (in the rat, so far) that a spinohypothalamic tract originates in spinal cord lamina I, V, and X, and projects directly to both the medial and lateral hypothalamus, apparently carrying both pain and non-pain stimuli from both visceral and somatic sources. There is also a pathway from ventrolateral medulla and noradrenergic nucleus A1 to the hypothalamus. The A1 neurons are involved in control of vasopressin release. The A2 and A6 noradrenergic nuclei also send projections to the hypothalamus. There is likely input to the hypothalamus from the NTS, parabrachial nucleus, and periaqueductal gray, but it is not certain if there are direct tracts (Cechetto, 1987). In addition to norepinephrine, evidence exists of involvement of the neurotransmitters cholecystokinin, CRF (corticotrophin-releasing factor), GABA, and substance P in these tracts (Cechetto, 1987; Waterhouse et al., 1991; Sawchenko et al., 1996).

Examples of other important specific afferent fiber tracts to the hypothalamus include the fornix mainly from the hippocampus, the stria terminalis and the ventral amygdaloid bundle from the amygdala, and the medial forebrain bundle, which originates in multiple places including the midbrain tegmentum, nucleus accumbens, and the septal nuclei. There is a thalamohypothalamic tract and a (prefrontal) corticohypothalamic tract. Most connections of the hypothalamus with other regions are reciprocal. Other inputs exist that are beyond the scope of this chapter, since the relevance of many of these to interoceptive processes has not been demonstrated.

Amygdala

Afferent pathways reach the amygdala mainly in the lateral region. Inputs come from the olfactory bulb, septal nuclei, thalamus, primary sensory cortex, medial prefrontal cortex, regions of the association cortex, cingulate cortex, and the subiculum (which in turn receives input from regions including the hippocampus and the entorhinal cortex). Evidence exists of direct inputs from the brainstem from both the NTS and the parabrachial nucleus (Cechetto, 1987), and at least indirect inputs from the raphe nuclei, periaqueductal gray, dorsal motor nucleus of the vagus, and the locus coeruleus. Reciprocal connections exist with the hypothalamus, the stria terminalis, the substantia innominata, and the temporal lobe (insular cortex and parahippocampal gyrus) (Chronister and Hardy, 1997).

Neuropeptides are the major neurotransmitters that have been implicated in these connections.

There are internal connections in the amygdala leading to output mainly from the central nucleus. The outputs affect stress hormones, sympathetic and parasympathetic autonomic function, arousal and attention, and specific emotional behaviors. LeDoux (1996, 1998) argued that there are two separate pathways going through the amygdala that mediate fear, one not involving the cortex that produces a rapid response and does not involve or require conscious awareness of the fear stimulus, and another that is slower and does involve the cortex along with conscious awareness. These outputs affect several functions associated with emotion and, thus, indirectly at least with interoception.

Limbic System

The hypothalamus and the amygdala are often considered part of the limbic system. Many of the afferent (and efferent) connections to these two structures are also considered parts of the limbic system (Mega et al., 1997). Other structures already discussed that are sometimes considered so are the ventral tegmentum and periaqueductal gray. The cingulate cortex and the prefrontal cortex (especially orbitofrontal), areas implicated in interoceptive function (see below), are occasionally included as well. Pharmacologically, the limbic system receives input from a wide variety of neurotransmitters including acetylcholine and the monoamines.

The hippocampal formation, including the subiculum and the dentate gyrus, has been an extensively studied part of the anatomical limbic system because of its essential involvement in memory and cognitive processes. Evidence exists for input to the hippocampus from the amygdaloid complex, implicating it in the processing of emotional information and thus indirectly in interoceptive processes. As noted above (Chapter 4 and Fig. 4–4), the hippocampus is included in the original Papez circuit.

Two other important regions of the limbic system are the septal nuclei and the striatal nucleus accumbens (Groenewegen et al., 1996). These structures are of potential relevance to interoceptive processes because of their involvement in emotional–motivational behavior, including rage reactions (septum) and drug self-administration (nucleus accumbens).

Cerebellum

The cerebellum is not considered part of the limbic system, but because of the fundamental relationship of classical conditioning to interoception and of the cerebellum to (at least some forms of) classical conditioning, it bears mention. There is evidence of involvement of the cerebellum in higher order behavior (Schmahmann, 1991), including emotional function (Ghelarducci and Sebastiani, 1997). Ascending afferent input to the cerebellum comes from the spinal cord, medulla, and pons, including the locus coeruleus and the raphe nuclei, as well as connections with the hypothalamus. Conditioning pathways for transduction of information from both the conditioned stimulus and the unconditioned stimulus pass through the brainstem, although not obviously involving either the NTS or the parabrachial nucleus. Evidence exists not just for involvement of the cerebellum in the conditioning process, but also for storage of the memory trace for this conditioning in the cerebellum (Beggs et al., 1999). Evidence for involvement of GABA in this process has been reported (Waterhouse et al., 1991).

CORTEX

Early theories of human psychology assumed that all the highest mental functions were associated with consciousness and that consciousness resided in the cerebral cortex. Pavlov believed that in studying the conditioned reflex he was studying processes that occurred and must occur in the cortex, whether it be in humans, dogs, or other mammals. Despite this, within the Pavlovian tradition itself, interoceptive processes (classical conditioning) were demonstrated and analyzed that did not appear to require conscious awareness in order to occur. On the other hand, the more recent tradition of study of interoceptive processes has focused on processes specifically related to consciousness. To answer the question, Can one feel one's heart beat?, it is necessary to be able to report what one is aware of. In other words, depending on the definition used (see Chapter 1), conscious awareness (and the ability to report this awareness) is or is not necessary for interoception. This issue begs the question, What role might the cerebral cortex play in interoception? That issue will now be addressed. It is not assumed that the cortex is the sole seat of consciousness in humans or other organisms. That assumption is almost certainly incorrect, or at least oversimplified. Both cortical and subcortical mechanisms are undoubtedly involved (i.e., necessary as well as sufficient). The question of the role of consciousness in interoception will be revisited in Chapter 13.

Frontal Cortex

As one ascends in the neuraxis it becomes harder and harder to recognize and identify sensory processes per se, as opposed to motor or association functions. This is especially true of cortical structures, and it is true for interoception as well as other sensory processes. At this level sensation per se, or even pure perception, might not properly be said even to exist. Thus, one must rely on identifying functions known to be associated with interoception that can be identified and associated with specific structures at this level. The most appropriate choice seems to be emotional–motivational processes. Therefore, regions of the cortex either directly related to visceral sensory processes or to emotional–motivational processes, and the regions that these areas connect to, will be reviewed.

Anterior Cingulate Cortex. The cingulate gyrus, or cingulate cortex, is considered part of the limbic system (see Devinsky et al., 1995; Benarroch, 1997; Mega et al., 1997). It was a part of the original circuit described by Papez. It is often divided into two large parts, the more anterior, which consists of Brodmann areas 24, 25, and 32 and has been associated with affective function, and a more posterior part that has been associated with more cognitive processes. Anatomically, these two regions are quite separate, with few interconnecting fibers. They are also different cytoarchitecturally; the anterior part has an agranular structure.

The anatomical connections of the cingulate can be subdivided not only into anterior versus posterior, but also into cortical connections versus subcortical connections. The cortical connections tend to be reciprocal. Cortically, the anterior part is connected mainly with the prefrontal cortex, both medial and lateral (although medial orbitofrontal is probably more prominent), and with the anterior region of the insular cortex. The posterior cingulate (and contiguous retrospenial cortex) is connected mainly with the parietal and occipital cortical areas, frontal eye fields, and some prefrontal connections.

The subcortical connections of the cingulate can be subdivided not only into anterior versus posterior connections, but into afferent versus efferent connections (although many of these are reciprocal as well). The afferent connections of the anterior cingulate include the anterior parahippocampus, inferior temporal region, amygdala, hypothalamus, periaqueductal gray (especially involved with defense reactions—flight, immobility), medullary autonomic nuclei, midline-anterior-dorsomedial thalamic nuclei, ventral striatum (especially involved with long-term memory for classical conditioning), and the septum. The efferent subcortical cingulate connections are to the agranular part of the insula, hypothalamus—paranentricular nucleus (PVN), amygdala, bed nucleus of the stria terminalis, thalamus, and regions of the brainstem including the periaqueductal gray, dorsal vagal nucleus, and ventrolateral medulla. Briefly, the posterior cingulate connects to the posterior parahippocampal gyrus, subiculum (hippocampus), and the dorsal striatum.

Generally, the functions of the cingulate have been variously described as *(1)* top-down control of sensory–attentional and cognitive processes, *(2)* management of executive functions—initiation, motivation, and execution, and/or *(3)* activation of visceromotor and skeletomotor processing. Specific functions of the anterior region include emotion—aggression, fear, startle, maternal/social—emotional learning, emotional vocalizations, nociception, and visceromotor (areas 25, 24) including autonomic nervous system activation via the NTS and the vagal dorsal motor nucleus. This area supports self-stimulation as well. The posterior region involves not only cognitive, sensory, and memory function, but some evidence indicates that emotional–cognitive interactions occur in this area. The connection of the posterior cingulate to the parahippocampal gyrus appears to include involvement in classical conditioning (Vogt et al., 1992; Maddock, 1999)

A more detailed, four-division anatomical–functional breakdown of the cingulate has also been described (see Mega et al., 1997). The infracallosal region (anterior, below the corpus callosum) mediates visceral functions. It is connected to the medial orbitofrontal cortex, rostral insula (for gustatory, olfactory, and alimentary functions), ventromedial temporal lobe, and amygdala. The supracallosal region mediates cognitive function. It is connected to the lateral orbitofrontal area. The medial region is involved with skeletomotor activity, and the posterior area mediates sensory processing connected to the perirhinal cortex, parahippocampal cortex, subiculum, and eye fields. Stimulation of the anterior area produces prominent autonomic and behavioral reactions. Evidence from imaging studies shows that anterior activation is produced by experimental emotional activation, including the induction of happy or sad affect in normal subjects as well as clinical depression and several of the anxiety disorders (see Whalen et al., 1998)

Medial Prefrontal/Orbitofrontal Cortex. The prefrontal cortex, generally defined, has been considered a major multimodal association cortex, involved in the so-called highest human traits, including planning, judgment, foresight, responsibility, and so on. More specific functions can sometimes be recognized in more specifically identified regions of the prefrontal cortex. Of potential relevance to interoception are the medial regions, especially the inferior or ventromedial (i.e., orbitofrontal) region (Ongur and Price, 2000).

In referring to the putative functions of the limbic circuit, which can include the orbitofrontal cortex, Trimble et al. (1997) speculated that the more medial brain structures, including this part of the overall prefrontal cortical area, carry ". . . information more closely related to internal states" (p. 124). Particularly in the orbitofrontal region this includes processing of emotional functions and might specifically include visceral sensory input as well as efferent feedback onto viscerosensory regions of the brainstem.

Consistent evidence has been found for involvement of the medial frontal cortex in autonomic functioning, including convergence of exteroceptive and viscerosomatic input (Cechetto and Saper, 1990; Benarroch, 1997). Stimulation of this region can produce vigorous autonomic effects such as effects on heart rate, blood pressure, and gastric motility, while lesions produce cardiovascular effects. Respiratory and cardiac activity as well as stimulation of the vagus nerve can produce firing of prefrontal neurons. Evidence exists for its involvement in conditioned fear responses.

Mega et al. (1997) described the orbitofrontal cortex as being part of one of two paralimbic belts. It is associated in function, in cytoarchitectural structure, and in phylogenetic development to the olfactory system and the amygdala. As already described, strong linkages exist between the orbitofrontal cortex and other limbic areas, especially the amygdala, rostral insula, ventromedial temporal pole, and the infracallosal cingulate gyrus (i.e., the part of the cingulate thought to be most closely associated with visceral function). The medial prefrontal cortex also has connections with the more lateral prefrontal regions. Morecraft et al. (1992) provided a detailed description of the afferents to the monkey orbitofrontal cortex, noting that this cortical region can be further subdivided into three separate regions, caudal to rostral, based on cytoarchitectural differences. Their general description of the connections is the same as above.

Somatosensory Cortex

The somatosensory cortex is divided into two broad areas, referred to as SI and SII (see Hendry et al., 1999). Both of these areas show somatotopic organization. The SII area is smaller than the SI and is near the lateral fissure, in close proximity to the insular cortex, a region of fundamental importance to interoception (see below). SII receives input from SI and also directly from the thalamus (ventrobasal complex). In addition to SI and SII, regions in the parietal lobes behind SI and SII are considered to function as either a unimodal or multimodal sensory association cortex. Evidence exists for input of impulses from visceral sympathetic stimulation into the sensory cortex, as well as evidence for visceral–somatic sensory convergence at the cortical level (Cechetto and Saper, 1990; Bruggemann et al., 1997). Evidence also exists for pharmacologic interaction of GABA and norepinephrine in the somatosensory cortex (see Waterhouse et al., 1991).

Insula

Several regions of the cerebral cortex, including the orbitofrontal cortex, the anterior cingulate gyrus (which is often considered cortex), and probably the somatosensory SII region and parts of the temporal pole, are at least indirectly involved in interoception by virtue of their involvement in emotional functioning. One region of the cortex appears to be most directly involved in visceral sensation, the insular cortex, which will now be described in some detail. It is also known as the Island of Reil, after the sixteenth century anatomist who described it. Additionally, it has also been called the central lobe or the fifth lobe of the brain. A number of especially useful reviews exist to which the interested reader is referred for further information (Augustine, 1985, 1996; Mesulum and Mufson, 1985; Cechetto, 1987; Cechetto and Saper, 1990).

Development. The insula is part of the cerebral cortex. It expands more slowly during early embryonic development than does the adjoining frontal, parietal, and temporal cortical regions, which eventually come to surround it, thus forming the lateral, or Sylvian, fissure. At birth the insula is buried by these other regions, which is why it has been called an island. In humans it has 5–7 sulci, which have fully developed by 32 weeks of embryonic development.

Anatomy. The fully developed insular cortex consists of 4–6 gyri. There is a central sulcus that divides it into an anterior part and a posterior part. The size and gyral–sulcal structures of the right and left insulae often show asymmetries, with the left often larger than the right. The insula is closely related to the contiguous orbitofrontal and temporopolar cortical regions, and it has been argued by some that these structures should be considered one anatomical complex. This region has been called paralimbic, based not only on macroscopic and functional linkages, but also because of its phylogenetic development and because it is a transition zone from allocortex to granular cortex. The insula is also closely related anatomically to the frontoparietal operculum and the primary olfactory cortex.

From a cytoarchitectural point of view, microscopic differentiation within the insula is not anterior–posterior, but rather more radial and concentric, with three different microscopic cortical structures—an innermost rostroventral agranular (peri-allocortical) region, an intermediate dysgranular (partially granular or periisocortical) region, and a posterocaudal outer granular (isocortical) region, based on the amount of myelinization, the presence of laminar differentiation, and the presence or absence of granule cells in layers 2 and 4. In the Brodmann system of numbering different cortical regions, the dorsocaudal granular region is no. 13, and the remaining area is subdivided into regions 14–16. Pharmacologically, the agranular and dysgranular layers are rich in acetylcholine, while the granular layer is less so.

Table 7–1. Summary of Cortical Connections of the Regions of the Insula

Dysgranular	Granular	Agranular	
Somatosensory	X	X	
Auditory	X	X	
Visual	X	X	
Gustatory	X	X	
Motor	X	X	
Association	X	X	
Olfactory–Limbic	X		X
Paralimbic	X		X

[Adapted with permission of the holder of the copyright from Mesulam and Mufson, 1985.]

Anatomic Connections. Afferent and efferent insular connections and circuitry are extensive and complex. Cortical connections are summarized in Table 7–1, subdivided by the three cytoarchitectural regions.

The connections of the insula to other cortical areas are extensive, including medial (orbitofrontal) and lateral frontal, temporal, and parietal, as well as cingulate gyrus. Afferent connections of particular relevance to interoceptive processes are especially prominent and important from the medial frontal lobe and anterior cingulate. Efferent connections from the insula go to frontal and parietal regions and the cingulate, and there are significant intra-insular connections. More specifically, there are connections to somatosensory areas SI and SII that appear to have a somatotopic organization. There are auditory connections to the agranular and dysgranular regions. There may be connections to visual cortex, but if so they are minimal. There are connections with gustatory and olfactory areas that mediate, for example, flavor sensitivity (Shipley and Geinisman, 1984). There are connections with the supplementary motor area. Finally, there are connections to higher order association areas, such as the prefrontal, orbitofrontal, temporal cortical, and limbic–paralimbic areas, already mentioned. As summarized in Table 7–1, connections to the agranular insular region are most prominent for the limbic and paralimbic areas, while the granular region connects with the other cortical regions, and the dysgranular has connections to all. Regional connections of these three cytoarchitectural insular areas to other cortical and subcortical regions are demonstrated in Figure 7–3.

There are significant subcortical as well as cortical insular connections. Connections with the basal nuclei include the lentiform nucleus, putamen, claustrum, and tail of the caudate. Subcortical limbic connections are most prominent for the amygdala and also include the hippocampus, piriform and entorhinal, peri-amygdaloid area, and olfactory tubercle and bulb. The amygdaloid connections are mainly with the basolateral nucleus, but connections with other amygdaloid nuclei also apparently exist. At least for the basolateral amygdala, connections are reciprocal.

There are reciprocal connections between the insula and dorsal thalamus. Mesulam and Mufson (1985) describe "widespread thalamic connections" in macaque monkey (p. 206), including all three cytoarchitectural regions. There are also connections to the hypothalamus, white matter regions, and brainstem areas, including reticular formation, raphe region, and locus coeruleus; the brainstem connections may be direct or polysynaptic. The anatomic connections of the insula to the various cortical and subcortical regions are highly consistent with its involvement in emotional processing (as well as many other sensory and motor functions). Its functional (i.e., physiological) connections (see below) show even more direct evidence of its probable involvement in interoception, especially with relevance to the putative role of the cortex in interoception and awareness.

Function. Anatomical evidence provides support for insular connections with regions implicated in emotional processing, and thus indirectly with interoception. Even more direct evidence exists for a functional link between the insular cortex and interoception, namely activation of the insula by visceral sensation in a somatotopically organized way. Insular function will now be reviewed. In addition to animal data, human data exist, including data from neurosurgical procedures and from people with epileptic foci in the insula (e.g., Ostrowsky et al., 2000). Anatomical studies often focus on connections by fiber tracts that do not involve synaptic connections. Functional studies, on the other hand, often study multineuronal, polysynaptic pathways as well as connections without intervening synapses. Table 7–2 contains a summary of insular

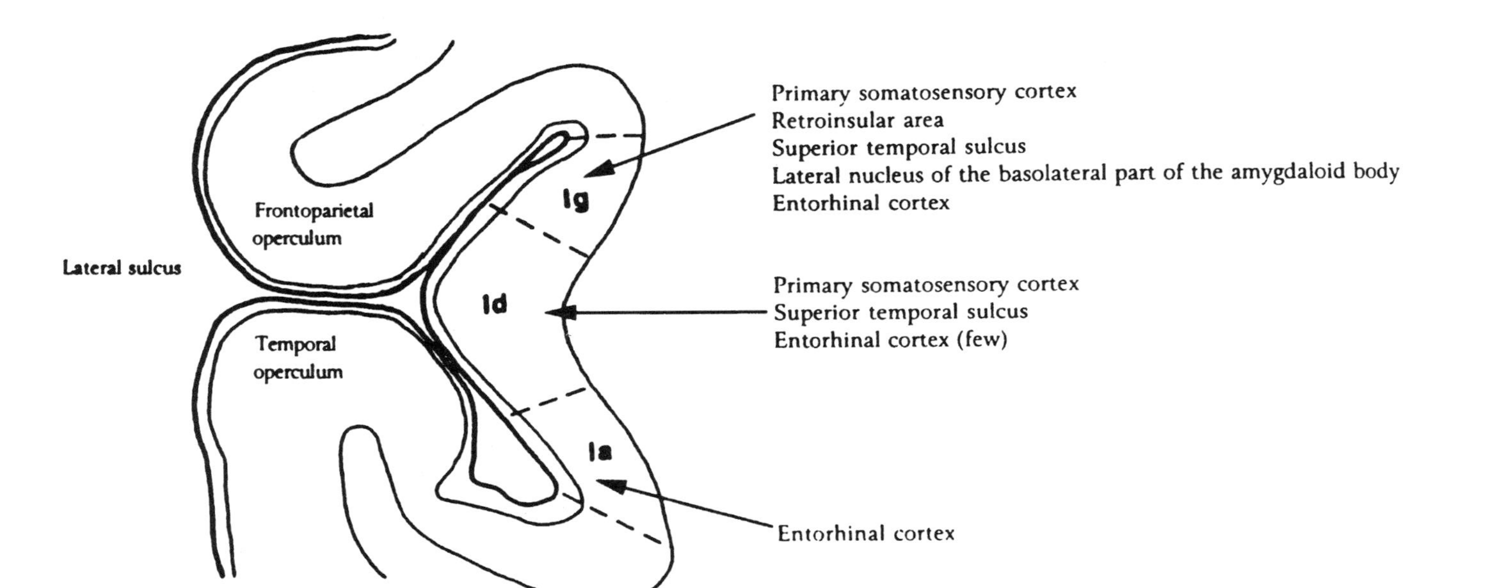

Figure 7–3. Topographical afferent input to different cytoarchitectural regions of the insular cortex. [From Augustine, 1996, reprinted with permission of the holder of the copyright.]

Table 7–2. Outline of Functions of the Insular Lobe

Visceral Sensation	Somatosensory
primary cortical gustatory	tactile recognition and recall
esophageal	asymbolia for pain
partial seizures	pseudothalamic pain
Visceral Motor (Autonomic)	Multifaceted Sensory
vomiting	feeding
cardiovascular	anatomic neglect
Limbic Integration	Motor Association
simple phobia	post-stroke motor recovery
	ocular movements
Vestibular	Memory Dysfunction
Language	Alzheimer's
language-related memory	Other
verbal component of working memory	selective visual attention
auditory processing of speech	

[Adapted with permission of the holder of the copyright from Augustine, 1996.]

functions, as known as of the mid-1990s. It is comprehensive in its general categories, but not in all functions listed.

The visceral sensory (afferent) functions of the insular cortex are the most relevant to interoception, of course, and will therefore be reviewed first. Other functions, including visceral motor (efferent), will also be reviewed. There is a clear somatotopic functional organization to the insula in the anterior–posterior direction, with gastrointestinal and gustatory function in the anterior region, cardiovascular (including baroceptive) and respiratory posterior, and some data indicating an intermediate chemoreceptive region (see Cechetto, 1987; Benarroch, 1997; Hanamori et al, 1998). There is also evidence for a nociceptive receptive function. This organization is based on a combination of electrical, chemical, and mechanical stimulation studies, in conjunction with neuroanatomical tract tracer studies and electrical recording studies (see Cechetto and Saper, 1990). Functional evidence exists for specific connections with several structures known to be involved in visceral sensation, the NTS and the parabrachial nucleus, via tracts through the lateral hypothalamus and parvocellular ventroposterior thalamic complex. (These tracts have reciprocal efferent connections as well—Kapp et al., 1985, including some involved in cardiovascular control—Ruggiero et al., 1987.)

Involvement of the insular cortex with efferent visceral function has been known for many years. Involvement in gustatory behaviors was recognized as early as 1900 (see Cechetto, 1987), including the more recent observation that a lesion in this area can eliminate a learned taste aversion (see Cechetto and Saper, 1990). As early as the 1920s, a relationship of this region to sham rage was recognized, although the regional relationship was not very specific (see Cechetto and Saper, 1990). As early as 1950, it was known that in humans electrical stimulation of the anterior insula produced sensations referable to the abdomen and the vagus nerve. Stimulation of the insula in the appropriate area, either electrically or chemically, produces changes in cardiovascular function, respiration, piloerection, pupil dilation, gastrointestinal activity, salivation, and catecholamine release (see Cechetto and Saper, 1990; Oppenheimer et al., 1992). Pharmacologically, not only is acetylcholine important in insular function, but glutamate has been identified in efferent projections (Butcher and Cechetto, 1998). Some evidence exists for hemispheric asymmetry in insular function. The right insula seems more involved in sympathetic function, while the left appears more related to parasympathetic function (see Benarroch, 1997).

Evidence exists for involvement of the insula in many other central nervous system functions (Table 7–2), including somatosensory function with special involvement of the contralateral side of the body, and convergence of multiple limbic activities indicative of important involvement in stress and other emotional processes (Krushel and van der Kooy, 1988; Allen et al., 1991). Of course, several of these other functions (e.g., motor association, memory) are implicated in emotion and therefore also in interoception.

CONCLUSION

Saper (1996) reviewed several of the anatomical issues already addressed, and links this anatomy to emotion. He agrees that emotion is a concept that is highly related to motivation (e.g., hunger, sexual arousal, p. 538). He also underscores the fact that emotion and cognition are closely allied and notes that emotion as it is normally understood could not exist without a motor, or efferent, component (somatic, autonomic and endocrine) as well.

While Saper agrees that older concepts of the limbic system are

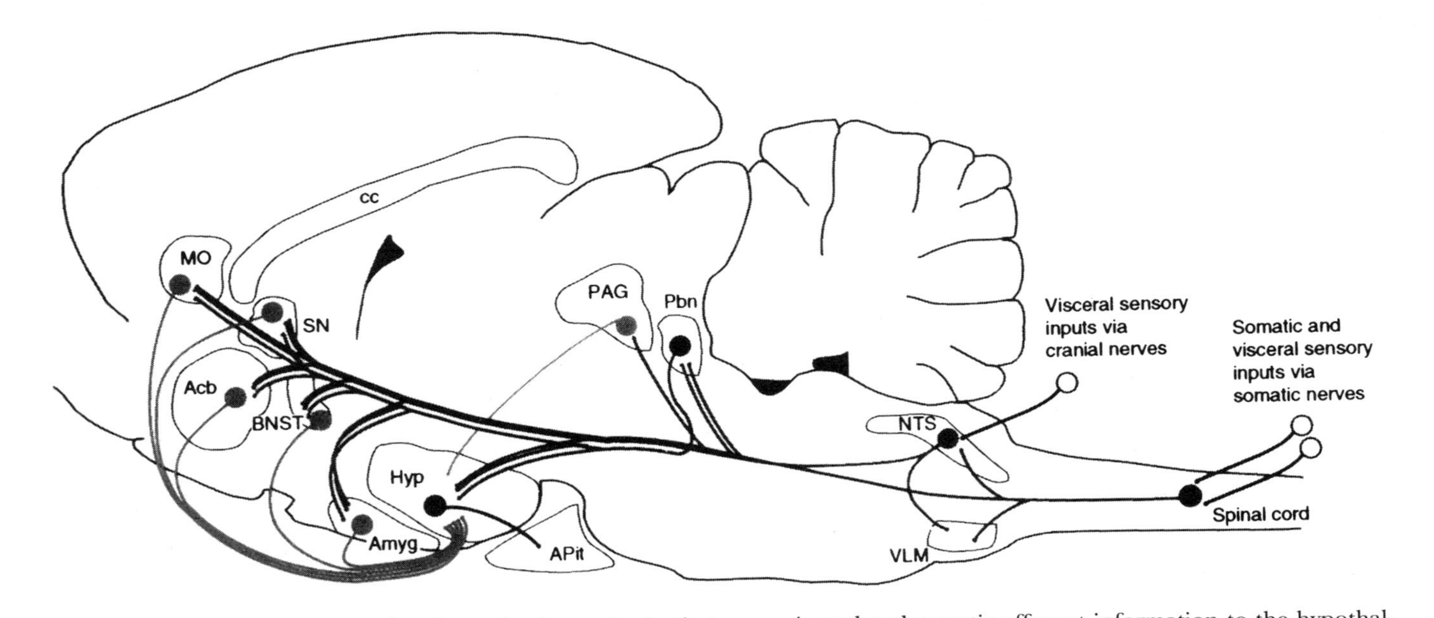

Figure 7–4. Illustration of pathways in the rat brain that carry visceral and somatic afferent information to the hypothalamus and limbic system. (See original for abbreviations of structures.) [From Burstein, 1996, reprinted with permission of the holder of the copyright.]

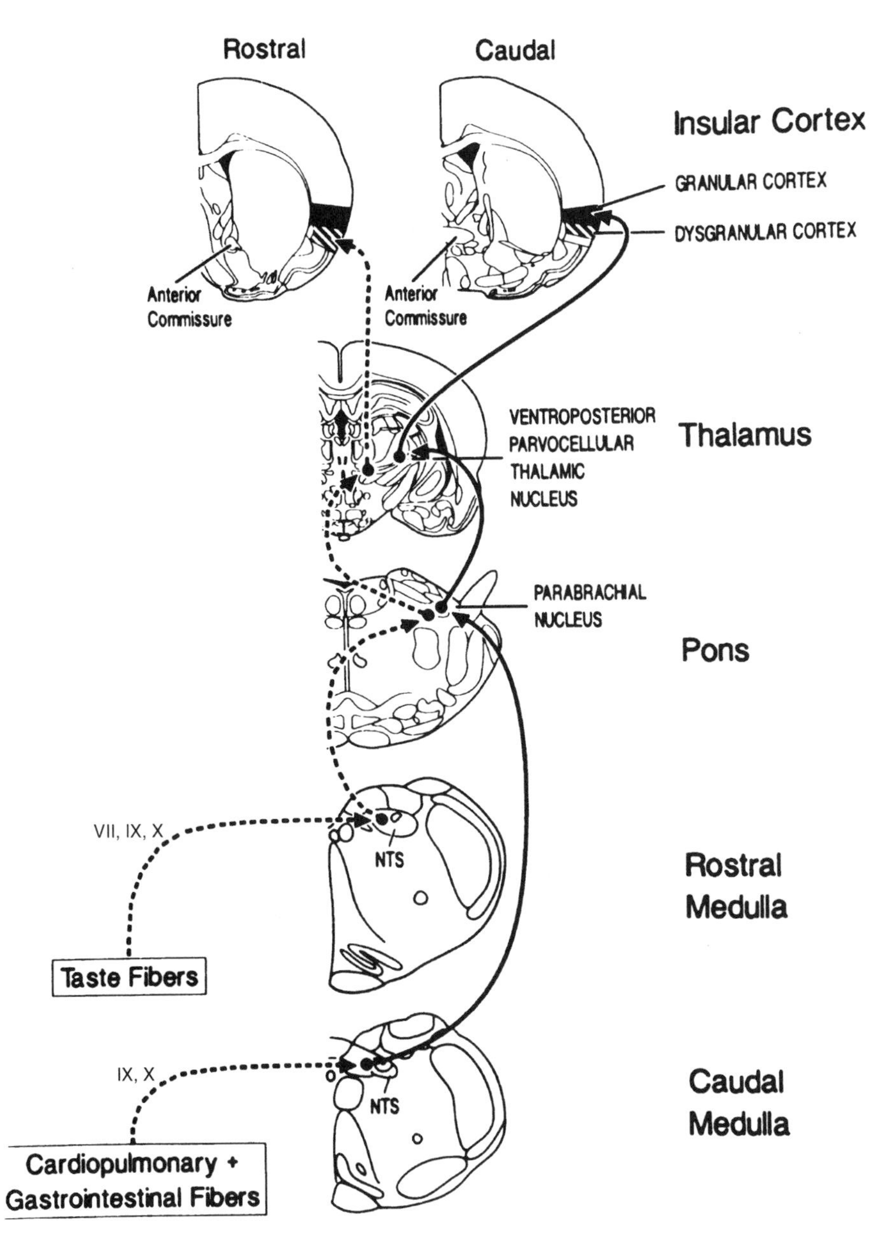

Figure 7–5. Summary of pathways carrying parasympathetic visceral afferent information from brainstem to cortex. [From Cechetto and Saper, 1990, adapted from Cechetto, 1987, reprinted with permission of the holder of the copyright.]

either incorrect (Papez's circuit) or too inclusive and therefore vague and nonspecific (MacLean's visceral brain), he also points out that the structures usually included in what is now referred to as the limbic system are in an ideal configuration to be involved in the mediation of emotional processes. He identifies a potentially important difference between emotion and cognition, in that emotion has more "momentum" (i.e., longer lasting effects, although, of course, memory also has long lasting effects). He does not mention the likely possibility that conditioning effects may be important in the persistence of emotional changes as well. As described above, aversive (fear) conditioning is often long lasting because it resists extinction. Its association with visceral changes, including visceral sensory effects, raises the possibility that conditioned visceral changes may contribute to the resistance to extinction, which of course links it to interoception.

Saper notes that subcortical mechanisms of emotion are very important. He argues that the long lasting effects of emotion may relate to the role of the ventral striatum in emotion and notes that human illnesses that have prominent motor abnormalities and are associated with pathophysiological changes in the striatum also have prominent emotional dysfunctions. If it is true that full awareness of emotional changes requires sensory information to reach the cortex, it suggests that any types of emotional changes that do not reach the cortex might occur partially or fully outside awareness. How does sensory information from the body, but outside awareness, affect emotion, cognition, and behavior? Is it a somatic marker that is "in the background," to use a concept from Damasio (1994)?

Overall, the neural science of visceral sensory systems shows the potential linkages between visceral sensation and various other essential psychobiological functions, such as motivation, emotion, control of autonomic function, and the emotional motor system (Holstege, 1992). Figure 7–4 summarizes the major sensory connections in the rat, while Figure 7–5 shows the major anatomic structures and their central nervous system locations in the human.

eight

THE CARDIOVASCULAR AND RESPIRATORY SYSTEMS

> I am Thane of Cawdor: If good, why do I yield to that suggestion whose horrid image doth unfix my hair and make my seated heart knock against my ribs, against the use of nature? Present fears are less than horrible imaginings.
>
> William Shakespeare, *Macbeth*

This chapter will address directly such questions as, Can one be aware of one's cardiac, vascular, and/or respiratory functions? How reliable and valid is such awareness? Under what circumstances does such awareness occur? Can it be learned?

Cardiac responses are of particular interest because there is a clear, discrete, relatively easily measured physiological response, although all visceral awareness studies share the problem that few, if any, independent criteria can indicate whether or not awareness actually occurred (e.g., evoked cortical EEG responses or functional imaging methods might serve this function). In subsequent chapters, awareness in other visceral organ systems, mainly the alimentary tract, will be addressed in the same way.

Research studies attempting to understand visceral sensation date back at least to early in the twentieth century (e.g., Boring, 1915). Current interest, however, was kindled largely by biofeedback investigations. Even though the biofeedback technique serves to externalize the visceral–autonomic functions, the possible role of visceral sensation in biofeedback focused the attention of investigators on visceral sensory processes per se.

Before reviewing the data that explicitly focus on these questions vis à vis the cardiovascular and respiratory systems, several methodological issues will be discussed. First will be the role that factors other than specific sensory processing of visceral afferent impulses play in subjects' reporting of visceral experiences. Pennebaker and Hoover (1984) made the distinction between visceral detection and visceral perception. "Perception implies the subject's use of both internal physiological and external environmental information in the perception of the visceral state. Detection connotes the subject's use of only physiological information—to the exclusion of all other factors" (p. 339). Second will be a discussion of methodological factors in measuring the ability of subjects actually to detect either individual heart beats or heart rate. These cardiac parameters will be discussed because methodology relating to the perception and detectability of cardiac action has been studied in more detail than other visceral–autonomic functions. Third, when a phenomenon is studied, it is highly useful to isolate it and study it in full force. However, much of the research reported here involved assessment of cardiac interoception with resting heart rate, not in full force.

VISCERAL DETECTION AND VISCERAL PERCEPTION

Pennebaker (1982) addressed the issue of the sources of the various factors that contribute to the occurrence of physical symptoms. Physical symptoms are not synonymous with interoception but are clearly related to it. Although his discussion goes beyond consideration of the cardiovascular and respiratory systems, the issues he raises will be reviewed here. They potentially apply to physical symptoms of all kinds. Two specific sources of physical symptoms he reviews, changes in blood pressure and changes in blood glucose, will be summarized later—for blood pressure later in this chapter and blood glucose in Chapter 11.

A symptom, indeed any perception, is a multiply determined event. The use of signal detection methodology, for example, highlights this fact by using a technique that separates detection from other factors (e.g., motivation) that determine whether an event is reported to have occurred. A fact that is essential to this issue, but often not explicitly mentioned, is the issue of intensity of the stimulus. The stronger the stimulus, the less likely it will be that other factors will influence reporting (and vice versa for weaker). For example, if the task is to decide if one's hand has been touched, bias from non-sensory factors might influence reporting when the stimulus is a small feather but is not likely to do so if the stimulus is a hammer blow. In the case of visceral sensory events, many events seem to be like the feather, but some certainly are not—pain from a myocardial infarction, respiratory distress from occlusion of the airway, the pressure sensation of a full urinary bladder. The more subtle the sensation, the more likely it is that other factors will influence reporting. Not only is reporting influenced by other factors, but actual detection is almost certainly affected as well. Visceral sensory information is embedded in the totality of afferent impulses into the central nervous system. The conscious experience is due to this totality. How much harder it seemingly is to extract the primary visceral sensory information from the complex that constitutes or comprises, for example, the emotion of fear.

As reviewed by Pennebaker, symptom reporting is a combination of encoding, awareness, and reporting. Each of these factors can be influenced by many other factors. For example, Pennebaker lists multiple cultural and demographic factors that have been shown to contribute—for example, age, sex, marital and socioeconomic status, racial and subcultural differences. Other factors such as cognitive attribution are also important. Of great importance is the fact that there seem to be individual differences in what symptoms individuals experience, as well as the fact that what might be the same experiences are labeled differently by different individuals.

Pennebaker addresses the role of perceptual processes in symptom production. He notes that expectations strongly influence symptom occurrence, and suggests that there may be competition between internal and external cues. The less information is coming from external sources, the more likely the individual will be to attend to internal cues. Expectations are learned and are influenced by the specific circumstance the organism is in. Different methods of interoceptive testing can be differentially affected (Phillips et al., 1999).

Can people validly report their internal states? Data that Pennebaker reviewed indicate that this question can best be answered individual by individual. Group data, it turns out, tend to obscure results. Among subjects, correlations between physiological states and symptoms are typically significant but not robust. When the same question is asked subject by subject, several findings are apparent: *(1)* some subjects are better than others, *(2)* different subjects label symptoms in different ways, *(3)* different subjects do well with different physiological state-symptom combinations—that is, this ability is not unidimensional, *(4)* different subjects may respond to different aspects of the physiological stimulus (e.g., which aspect of heart action—electrical, mechanical, etc.—is a given subject responding to?), and *(5)* most importantly, although the individual approach reveals that subjects frequently show inaccurate judgments, sometimes the associations are quite robust, including reliability over time. There is some suggestion that situational factors can influence this ability—for example, differences in arousal state. Concerning symptom self report, not only is it apparent that there are differences among people, but the ability to verbally report visceral sensations (to put it into words) is probably generally difficult. The experiences themselves are typically vague and faint. Furthermore, for some visceral functions, there seems to be localized representation in the non-dominant—that is, non-verbal—hemisphere of the brain (see below).

At the very start of this book the issue of the potential linkage between bodily sensations and emotion was raised. Pennebaker discussed this issue in his book as well. Two questions were addressed: *(1)* Are reports of emotional experiences correlated with reports of physical symptoms?, and *(2)* Are the reports of physical symptoms correlated with actual physiological changes? Concerning the first question, both among-subject and within-subject analyses were described. Various emotions such as happy, tense, angry, sad, and guilty were correlated with such physical symptoms as racing heart, tense muscles, upset stomach, warm hands, and dizziness. Using racing heart as an example, for the among-subject data, there were significant correlations with tense, sad, guilty, and happy (negative). For within-subject data, only tense showed a significant individual correlation. Thus, for racing heart—and for symptoms in general—significant patterns were seen, and these patterns were different for within-subject versus among-subject correlations. Conversely, subjects given symptom clusters were reliably able to identify associated emotional experiences.

In contrast to the significant associations noted above, when the question is asked, What is the linkage between either symptoms or emotional experiences and actual physiological changes?, many studies have found that the correlations are typically weak. The data cited in this and the prior paragraph indicate that subjects associate, at a perceptual or cognitive level, physical symptom complexes with particular emotional experiences, but specific physiological changes do not correlate very well with the more complex perceptual–cognitive states. At least two possible reasons exist why this might be so. First, the physiological changes are only a part of the overall states, and as such the correlations are weak. Second, the wrong physiological changes are being monitored. For example, if a person says, "I felt my heart beating rapidly," what physiological change(s) was the subject actually feeling? Was it heart rate (chronotropic effect)? Was it strength of the heart beat (inotropic effect)? Was it blood flow in the aorta, the carotid arteries, or elsewhere? Was it the heart beating against the inside of the chest cavity, as Shakespeare's Macbeth described almost 400 years ago. (Shakespeare knew 400 years ago that visceral sensations may be mediated by skeletal sensory receptors—see epigraph.) Was it some or all of these, and/or something not yet known—release of a substance from the heart such as atrial natiuretic peptide (Levin et al., 1998) correlated with heart action that produces detectable sensations? If the physiological contributions to emotion, or to cognition more generally, are only one part of the overall complex of inputs, even if an important one, it should perhaps not be surprising that the correlations observed are not large, especially in circumstances when the emotional state is either relatively mild or is not one of the emotions usually associated with visceral–autonomic activation. That does not mean, however, that symptoms and specific physiological changes have not been observed and reported, as described below.

Demographic and cultural factors affect symptom reporting. Personality factors appear to have an influence as well. Pennebaker highlights two such variables that appear to be associated with symptom reporting—attention to body function and locus of control. Additionally, relationships were found for symptom reporting with health-related behaviors, eating patterns, interpersonal behaviors, and self-image.

Much additional research has addressed questions like those described above. This work described by Pennebaker gives a reasonable sampling of the kinds of factors that have been studied. The overall conclusion from Pennbaker's review highlights the fact that cognitive–perceptual states are very complex and that symptom reporting is determined by many factors. Furthermore, different ways of assessing visceral perception—for example, self report, heart beat tracking, and signal detection—do not necessarily correlate among themselves, or over time (reliability), or with actual heart action (Pennebaker and Hoover, 1984). Nonetheless, visceral–autonomic sensory afferent information is likely to play a significant role in visceral perception.

METHODOLOGICAL ISSUES

Two common but somewhat contradictory beliefs exist about an individual's ability to sense the action of the heart and other visceral organs. The first belief is that it is very hard, or even impossible, to do in a valid and reliable way. The second is that one can feel such states as stress or fear, and that a major part of that awareness involves cardiac sensations. People will say, for example, "I felt my adrenaline pumping." There are several things to note about this statement. First, it commonly refers to a constellation of physical and cognitive experiences that includes cardiac awareness. Second, this constellation is consistent with a larger general pattern indicative of autonomic sympathetic activation. Third, as stated literally, the statement is almost certainly usually incorrect. Anxiety reactions and typically encountered psychological (and severe intense physical) stresses usually do not raise circulating epinephrine (adrenaline) levels high enough to be sensed (Cameron et al., 1990a; see Fig. 2–2). It seems more likely that the overall sympathetic discharge pattern is sensed rather than the effects of adrenaline per se.

Most of the research designed to determine the ability of individuals to sense the action of the heart has two characteristics. The large majority has been done in humans, to a considerable extent because verbal reporting can only be done with humans, even though the technology to answer such questions in non-human animals is available. Second, the majority of this research has been done in the unstimulated state (i.e., in the resting condition), although some studies involving manipulation of cardiovascular activity have been described. A major advantage of doing resting state studies is that various potential sensory–perceptual distractions can be avoided. A major potential disadvantage is that studying subjects under only basal conditions runs the risk of not detecting interoceptive abilities apparent only in non-basal situations.

Reed et al. (1990) and Jones (1994) reviewed methodological issues related to interoception, focusing on cardiac awareness. Jones provided an extensive review of methodological issues, while Reed et al. provided a shorter summary. The number of studies that have addressed this question has not been that large. For example, Jones lists fewer than 50 published papers and fewer than 60 abstracts that describe research not described in the published papers over a period of approximately 20 years.

In assessing the ability to sense or detect cardiac action, an important question is, What is being detected? Possibilities include chronotropic versus inotropic cardiac effects, blood flow in large vessels, stimulation of somatic receptors in the chest wall, hormonal factors, and/or other as-yet-unknown variables. In addition, investigators have separated the question, Can the subject sense heart rate? from the question, Can the subject sense individual heart beats? Separate methods have been used to attempt to answer these two questions.

As noted by Jones, an issue of concern is the fact that some of the methods found to be least valid and least reliable are still being used, especially by investigators focusing on clinical questions and on clinical populations. As noted by Reed et al., even the basic-science-based investigators often have not paid sufficient attention to such issues as psychophysical scaling and standard reliability requirements. Another issue of major concern for investigators is guessing. Most adults know that the resting heart rate is somewhat more than one heart beat per second. Any valid measure of a subject's ability to sense heart rate will need to exclude this factor.

Reed et al. divided the procedures that have been used to assess cardiac interoception into three categories: questionnaires, tracking techniques, and discrimination tasks. Tracking and discrimination tasks will be separately discussed. Suffice it to say about questionnaires that studies usually have found little relationship between questionnaire measures and measures of cardiac awareness that are closely tied to actual physiological changes. Based on what has already been said, that is not surprising. The questionnaires measure perception, not detection, that is, the whole sensory–perceptual–cognitive complex that determines the end result of self ratings.

Under tracking techniques, Reed et al. identified self-report (i.e., verbal report) and behavioral procedures. Both of these apply to assessment of the ability to sense heart rate (i.e., a train of heart beats rather than an individual beat). Guessing is a problem with this method, although (based on studies they reviewed) if the duration of tracking is varied without informing the subject of each trial's length, this problem can be reduced somewhat. Reed et al. concluded ". . . self-report . . . show promise. . . . However . . . findings are neither conclusive nor definitive" (p. 274). Interestingly, changing the conditions of arousal (presumably thereby modifying heart rate itself) improved performance.

Behavioral tracking typically involves requiring a subject not to count heart beats but rather to perform a behavioral task such as a button press when each heart beat is sensed. An advantage of this procedure is that with proper technology each motor response can be paired with a particular heart beat. Here, again, guessing is a potential problem, as is reaction time and the possibility that the motor response in some way interferes with task performance (which also could be true for the self-report counting task). Also, if an attempt is being made to assess the accuracy of sensing actual heart beats by counting as correct only motor responses that follow a specific component of cardiac action by no more than some fixed temporal amount, a potential bias is introduced. This procedure requires the investigator to make an assumption about what (i.e., which aspect of the cardiac cycle and/or vascular mechanical response) the subject is sensing. Despite possible individual differences, the investigator is likely to make the same assumption for every subject. If the investigator is wrong, performance will appear to be less correct than, in fact, it is for every subject for whom the incorrect assumption is made.

In order to address some of the problems of the motor behavioral tracking task, some investigators have observed that it is temporal variability of response latencies that is most important. Assuming a consistent response latency, this would remove any effects of different subjects responding to different components of the cardiovascular events (as long as it was the same event each time). With this procedure, if the subject responds accurately, latencies will show only minimal variability, while random responses will lead to more variance in latencies. Various ways of calculating this variability factor have been used. Adjustment for heart rate was required, and it should be noted that a variability in the underlying heart rate itself exists due to respiratory effects mediated by the vagus nerve. Using distributional measures was of some benefit. Overall, Reed et al. concluded ". . . although tracking tasks provide a great deal of face validity, they contain enough ambiguity and lack of precision to render them dissatisfying. . . . Behavioral tracking tasks . . . are certainly more satisfactory than the self-report tasks . . ." (p. 279).

Because the tracking tasks all have problems, discrimination tasks were developed. In the discrimination tasks, subjects are presented with external signals that have fixed but different relationships to heart beats. Subjects are required to identify the external signal that bears the experimentally defined relationship to either heart beats or heart rate. Reed et al. concluded that heart rate (as opposed to individual heart beat) discrimination tasks were not clearly valid and should therefore be avoided.

Concerning heart beat discrimination, Reed et al. described a procedure first used by Brener and Jones and then modified with improvement by Whitehead and colleagues. In the procedure by Whitehead et al., subjects received stimuli that were either 128 ms or 384 ms after the R-wave of the ECG. The task was to associate one of the stimuli (128 ms) with the cardiac event and not the other (384 ms). This task avoided many of the confounding issues of other methods. However, as Reed et al. describe the outcome of this study, "Now that they had an apparently valid measure of cardiac perception, there was nothing to measure; few people actually were able to perceive their heartbeats" (p. 282). Or, to state it another way, when all the extraneous or artifactual sources of information as to actual heart action were removed from the interoception task, nothing was left. On the other hand, Jones reviewed data later collected by Clemens that indicated that subjects could perform well with the procedure of Whitehead et al. using the same delays (see Jones, p. 85).

Later investigators concluded that the demands of the procedure used by Whitehead and colleagues might have provided an incorrectly negative conclusion because the task was too hard. This procedure required the subject to make a discrimination of one-quarter of a second. Furthermore, potential experimenter bias was present in this design. It assumed that the component of the overall cardiovascular event to which the subject was responding was the R-wave (ventricular systole). While this assumption has face validity (ventricular systole—the squeezing of the heart to force out the blood—as the event that produces the discriminable stimulus) as already stated above, this might not be correct for at least some subjects—and probably is not.

Later improvements in study design have been made. These included *(1)* making the incorrect stimulus more clearly incorrect because it did not have a fixed relationship to the cardiac cycle, as did the correct stimulus; *(2)* placing the correct stimulus later after the R-wave, at 200–300 ms, near the time when most subjects who are able to actually perceive heart action experience it as occurring; and *(3)* giving subjects more than two choices of a right and a wrong interval (under the assumption that consistent choice of any one interval indicates valid responding, but allows for different subjects to respond to different components). As of the late 1980s, when Reed et al. prepared their review, they concluded that this design choice appeared to be the best.

Jones also reviewed methodological issues in more procedural detail than Reed et al. Jones also divided the various methods into the two large initial categories of tracking, or counting, paradigms versus discrimination, or signal detection, paradigms, and agreed that, as a measure of interoceptive detection of actual physiological cardiac events, tracking, or counting, paradigms were inferior.

Jones distinguished several different discrimination paradigms, and he noted that not only might respiratory changes affect results, but some of the paradigms were liable to subject manipulation by voluntary modification of respiration, even if not done intentionally. For example, breath holding as a means of removing breathing motions and noises as distracting stimuli could influence results. Another important factor noted by Jones was knowledge of results (i.e., feedback training). Some of the paradigms and experiments included it, and others did not. For at least some studies, knowledge of results improved performance, sometimes quite substantially.

Jones reviewed studies that compared the performance of the same subjects in two different procedures. For example, two studies are described that compared the Brener–Jones versus the Whitehead paradigm (see the original Jones article for more detail on how these paradigms are the same and how they differ). Both studies concluded that these procedures tap different subject behaviors, and the subjects' results on one did not correlate significantly with results on the other. Other paradigm comparisons also indicated that the subjects' performance often did not correlate across procedures. The comparison study by Pennebaker and Hoover referred to above arrived at a very similar conclusion. The unavoidable conclusion is that, despite the goal of all these procedures to measure subjects' ability to sense cardiac action, they do not measure the same thing. Tracking procedures, and especially questionnaires, seem to measure, at least to some extent, the totality of perception and not just detection. It is not clear whether and to what extent larger perceptual factors might affect the discrimination procedures, and among the discrimination procedures, what visceral–autonomic afferent impulses from which cardiac (and, possibly, vascular) interoceptors are affecting which paradigm. There is no obvious gold standard paradigm. The results described below, although generally indicative of the validity of cardiac interoceptive processes, must be interpreted with this caveat in mind.

THE CARDIAC DATA

It is not the goal of this section to provide critical commentary on the studies and data described. As with most other fields, at the "cutting edge" substantial uncertainty exists about some of the findings, and in many cases replication has not been reported. This field has special methodological difficulties, as described above. That should not lead the reader to be overly critical but should prompt circumspection in interpretation of findings. The reader should, however, not allow himself or herself to be unable to see the forest for the trees. While it has long been the given wisdom that, with the exception of pain, one generally cannot feel the functions of one's visceral organs, the bulk of the data does not support a conclusion this conservative. For the cardiovascular system (and for other visceral organ systems such as the gastrointestinal system), under at least some circumstances at least some individuals clearly do experience visceral–autonomic sensory awareness. Thus, for all the visceral organs, the questions of interest are:

(1) Does the organ in question produce visceral sensory afferent impulses?
(2) What role do such impulses play in visceral function, including homeostatic processes?
(3) Do these impulses contribute to complex motivational states such as hunger or fear, and/or do they contribute to cognitive processes?
(4) Do these impulses reach awareness?
(5) Do these impulses affect behavior, and if so, how?
(6) What is the relationship between the answers to questions 3–5, and the physiological functioning of the central nervous system?

The first two of these questions are in the domain of sensory physiology, and although of great importance, it is not the main focus of this book. The main focus concerns the remaining four questions. The reader will find that studies to answer these questions are often lacking, although some of the studies that addressed especially questions 3, and to a lesser extent 5 and 6, have been discussed in several of the previous chapters. What will be shown is that a simple, Can one feel one's heart or gut, and so on?, is a question to which the answer is already known to be "yes," and more sophisticated questions should now be asked.

This chapter will review data related to the heart, vascular system, and respiratory system. Studies of cardiac interoception in normal subjects will be presented first. Then blood pressure and respiratory functions will be discussed, followed by a review of studies of interoception in association with various abnormal states. Finally, the potential role of interoceptive processes in stress and neurocardiology will be briefly discussed. The data reviewed will in large part address question 4. The specific factors that have been studied in normal subjects are outlined in Table 8–1.

Much of the original interest in determining if subjects could demonstrate awareness of cardiac or other visceral functions arose from biofeedback research, as already noted. One of the original investigators was Brener (1974, see also 1981), who hypothesized that the ability to gain control over visceral function necessarily involved the ability to sense visceral function. Carroll (1977) and Gannon (1977) reviewed data linking cardiac perception to cardiac control.

Even before these investigations in the 1970s, a few relevant

Table 8–1. Variables Related to Cardiac Interoception

Variables
Subject Variables
age
weight and percent body fat
sex (including control for body fat)
fitness
cardiovascular dynamics
symptom proneness (including diagnosis: anxiety, etc.)
Methodological Variables
rest versus exercise (versus pharmacologic stimulation)
posture (body position)
emotional induction (arousal, stress, etc.)
knowledge of results (feedback)
test–retest (reliability)
external stimuli
number of training trials
training before study
magnitude of change in heart action
Outcome Variables
cross-modality correlations
potential mediating factors
what is the stimulus?
somatic (and proprioceptive?) changes
Central Nervous System
EEG
somatosensory cortex
frontal cortex
laterality
imaging findings

[Adapted with permission of the holder of the copyright from Jones, 1994.]

studies had been reported. For example, Fisher (1967) determined whether the degree of awareness of specific organs was correlated with actual levels of activation. Two studies were performed, one involving both heart and stomach function and a second involving heart only. Electrographic activity was used as the measure of actual activation, and measures were made while resting and (in one of the experiments) in response to a noise stress as well. Significant positive correlations were observed for both resting and stress measures for both heart and stomach activity.

Lacroix (1981) provided a criticism of the hypothesis that visceral sensation was necessary to visceral conditioning and argued that afferent processes play, at most, a secondary role in gaining control. The resolution of this issue is unclear and is further clouded by the problems that occurred in replication of the biofeedback studies involving muscle paralysis described above. In other words, the questions of the need for visceral afferent information and/or the need for somatic afferent information in learning control of the function of visceral organs, not to speak of the role of skeletal voluntary motor functions in changing these functions, are all intertwined. Some reported data indicate that perception of heart activity and the ability to demonstrate control of heart rate are not linked (e.g., Young and Blanchard, 1984; DePascalis et al., 1991). Sometimes overlooked is the possibility that visceral–autonomic awareness is not necessary to learn visceral control but that visceral afferent impulses play an essential role nonetheless. That is, in order for the central nervous system to gain effective control of the functioning of some visceral organ, it must be able to monitor the functional status of the organ (i.e., there must be visceral afferent information). This monitoring logically need not necessarily involve awareness.

Subject Variables (Individual Difference Factors)

The first group of factors to be considered in normal subjects are subject factors—individual differences. Several differences exist among subjects that have been shown to relate to awareness of heart activity. These include age, weight and percentage of body fat, sex and sex controlled for lean body weight, physical fitness, factors related to cardiac function, and proneness to report symptoms (Table 8–1).

A methodological point related to this individual difference issue is that in some studies investigators have enrolled subjects on the basis of presumptive or demonstrated ability to perform the cardiac awareness task at above-chance levels. This has been done, for example, by studying only men, since men as a group perform better than women (see below), or by pretesting subjects and only including those who perform above chance. Such a tactic has advantages—only by studying subjects who can do the task can one understand how the task is done, and disadvantages—lack of generalizability and no opportunity to learn why those who cannot do the task are unable to perform it.

Sex. One of the factors that has been reported to show individual differences is sex (reviewed by Katkin, 1985; Jones, 1994, 1995). Specifically, several studies have reported that men performed more accurately than women. This finding has been true not only of cardiac function but also, for example, of gastric contractions and GSR. On the other hand, not all studies have reported this finding, and a few reports showed small advantages for women. Factors that have been proposed as possible sources for male advantage are fitness differences (see below), training effects and knowledge of results (men showed more improvement across trials—see below), and body fat composition, or leanness (see below). Jones (1995) observed that the advantage men might have is more apparent in studies that require awareness of individual heart beats instead of ongoing heart rate. Studies not cited by Jones (Ludwick-Rosenthal and Nuefeld, 1985; Harver et al., 1993) also demonstrated that men performed more accurately than women.

Weight and Body Composition. Jones (1995) reported three studies that examined the effects of body composition, that is, percentage of body fat, on the ability to perceive heart action. The first study was a retrospective analysis that found that lower body mass index (leanness) was associated with better perception. In a second study, which also was designed to assess effects of body position on awareness, percentage of body fat had the strongest association with heart beat awareness.

A study by Rouse et al. (1988) was designed directly to assess the effects of sex and body composition. The Whitehead procedure, described above, was used. Four equal-sized groups—lean and non-lean men and women—were studied. Percentage of body fat was determined. Body composition was strongly significant but sex was not, and there was no interaction. In other words, when body fat composition and sex effects were assessed independently, body composition was the relevant variable and the effect of sex disappeared. Lean men and women group both performed above chance on awareness, while both non-lean groups did not. It should be noted that this study did not provide knowledge of results. Thus, it did not test the question of whether men do better when knowledge of results is provided. One possible explanation for the observed results is that non-visceral somatic sensory receptors play a role in cardiac awareness (e.g., receptors in the chest wall), and that body fat damps the sensations these receptors otherwise can sense.

The Rouse et al. study produced a clear and robust result. Unfortunately, another study calls these results into question. Using a somewhat different procedure and signal detection analysis, Gardner et al. (1990) found that both obese and non-obese subjects demonstrated cardiac awareness, and no difference was found between either obese and non-obese subjects or between men and women.

Age. Dickerson and Jones, in a study described in Jones (1994), assessed the effects of age on awareness. They reported that the best performance was observed in adolescents and young adults. They hypothesized that this might be due to leanness, since percentage of body fat is greater in older individuals. Consistent with this presumptive effect of age, they noted another study done in older adults that found poor performance in both older subjects who were post myocardial infarction and matched older subjects who were not. Another possibility also suggests itself. It is known that adrenergic autonomic nervous system function changes as individuals age (e.g., Cutler and Hodes, 1983; Ebstein et al., 1985). Since it has been shown that myocardial function, in part under adrenergic sympathetic control, influences awareness (see below), changes in myocardial function may contribute to the age effect.

Fitness. The hypothesis has been offered that physical fitness may affect cardiac awareness. This may occur because the exposure to changes in heart rate associated with physical training sensitize the person to the feelings of changes in heart action and/or because the trained heart functions differently. For example, the trained heart typically has a stronger inotropic effect—it beats harder and ejects more blood per beat. Several studies have been reported that address this question.

One study reported by Jones (1994, 1995) found that more fit subjects were more aware, but that this effect was only true for men. During exercise, awareness was increased. In contrast to the first study, a subsequent study did not find differences across fitness groups when subjects were at rest but did find again that exercise improved awareness. A third study also found exercise improved awareness. While these studies do not specifically address the hypothesis offered in the paragraph above, they do demonstrate two things. First, there is contradictory evidence as to whether fitness training improves cardiac interoception, but the data are consistent that the action of the heart is easier to sense when it is increased. These studies do not answer the question of how. This finding relates as well to the issue of whether myocardial factors influence awareness, and it relates to methodological issues as well—how to increase awareness.

Cardiovascular Dynamics. Studies of awareness of cardiac action are designed to be linked to the action of the heart itself, typically the heart's ventricular contraction. Data collection is typically time-locked to the R-wave of the ECG (electrocardiogram), which is the electrical manifestation of ventricular contraction.

First, it is logically possible that the perceptible stimulus is not mechanical. As reviewed above in Chapter 5, in addition to mechanoreceptors, other types of interoceptors have been identified, such as osmoreceptors. However, the timing of the cardiac-generated stimulus would appear to exclude the possibility that chemoreceptors, osmoreceptors, or thermoreceptors could be involved. Are there, for example, electroreceptors that could sense the electrical events in the heart? That possibility seems unlikely because such receptors apparently have never been identified, but neither does it seem that such a possibility has been seriously considered or studied.

The type of receptor that seems most likely to be involved in cardiac interoception is mechanoreceptors. Where might such receptors be? They could be in the heart itself, or elsewhere. Two general possibilities for being located elsewhere need to be considered. The first possibility is in the blood vessels, especially the great vessels near the heart such as the aortic arch. The second possibility is the other tissues near the heart that could be affected by the mechanical action of the heart. The main consideration is the non-visceral somatic touch, or pressure, receptors in the thorax, especially on the inside of the anterior chest wall.

It is logically possible that more distant receptors could be involved, but fairly good reasons exist to doubt this. First, timing of the perception of the cardiac event in relationship to the contraction of the heart indicates that whatever change is being sensed is near the heart, because there is not sufficient time for the mechanical stimulus to be propagated very far from the heart. Second, a drop-off of stimulus intensity occurs as the mechanical stimulus goes farther from the heart, which also implies that any mechanical stimulus able to be sensed is more likely able to be sensed near the heart in the thorax.

Several studies have attempted to identify the cardiodynamic changes that generate the stimulus. These will be discussed below, because these studies have typically involved experimental manipulations. It should be noted that differences among subjects before any experimental manipulation are likely to be important as well. Several have already been discussed. Individuals who are more fit seem to be more aware of cardiac action. This could be because the cardiodynamics of their hearts are different—they typically beat more slowly and have greater contractility (inotropic effect), and they may experience a training practice effect. They also typically have less body fat, as do young men in comparison to young women, a factor that has been shown probably to be relevant. Greater weight and greater body fat, in particular (body fat has fewer, if any, relevant receptors), could have a dampening effect on stimulation of mechanoreceptors. Genetic factors that contribute to both weight and body composition are probably relevant, although there seem to be no studies of familial correlations in cardiac interoceptive ability.

Methods to assess potentially relevant aspects of cardiac and local thoracic vascular function are available and have been used to study this issue. The main methods have been impedance cardiography and measurement of systolic time intervals. Additionally, pharmacologic methods can be used. For example, this author has successfully used infusions of isoproterenol, a substance that stimulates beta-adrenergic receptors in the periphery only (it does not cross the blood–brain barrier) and has both positive chronotropic and inotropic effects on the heart in combination with positron-emission tomographic (PET) human brain imaging to identify regional brain changes associated with cardiac interoception.

Symptom Proneness. The tendency to report symptoms is, of course, associated with the presence of some illness or dysfunction. Beyond that, however, there appear to be substantial individual differences in propensity to report symptoms. This issue was addressed above, in the first section of this chapter, and it will be addressed again later in this chapter. Suffice it to say here that some syndromes are probably more likely to be associated with awareness of cardiac activity than are others. Personality factors may play a role. Two specific disorders that will be discussed in some detail in subsequent chapters are diabetes mellitus, with associated hypoglycemia and autonomic neuropathy, and panic disorder. One underlying physiological mechanism that is linked to cardiac awareness in these disorders is change due to adrenergic sympathetic nervous system changes. Last, stress and cardiac awareness will be discussed briefly later in this chapter. Much of this topic could be subsumed under the broad rubric of psychosomatic processes.

Methodological Factors (Independent Variables)

Posture and Body Position. Posture affects cardiac awareness. Individuals are least aware when standing, most when lying down, and intermediate when sitting (or lying at an angle, as on a tilt table). This appears to be true because postural change is a method of producing cardiodynamic changes created by gravity. Venous blood pooling patterns change with postural change and affect cardiovascular dynamics. Studies have addressed effects of controlled postural change either by varying usual posture—standing versus sitting versus lying—and by using passive tilt.

Jones et al. (1987) used the Whitehead procedure to compare heartbeat awareness in seated versus standing subjects. Only subjects who performed at 70% or better on a pre-test were studied. Awareness was significantly better in the seated position than in the standing. The investigators found no differences between positions for mean blood pressure or respirations but found that heart period (inverse of heart rate) was related.

In contrast to this result, Ring et al. (1994) manipulated posture, and thereby cardiac stroke volume, by means of a passive tilt table. They used a different heart beat detection procedure, and the subjects apparently were not pretested for accuracy. The procedure they used was one that has become very popular and by some estimates is the best. It involves subjects' choosing stimuli from several temporal choices that most closely approximate a sense of simultaneity with subjective awareness of heart beat. A non-random distribution of choices is considered to be evidence of awareness, unconfounded by many of the problems inherent in other procedures (e.g., the difficulty of the temporal discrimination in the Whitehead method). Also in contrast to the earlier results, this study did not find an effect of modification of posture on awareness. It is not clear to what extent the substantial methodological differences might have led to the opposing study outcomes. These authors point out that several other studies have shown inconsistent methods and results, as well.

Rest versus Exercise. Early studies examining the relationship between rest and exercise and cardiac awareness were reviewed by Jones (1994, 1995). One study found that prior to bicycle exercise preexisting fitness level affected awareness and that exercise improved awareness in fit and non-fit groups. In another study the presumptive anxiety and associated change in cardiac action attendant to preparation for a public speaking task improved awareness, but deep knee bend exercise improved these subjects' scores even more. A third study found that walking on a treadmill improved awareness. The first two studies assessed subjects immediately post-exercise, while the third made the assessment during the walking task, when heart rate was at least 120 beats per minute. In a study not reviewed by Jones, Pennebaker and Lightner (1980) reported that subjects who heard distracting sounds during treadmill exercise reported less fatigue and fewer symptoms than did those who heard their own breathing sounds amplified. Although not focused specifically on cardiac awareness, these results indicate that focusing one's attention on internal cues during exercise makes one more aware of them. Kollenbaum et al. (1996) demonstrated that subjects could reproduce heart rate, a myocardial metabolism parameter (and a respiratory parameter), on a subsequent occasion after training at specific intensities. They concluded that this was evidence that subjects could interocept the visceral sensations produced in the first session in order to reproduce it in the second. Katkin (1985) speculated that cardiodynamic and exercise-related increases in cardiac awareness were likely to be due in significant part to adrenergic factors but only offered indirect evidence at that time.

A more recent study by Schandry and Bestler (Schandry et al., 1993; Schandry and Bestler, 1995) touches on a number of the issues discussed above and provides specific and detailed information on the nature of the source of the heart beat stimulus. It combines changes in posture and in exercise intensity and uses a detection task for heart rate instead of individual heart beats, with no feedback of results. Because of the conclusions it reaches, it will be described in some detail.

Subjects were pretested, and those with perfect heartbeat perception at rest were excluded because of the ceiling effect—no further improvement would be possible. Subjects were required to exercise on a bicycle ergometer that was attached to a tilt table. Tilt table angles between upright (90°) and horizontal (0°) were varied parametrically in 15° increments, along with four different exercise levels including no exercise up to 75 watts. Three baseline assessments were interspersed to demonstrate that learning and order effects did not account for the results, such that the total number of trials for each subject was 31. The order was the same for all subjects, as is illustrated in Figure 8–1. Impedance cardiography and systolic time intervals were determined and used to derive several measures that reflected the following variables, in addition to heart rate: stroke volume, left ventricular contractility, and left ventricular ejection time. A final derived variable, "momentum" of the blood leaving the heart, was the quotient of stroke volume and left ventricular ejection time. A hypothesis underlying the study was that the

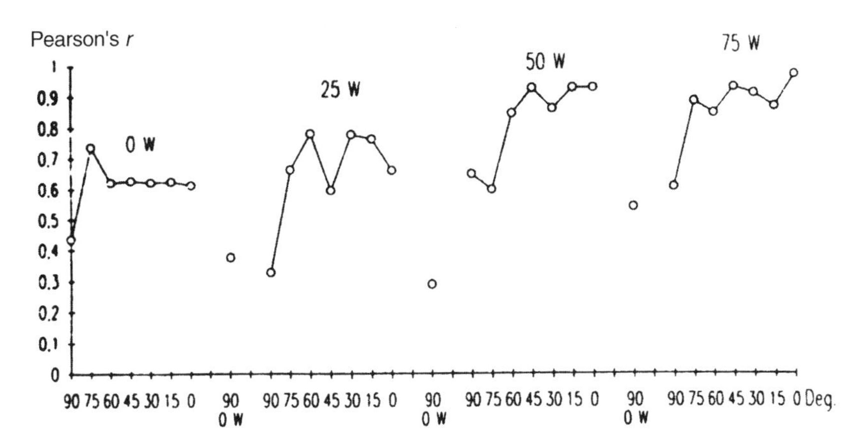

Figure 8–1. *X*-axis: angle of tilt (0 = horizontal, 90 = vertical) for four different levels of exercise (W = watts). *Y*-axis: cardioreceptive sensitivity (Pearson's *r* correlation between counted and measured heartbeats). Variables interact such that accuracy is increased by exercise and by horizontal posture. [From Schandry et al., 1993, reprinted with permission of the holder of the copyright.]

actual stimulus for heart action awareness must be the stimulation of mechanoreceptors or pressure receptors in or near the heart, and thus variables such as stroke volume, contractility, and "momentum" would be closely associated with awareness.

The results of this study are shown in Figure 8–1 and Table 8–2, which are reproduced from Schandry et al. (1993). In Figure 8–1, correlation coefficients are represented on the Y-axis. This "cardioceptive sensitivity" variable is the correlation of actual heart rate with subjective reported heart rate across the groups of subjects. The points on the graph represent the "cardioceptive sensitivity" correlation coefficient for each individual combination of tilt and exercise intensity. It can be seen that the coefficients increase (i.e., the correlations become stronger) both as exercise intensity increases (W-watts) and as tilt goes from upright to supine (0 to 90). In other words, this study replicated earlier findings that posture and exercise both substantially affect cardiac interoception. The correlations between actual and subjective heart rates range from approximately 0.30 to approximately 0.95.

Table 8–2 presents the data showing the associations between the physiological measures and "cardioceptive sensitivity." It should be noted that "cardioceptive sensitivity" is itself a correlation coefficient. Therefore, the correlational data in this table represent a correlation of a correlation with another variable. These findings must be interpreted with this somewhat unusual data analysis methodology in mind. Assuming that the analysis is appropriate, these

Table 8–2. Correlations of Cardioceptive Sensitivity

Heart Rate	Stroke Volume	Heather Index*	"Momentum"
Cardioceptive Sensitivity (r)			
0.49†	0.59‡	0.58‡	0.67‡
Cardioceptive Sensitivity (% variance explained)			
24%	35%	34%	45%
Cardioceptive Sensitivity §			
	0.58‡	0.37‡	0.53‡

* Heather Index: measure of left ventricular contractility.
† $p < 0.05$.
‡ $p < 0.001$.
§ partial r: r after removal of heart rate effect.
[Reprinted with permission of the holder of the copyright from Schandry et al., 1993.]

results demonstrate that awareness is more closely associated with measures of heart action related to the inotropic effect of the heart rather than with the chronotropic effect. These results agree with the investigators' hypothesis.

Some of the related observations in the 1995 article and some mentioned in the 1993 article will also be briefly described here. The most common threshold for exercise-induced cardiac awareness is at a heart rate of approximately 100 beats per minute, perhaps not coincidentally the high end of the normal adult resting heart rate. Consistent with other research, the increase in inotropic effect, and thus the increase in cardiac awareness is primarily adrenergically sympathetically mediated. Also consistent with earlier studies, physically fit subjects were more aware of cardiac action, both at rest and at each level of exercise tested. The overall conclusion from this research: It is the strength of the heart beat, not heart rate, that is the more important determinant of the sensory stimulus.

Training Before Study and Number of Training Trials. It is clear that subjects, for a variety of possible reasons, come to the experimental situation with substantial differences in ability to sense cardiac function. This individual difference is undoubtedly true for visceral–autonomic detection generally. (The question of cross-modality ability will be discussed below.) This fact raises the issue of prior experience, and thus, of training (i.e., learning). It has already been observed that fitness probably plays some role in cardiac awareness, and fitness perhaps contributes to the advantage these individuals have.

In the research setting, training can be thought of in two ways. The first is the prior experience that individuals have before any structured experimental protocol is initiated, and the second is any ongoing learning or benefit that subjects accrue during the study. The relevance of the second factor depends on the duration of the study. Some protocols involve only a small number of trials, while others involve many. Those that involve many trials are generally just the ones that appear to require many. In other words, opportunity for learning and task difficulty are often confounded.

There seems to have been no systematic use of pre-training in order to optimize subject performance during data collection, although it would seem to be of potential value in at least some circumstances. For example, it has been observed that some of the techniques used, especially those that do require many trials, lead to boredom on the part of the subjects, thus introducing the potential for unreliable data and bias (later trials less reliable than earlier). Training prior to study, perhaps on a separate day, would be of value. Furthermore, in studies where called for, a stable baseline could be obtained, and novelty effects and/or misunderstanding of task demands could be minimized.

One form of potential training effect that has been studied is knowledge of results, or feedback (see Katkin, 1985). It was found that feedback training was beneficial, but only for a subgroup of subjects (men who had labile skin conductance). In another study described in Katkin, improvement during a study, over blocks of 40 training trials per block, was demonstrated for men, while women showed no improvement. Thus, there was again an interaction of training by sex. In another study, reported by Jones (1994), a study involving 50 trials was analyzed in blocks of 10 trials, and performance in subgroups was correlated with overall performance. A monotonic increase occurred in the correlation across trials, from +0.78 for the first 10 trials with all 50, to +0.97 for the first 40 with all 50. (It might have been more interesting, and reduced confounding, to present correlations of each block of 10 with all 50, and/or each block of 10 with the other 40.)

External Stimuli and Knowledge of Results (Feedback). It has already been noted that, historically, the interest in interoceptive processes arose in large part as an outgrowth of the interest in biofeedback and learned control of visceral function, even though the question of the need to be able to sense visceral function in order to gain control over it has not been fully resolved. The reverse question is also of interest: What role does feedback, that is, knowledge of results, have on interoception?

Knowledge of results and external feedback of actual heart action are both forms of external signaling. Tracking methods of cardiac interoceptive assessment do not involve signaling of actual heart beats, while discrimination procedures by definition include true and false signals, with various different temporal relationships between actual heart beats and the true feedback signal. In the simultaneity procedure, the true signal is defined by the subject from among several. (Full descriptions of these paradigms are beyond the scope of this volume but can be found in Jones, 1994.) In Jones's

1994 review, his list of studies published mainly in the 1980s indicates that almost 40% of these studies included knowledge of results.

Grigg and Ashton (1982) performed two experiments in which knowledge of results was compared to no knowledge. The second experiment included a third group that received external feedback of actual heart beats but no knowledge. Results demonstrated that the group that received knowledge was statistically best in both studies. The external feedback group did significantly better than the group that received no feedback or knowledge, but performed significantly less well than the knowledge group.

It was noted above that knowledge of results was beneficial in the study described by Katkin (1985). Jones (1994) pointed out that some investigators commonly provide knowledge of results and others do not, and in those studies in which feedback has been used, substantial improvements are usually seen.

Brener and Ring (1995) described a study in which the amount of information (feedback) was experimentally manipulated. The procedure they used, the simultaneity task described above (Ring et al., 1994), allowed for a potential trade-off by the subject in amount of external information provided and amount of effort (inspections) the subject used. This is not exactly the same as knowledge of results, but it does relate to the subjects' performance with different levels of information. It was found that subjects could partially but not fully compensate for less information with more effort, again indicating that information related to performance improves performance. In summary, both knowledge of results and external heart beat feedback affect performance, and knowledge may have a stronger effect.

One study used a stimulus that might be considered a combination of external and (in a sense) internal. Weisz et al. (1988) had subjects focus on themselves as an external stimulus by looking in a mirror. They found that this self-focused attention improved performance on a discrimination task but not a tracking task. They proposed that these results improved performance on a task more directly related to heart beat detection (discrimination task), but not on a task (tracking) that is believed to be less reliable and more affected by subject beliefs (with beliefs not affected by the mirror).

Validity and Reliability. Reliability and validity are two of the hallmarks of testing. Whatever a test tests, it should test it every time the it is administered; that is, results should correlate—reliability.

Also, whatever a test claims to test, it should actually test that—validity. Are the paradigms used to test cardiac interoception reliable and valid? One type of reliability that has been assessed to some extent in this context is test–retest reliability. Pennebaker and Hoover (1984) found that a self-report method and a signal-detection method of assessing cardiac awareness were reliable over a one-week time period, but a heart beat tracking method was not. Brener and Ring (1995) cited studies that agreed with this conclusion with reference to the discrimination (detection) method, as did Jones (1994). In summary, however, reliability testing for most paradigms has been very inadequate.

Validity is also essential. Does a test measure what it claims to measure? For example, in the case of cardiac interoception, if a correlation between a subject's reports of heart beats and the actual heart beats is strong, it might nonetheless be the case that the subject is not sensing heart beats (or, more likely, is doing so only indirectly). As has already been discussed, what subjects might be sensing is the surge of blood flow into the aorta or even the beating of the heart against the anterior chest wall. (Note: If receptors in the anterior chest wall were the mechanism of sensation, lying supine—on one's back—would tend to move the heart away from the anterior internal wall of the chest and thus diminish performance, but in fact the opposite happens. Subjects perform better in the supine position, making this mechanism less likely as a necessary condition. Might there also be receptors in the posterior chest wall?)

In one sense, demonstration of above-chance correlations between actual cardiac events and reports is a measure of validity, and thus those paradigms that work—that do show significant correlations—are by definition valid. In another sense, more should be demanded. For example, cross-validation studies have indicated that results from different paradigms do not correlate with one another, even in the same subjects (e.g., Pennebaker and Hoover, 1984). Even if two paradigms both perform above chance, and even if they do so in the same subjects, if the results of the two paradigms do not correlate with each other, they are not measuring the same thing, and both cannot be valid. More work is needed in this field concerning both reliability and validity.

Emotional Induction. Given the interest that started more than a century ago in the relationship between emotion and visceral–autonomic sensory functioning, it is surprising that little research has been done on this question (see Ferguson and Katkin, 1995). The apparent lack of awareness of the visceral organs might have raised doubts about the likelihood of any relationship being found. There has not been a complete lack of interest, however. Studies have typically been of two kinds. First, self-ratings of somatic symptoms have been very common in studies of a variety of medical and psychiatric disorders, especially those in which symptoms have been specifically referable to visceral organs (e.g., diabetes mellitus, anxiety disorders). Second, the interest in interoceptive processes themselves has prompted some return to the Jamesian question more than 100 years after it was initially posed.

The issue of somatic symptoms and interoceptive processes associated with various disorders will be dealt with later in this chapter and in subsequent chapters. It should be noted that self-report ratings, which were most commonly used in these studies, as already described, appear more clearly to address visceral perception than detection. That is, these are ratings not just of visceral sensory processes, but of cognitions and beliefs, as well, and should be interpreted with that fact in mind.

Schandry (1981) assessed subjects on state anxiety and emotional lability and found that good perceivers on a task involving detection of heartbeats did not differ in heart rate detection but had higher scores on self-assessments of emotion (i.e., were more emotional) than poor perceivers. Katkin et al. (1982) found that electrodermally labile subjects could learn a heart beat discrimination task better than electrodermally stabile; lability is thought to be a marker for increased arousal or emotion. Using a tracking task, Ludwick-Rosenthal and Neufeld (1985) also found that good perceivers had higher state anxiety than poor but were less emotionally expressive. Using visual stimuli with affective content to induce arousal, Katkin (1985) reported that negative (disgusting) slides showing mutilated car accident victims improved performance on a discrimination task despite lowering heart rate. He speculated (but did not provide evidence) that this was due to an increase in inotropic effect. He also described data indicating that good perceivers rated themselves as more disturbed by the negative slides. Using a mental arithmetic stress test related to emotion or arousal, Eichler and Katkin (1994) reported that subjects who had shown higher cardiovascular reactivity to the stressful task were more accurate on heart beat discrimination. There was some tendency for good and poor perceivers to differ on tests of cardiac function, but it was relatively small.

Jones (1994) discussed the data reported by Katkin, above, as well as described an additional study that demonstrated that a public speaking task (which presumably raised anxiety levels) improved detection. He described a study by Davis et al. that did not agree with Katkin's results. More recent studies of people with anhedonia by Ferguson and Katkin (1996) and during everyday life in different groups by Myrtek and Brugner (1996) also were generally negative studies. In summary, two conclusions are apparent: *(1)* Evidence exists that emotion or anxiety can affect cardiac interoception, although not all studies found this. *(2)* It is unclear whether emotion and anxiety improve detection by a change in cardiovascular dynamics and/or by a change in other factors that contribute to detection (e.g., attentional processes).

Outcome Variables (Dependent Measures)

Cross-Modality Correlations. As has become clear, it is likely that all the paradigms and methods proposed to assess cardiac awareness do not measure the same thing. Self-report measures appear to measure the totality of visceral perception, while discrimination tasks of individual heart beats appear to come closest to actual detection of heart action separate from other factors, and tracking tasks are probably intermediate—less influenced by other factors than self-reports but more so than discrimination tasks. Even within the realm of detection of cardiac action, also as already discussed, the nature of the stimulus is not certain and could vary among tasks. So far, the mechanical (and pressure) effects of the contraction of the heart and/or blood flow into the aorta seem to be the best candidates, but that is not certain. Furthermore, lacking appropriate reliability and validity data, the question can be asked for at least some of the procedures, Do they consistently assess anything? That question is harsh and is overly pessimistic about the state of the field, but some carefulness and constructive doubt is called for.

Jones (1994, pp. 111–135) reviewed studies that performed more than one protocol and determined correlations of performance among them. It has already been observed that Pennebaker and Hoover (1984) found poor intercorrelations among the three different types of tasks. Note that finding that task results correlate with one another demonstrates only that they are measuring either the same thing or different things that themselves are closely correlated. Such intercorrelations do not reveal what the tasks are measuring (i.e., do not directly address the validity issue). The review by Jones is too long to discuss in detail here. In general, even for tasks that were heartbeat discrimination tasks, the correlation among tasks was typically weak. This appeared to reflect at least partly the difficulty of the tasks (i.e., tasks that produce near-random performance are not going to be correlated with one another), and knowledge of results (which improved performance) seemed to make some difference. Yet even with knowledge of results, correlations were not strong. This lack of correlation appeared to be due, at least in part, to the fact that subjects tended to use different performance strategies with different protocols.

What Is the Stimulus? What is the stimulus (stimuli) that the individual is responding to? Two broad categories of possibilities exist. First, there are sensory impulses coming from the heart or near it (cardiovascular dynamics). It could involve the inotropic or chronotropic effects of the heart, mechanical and/or pressure receptors from the blood vessels in the thorax, and/or stimulation of somatic receptors in the chest wall (or some other mechanism). Data have already been presented supporting the involvement of the inotropic effect of the heart, but other mechanisms cannot be excluded. Of note, isolated irregular heartbeats occur commonly in hearts that are otherwise functioning normally in both men (Brodsky et al., 1977) and women (Sobotka et al., 1981). For unclear and unstudied reasons, some occurrences of these irregular heartbeats produce conscious sensations, but most do not. But other factors cannot be excluded, either, as contributing to reports of awareness. These are non-cardiovascular contributions.

The non-cardiovascular factors were reviewed above. As further support that such factors are important, additional data will be reviewed here. Some factors beyond cardiovascular dynamics must exist that can account for the differences in results among different types of tasks and the poor correlations among methods, despite the fact that they each individually often demonstrate above-chance performance.

Pennebaker (1981) reported that in a study with a heart rate tracking task and also external tasks not directly related to actual heart rate, heart rate judgments were significantly affected by the external tasks. In the study described above by Weisz et al. (1988), observing oneself in a mirror, a task that has potential mixed aspects of internal and external characteristics, improved heart beat tracking. Ring and Brener (1996) reported that beliefs about heart rate affected performance on a tracking task, and actual heart rate affected beliefs. That is, both factors interacted to influence performance. Proprioceptive function has apparently not been as systematically studied in this context. It is assumed that the change in cardiodynamics associated with change in posture accounts for the improved performance, with supine most and standing least effective, but other factors may be present as well. Somatic sensory receptors beyond the thorax may also contribute, although the timing of the blood pressure pulse wave and the decrement in magnitude of the wave make involvement of these receptors unlikely and/or potentially less important. Cardiac interoception as narrowly defined applies to detection only, but in the real world interoception occurs invariably in the larger context of all these factors.

Central Nervous System. As already noted, for many years it was believed that the cerebral cortex was the region of the brain in which awareness and consciousness resided. That belief is still prevalent, although it is now clear that it is an oversimplification. It is not clear what level of awareness can exist without a cortex (e.g., the experience of at least some forms of pain), but it is clear that the cortex alone is not the sole mediator of such functions. Despite this, it seems likely that the cortex would be very involved in sensory awareness, including visceral awareness. Using that logic, investigative groups have studied central nervous system function involved in cardiac interoception with use of the EEG. The EEG measures changes that are affected by subcortical processes, but the electrical signals that make up the EEG signal come mainly from the cortex.

Before addressing the question of the possible association between cardiac awareness and the EEG, a broader question should be asked: Is there any evidence for changes in the EEG correlated with cardiovascular function, with or without awareness? The answer to the question is "yes." Evidence exists of EEG changes associated with hemodynamic changes dating back more than 50 years. The most extensive body of research related to this issue involves the function of the baroreceptors. It has been shown that baroreceptor stimulation produces cortical inhibition. For example, Vaitl et al. (1996) showed that gravity-related hemodynamic changes increase delta and theta frequencies in the EEG. For reviews of this topic, see Vaitl and Gruppe (1995), Velden and Wolk (1995), and Elbert and Rau (1995).

In addition to the research with baroreceptors, studies have specifically addressed the question of whether there is any association of EEG changes with cardiac action. For example, Walker and Sandman (1979, 1982) found that visual evoked potentials (a particular kind of EEG signal related to information processing) were affected by normal changes in heart rate (and carotid pressure—the locale of baroreceptors) and that this effect differed between cerebral hemispheres (see further discussion below). Koriath and Lindholm (1986) reported that EEG waveforms locked to the cardiac cycle differed from waveforms that were random with respect to the cycle, and Koriath et al. (1987) found that changes in cardiovascular function produced by exercise were reflected in changes in cortical function. These changes may be due to pressure changes (i.e., baroreceptor effects) rather than effects specifically due to heart action, but that could, of course, be true of cardiac interoception as well.

Two investigative groups have been most active in investigating the relationship between EEG changes and cardiac awareness. One group has been most interested in possible changes in the somatosensory cortical region, and the other has been most interested in the frontal cortex. Additionally, both groups have been interested in potential laterality effects, that is, that cardiac awareness might be more closely related to changes in one of the cerebral hemispheres. Descriptions of these somatosensory and frontal cortical region studies can be found in Schandry et al. (1986), Weitkunat and Schandry (1990), Katkin et al. (1991), Montoya et al. (1993), Jones (1994), and Schandry and Montoya (1996). Additional relevant references include those focused on laterality effects (see below).

Under the theory that cardiac sensation is actually mediated by somatosensory mechanoreceptors, Jones and colleagues assessed the relationship between heartbeat and evoked potentials over the anterior parietal somatosensory regions in the right and left cortex. Subjects were classified as "high aware" versus "low aware." High awareness subjects showed a different pattern of evoked potentials than did low awareness subjects in both the C4 region (right hemisphere—upper curves, Fig. 8–2) and the C3 region (left hemisphere—lower curves, Fig. 8–2). Furthermore, overall, the right hemisphere showed a more robust response than did the left for both groups.

Schandry and colleagues studied cortical evoked potentials in the frontal regions in good perceivers and poor perceivers. This group found that the form of the evoked potential wave changed when subjects were instructed to attend to their heartbeats. In one study these investigators studied the effects of financial motivation (Figure 8–3). It was reported that for good perceivers, but not for poor perceivers, motivation changed the waveform and that the observed effect was more prominent in the right hemisphere. In another study, this group reported that in the frontal region of the right hemisphere only, a significant correlation was found between one aspect of the waveform and the awareness score. Thus, the data from the evoked potential changes from the studies these two groups performed document brain electrical changes in both the frontal and somatosensory cortical regions that were related to cardiac awareness and were more prominent in the right (i.e., so-called nondominant) hemisphere.

Laterality. For both the frontal and somatosensory cortical regions, evidence was found in the studies cited above of hemispheric laterality differences in brain function changes associated with cardiac awareness. A number of additional studies have directly addressed the question of possible laterality differences in cardiac interoception. These laterality studies, of course, also relate to brain changes and interoception more generally.

Davidson et al. (1981) reported that finger-tapping in the left hand of right-handed people was more closely coordinated with the R-wave of the ECG than tapping in the right hand, indicating that this coordination was more closely related to the right hemisphere. The studies by Walker and Sandman (1979, 1982) discussed above found that the relationship between evoked potentials and cardiac activity related to cortical function in the right hemisphere, but less so in the left. Yoon et al. (1997), using unilateral carotid amobarbital infusions to inactivate one or the other hemisphere, found evidence of cardiac sympathetic autonomic control mainly lateralized to the right hemisphere. Katkin and Reed (1988) reviewed the research indicating that heart beat perception is associated with the right hemisphere. Drake (1986), however, discussed data indicating an

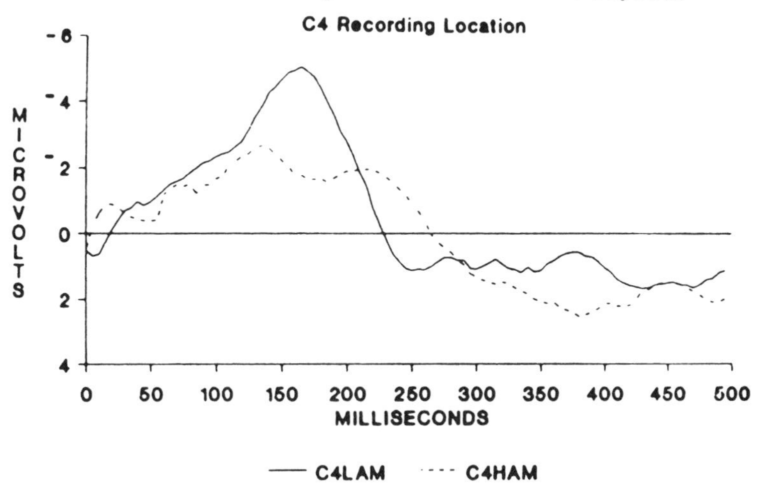

(Time zero = upstroke R-wave)

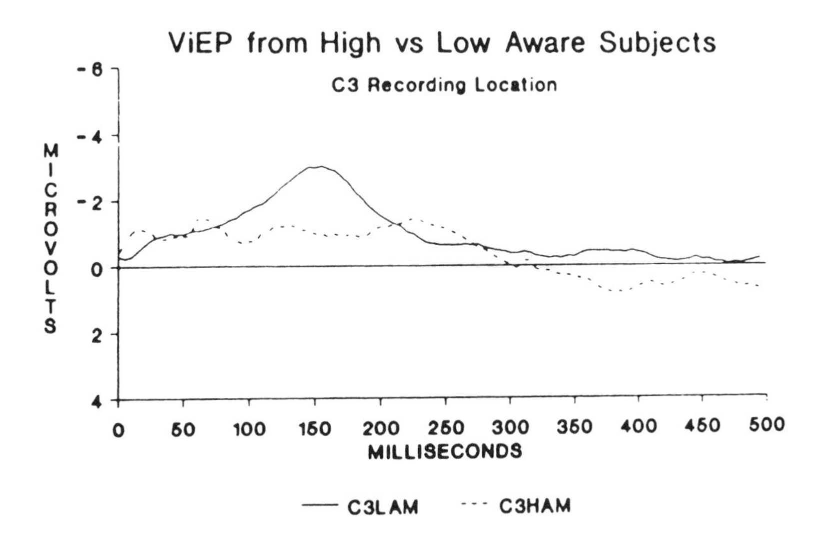

(Time zero = upstroke R-wave)

Figure 8–2. Cortical evoked potentials for high (HAM—dashed line) and low (LAM—solid line) aware male subjects in the right (C4) and left (C3) somatosensory cortices. Patterns were significantly different for HAM versus LAM subjects. [From Jones, 1994, reprinted with permission of the holder of the copyright.]

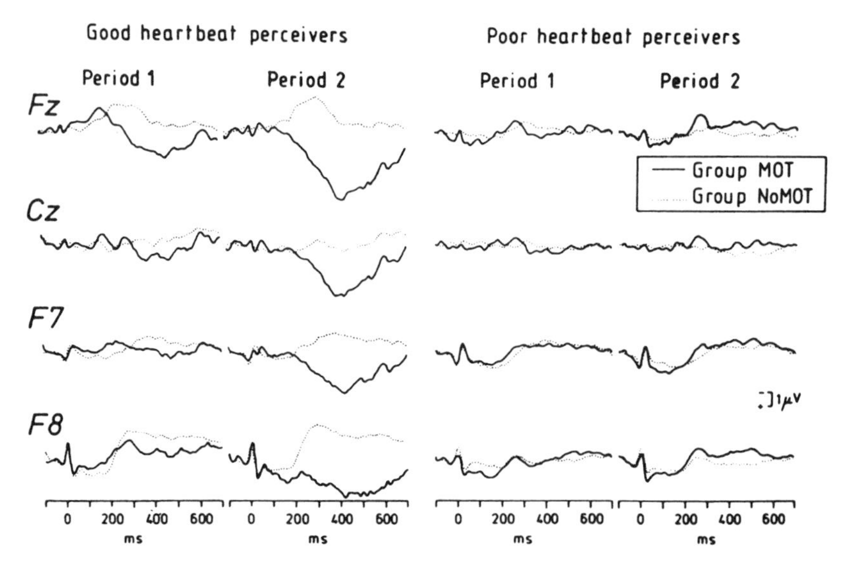

Figure 8–3. Effects of differences in motivation on heartbeat evoked potentials in good and poor heartbeat perceivers. Good perceivers were different from poor, especially in the frontal regions, with motivation influencing results. [From Weitkunat and Schandry, 1990, and Jones, 1994, reprinted with permission of the holder of the copyright.]

association of tachycardia with left hemisphere activity, and Natelson (1985) commented as well on functions related to the left hemisphere.

Direction of initial eye movements associated with task demands has been claimed to be an indicator of cerebral hemisphere function. Montgomery and Jones (1994) found that individuals who performed well on a heartbeat detection task were more likely to make left lateral eye movements, indicative of right hemisphere preference. In agreement with Montgomery and Jones, Hantas et al. (1984), also using the conjugate lateral eye movement test, found that right-movers performed at chance level on a cardiac awareness task while left-movers performed above chance. Weisz et al. (1990) found the same thing, and Weisz et al. (1994), using a monocular viewing task, observed consistent evidence of right hemisphere dominance in cardiac perception. Last, Violani et al. (1996) replicated this evidence, including finding the same advantage in cardiac awareness in left-mover female subjects. To summarize, the results of the studies of hemispheric localization of cardiac awareness, and more broadly of central nervous system cardiac control (at least sympathetic control), are remarkably consistent in documenting involvement of the right cerebral cortex in cardiac interoception.

Imaging Findings

Functional brain imaging studies are likely to play a very large role in furthering understanding of interoceptive processes and brain function. This has already occurred in study of the gastrointestinal tract and will be described in that chapter, along with a brief introduction to imaging techniques as background to that literature. A small amount of brain imaging related to cardiac function and interoception has also been completed. Here will be described two studies completed in our laboratory, and more data will be described below.

The first study (Cameron et al., 2000) investigated the effects of intravenous yohimbine on cerebral blood flow in normal human subjects using PET. Yohimbine is an antagonist of the alpha2-adrenergic receptor and functions as an adrenergic stimulator in the brain because the predominant function of these receptors in brain is inhibitory (i.e., inhibition of an inhibitory receptor leads to activation). It was demonstrated that activation, relative to the whole brain, occurred in areas known to be associated with visceral sensation, including the insular cortex, frontal cortex, and thalamus. Anxiety and cardiac self-reported symptoms were associated with changes in the frontal region.

In a subsequent study (Cameron and Minoshima, under review), intravenous isoproterenol, an agonist of the beta-adrenergic receptor that produces increases in both the inotropic and chronotropic effects of the heart but does not cross the blood–brain barrier into the brain, was given to normal subjects, and changes in cerebral metabolism (brain glucose metabolic changes) were measured with PET. This study included mostly women and mostly right-handed subjects. Including all subjects, isoproterenol activated a midline brain region representing the medial portion of the cingulate gyrus and the truncal region of the somatosensory cortex on the left side of the brain. After excluding possible sex and handedness effects by analyzing only the subgroup of right-handed female subjects, significant activation of the right insular cortex was found. The thalamus and frontal lobes were not significantly activated in this study, but both studies were consistent in demonstrating insular activation in association with administration of substances that produce adrenergic sympathetic activation in the periphery and central nervous system.

A study by King et al. (1999) addressed a very similar question with the fMRI functional imaging methodology. They used maximal inspiration, the Valsalva maneuver, and isometric handgrip to stimulate changes in heart rate, blood pressure, and respiratory function. Activation by these procedures occurred in the insular cortex, posterior thalamus, and medial prefrontal cortex. They found that stimulation of taste with a sucrose solution also activated the insular cortex more inferiorly than occurred with cardiopulmonary stimulation.

Some studies have addressed the question of possible cardiac interoceptive abnormalities. Those that have been done in primary cardiovascular disease will be described later in this chapter, and some others (for example, possible cardiac interoceptive changes in the anxiety disorder panic disorder) will be discussed later. First, however, interoceptive changes associated with blood pressure and respiration will be reviewed.

BLOOD PRESSURE AWARENESS

There are far fewer studies of awareness of blood pressure change than of cardiac change, although to the extent that cardiac interoception is actually awareness of blood flow out of the heart, cardiac interoception is related to blood pressure changes. This possibility is supported, at least to some extent, by the observation that blood pressure affects cardiac perception. People with elevated systolic blood pressure were better heartbeat perceivers than were people with normal systolic blood pressures (O'Brien et al., 1998).

In an early study Tursky et al. (1982) used blood pressure cuff inflation to demonstrate that subjects reported maximum subjective pulsations at the cuff site when the cuff pressure equaled mean arterial pressure, although in a different study subjects could not reproduce exercise-induced blood pressure changes (Kollenbaum et al., 1996). Guglielmi and Roberts (1994), in a biofeedback study, showed that subjects could learn to produce skin temperature differences between hands (most likely at least partly due to blood flow differences). Awareness per se was not the focus of this study.

McFarland and Kennison (1989) showed that such hand (finger) temperature changes are sometimes related to emotional state, and that can reflect cerebral hemisphere function differences. Vidergar et al. (1983), in another biofeedback study designed for subjects to learn to increase or decrease blood pressure, demonstrated that significant correlations existed between actual blood pressures and subjects' estimates of their blood pressures. Greenstadt et al. (1986) found that without knowledge of results subjects were unable to discriminate blood pressure but were able to do so when knowledge was provided.

Similar to cardiac interoception, blood pressure interoception is not simply a matter of detection. Broader perceptual factors play an important role. Pennebaker et al. (1982) found that systolic blood pressures correlated significantly with one or more symptoms in three-quarters of normotensive subjects but that substantial individual differences existed in which symptoms were involved. Factors of sex and variability of blood pressure were involved. In a subsequent study Pennebaker and Watson (1988) found that all subjects in all four groups—unmedicated hypertensives, medicated hypertensives, normotensives, and hypotensives—could estimate systolic blood pressure at greater than chance levels, two-thirds showed a significant correlation with at least one symptom, but that "overall, BP [blood pressure] beliefs were largely inaccurate" (p. 309).

In partial contrast to the above, Muller et al. (1994) reported that a much higher percentage of hypertensives than normotensives demonstrated significant with-subject correlations of actual systolic blood pressures with symptoms. Rostrup and Ekeberg (1992) showed that in people who actually were not hypertensive, belief that they were hypertensive influenced actual blood pressure (and heart rate) responses to a cold pressor test, plasma adrenaline responses, and associations with a number of personality factors. In ambulatory assessment studies, Baumann and Leventhal (1985) reported that some subjects did show some ability to sense actual blood pressure, but perceived symptoms and moods were stronger predictors of blood pressure estimates than were actual blood pressure readings. Brondolo et al. (1999) reported very similar results. Fahrenberg et al. (1995) found that there were no significant within-subject correlations between systolic blood pressure ratings and actual blood pressures, but there were associations between blood pressures ratings and subjective feeling states.

AWARENESS OF RESPIRATION

Dales et al. (1989) reported that respiratory symptoms such as cough and dyspnea were associated with such symptoms as anxiety and depression. Evoked potentials were associated with the sensations produced by both inspiratory and expiratory occlusion (Harver et al., 1995). Harver et al. (1993) showed that subjects could detect changes in inspiratory resistive load and that men were better than women at this task. Van den Bergh et al. (1997) described conditioning of respiratory symptoms using odors as the conditioned stimuli.

Respiratory interoceptive effects were examined in individuals with asthma. Schandry et al. (1996) found that measured peak respiratory flow rate was correlated with one or more identifiable symptoms in approximately half of individuals with asthma. Dahme et al. (1996) found considerable variation in asthmatic individuals' ability to discriminate different levels of airway resistance and that, in general, asthmatics were less sensitive than non-asthmatics. Meyer et al. (1990) found that asthmatics' ability to discriminate different exercise-induced respiratory loads was not very accurate. Studying children and adolescents with asthma, Reitveld et al. (1999) found that the subjective experience of breathlessness increased with exercise, especially if it was coupled with a skin itching sensation, and the strongest correlations with breathlessness were not for changes in exercise-related lung function or for asthma severity, but rather for general symptoms and worrying. In general, data are insufficient to draw any consistent conclusions about respiratory interoception.

PATHOLOGICAL CONDITIONS

It is well known that emotion can influence cardiac rhythm (e.g., Natelson et al, 1989; Wolf, 1995) and that stress and myocardial ischemia and infarction can affect the autonomic control of heart function (e.g., Cohn, 1989; Zipes, 1990; Ewing, 1991). The question of whether there are abnormalities of cardiac interoception in people with various cardiac pathologies is a natural one. As one editorialist asked, "Is the heart a sensory organ?" (Abboud, 1989, p. 390). Disorders that have been studied include anginal pain and myocardial ischemia, interoceptive processes in people with cardiac pacemakers or who have undergone cardiac transplant, and interoceptive function associated with personality function, such as so-called Type A behavior (Essau and Jamieson, 1987) and cardiac neurosis (Myrtek et al., 1995; Schonecke, 1995).

Considering people with pacemakers, Windmann et al. (1999), using a heartbeat tracking task, found that belief about present heartbeat influenced reporting on the tracking task but that actual ability to perform the task was fairly weak. Considering cardiac transplant, it has been shown that sensory re-innervation of the transplanted heart, including experience of angina, does occur in at least some patients (Stark et al., 1991), although lack of stress-induced heart rate reactivity after transplant also indicates that re-innervation can be minimal for up to one year post-transplant (Shapiro et al., 1996). Despite this, approximately one-third of transplant patients showed heart beat awareness approximately three months after transplant, a greater percentage than was observed in a general medical control group (Barsky et al., 1998), indicating that this awareness did not require significant cardiac re-innervation. Essau and Jamieson (1987) found that people with Type A personalities systematically overestimated their heart rates, and Myrtek et al. (1995) and Schonecke (1995) both reported that individuals with cardiac neurosis were not more accurate than were control subjects in cardiac awareness.

Jones et al. (1985) studied heart beat awareness in individuals who had had myocardial infarctions. They found that these subjects were not better than normal subjects, and might have been somewhat worse. Jones (1994) discussed this issue and reviewed other studies that had addressed this question. He noted the high rate at which post-infarction patients complain of experiences referable to the heart, such as palpitations. He also noted that studies of these patients must control for the fact that medications these people often take could affect awareness (e.g., beta-adrenergic blocking agents). He described a study by Davis and colleagues that found that patients showed poorer performance. This result was interpreted as relating in part to age. Another study, not mentioned by Jones (Umachandran et al., 1991), also found that awareness decreases with age.

A number of groups have compared individuals with silent myocardial ischemia to people with symptomatic ischemia. Deanfield et al. (1984) and Rozanski et al. (1988) demonstrated that mental stress could produce ischemia which is silent approximately one-third of the time. Barsky et al. (1990) reviewed various potential reasons for differences across individuals in the occurrence of silent ischemia. Freedland et al. (1991) argued that ischemia was silent in some individuals because of a general decrease in pain sensitivity and bodily sensation in those individuals. Droste et al. (1988, 1995) reviewed the issue of silent ischemia. It was noted that substantial differences exist in ischemic sensitivity. This sensitivity varied across both people and situations. Within-person sensitivity appeared to be consistent between laboratory and natural conditions. This reviewer provided data indicating the involvement of endorphin release in sensitivity to ischemic pain.

Investigators have begun to use brain imaging methods to determine which brain regions are associated with anginal pain. Rosen et al. (1994) found that dobutamine-induced angina was associated with activation of cortical regions. Rosen et al. (1996) reported that the thalamus was activated in both symptomatic and silent ischemia, while frontal cortical regions were activated only during symptomatic painful ischemia. To summarize, the theory that people with pathological cardiac function would be more sensitive to cardiac function has generally not been supported, but such expectations have tended to modify cardiac perception and symptoms. A few physiological correlates in brain and periphery have been identified. This kind of research has just started, however, and conclusions could easily change as more data become available.

NEUROCARDIOLOGY AND STRESS

So far, the emphasis in this chapter has been on cardiovascular and respiratory (mainly cardiac) awareness, in part because awareness of cardiac function has been a focus of a substantial amount of research. Interoception entails more than overt awareness. As had already been noted, it includes visceral sensory impulses and all the effects that these impulses have, including but not limited to conscious awareness. This volume has emphasized those effects that go beyond sensory physiology and homeostasis per se to involvement in complex motivational and cognitive states and influences on behavior.

Although it has been acknowledged that functions of afferent information from the heart to the brain are important to a full understanding of, for example, the roles of neural control of cardiovascular function during exercise (Stone et al., 1985; Koriath, 1989) and of behavioral factors in cardiac dysrhythmias (Natelson, 1985; Wolf, 1987), even to this day not enough systematic research has addressed this question. For that reason, the research that will be reviewed here mainly involves issues related to efferent pathways from the brain to the heart. Only further studies will reveal what essential roles afferent pathways and interoceptive processes play.

The role of the central nervous system in cardiac dysrhythmias and sudden death has generated substantial interest and research activity. This research is too extensive to review here, but see, for example, Lown (1982, 1987), Herd (1984), Verrier and Lown (1984), and Verrier (1987). Several groups have examined mechanisms that modify proneness to dysrhythmias. Aversive conditioning has been shown to decrease stability of electrical conduction. Psychological stressors and emotional states raise the likelihood of fibrillation, as does increased sympathetic function (e.g., Brodsky et al., 1987). Naturalistic studies suggest that approximately one in five instances of ventricular dysrhythmias in patients with pre-existing cardiac disease is related to stress. Blockade of efferent impulses from the brain during these heightened emotional states or decreases in sympathetic tone can diminish this tendency. Genetic factors probably play a role. Mechanistic studies have implicated processes within the heart (e.g., Samuels, 1987; Cechetto et al., 1989) and processes within specific regions of the brain such as the thalamus and frontal cortex (Verrier, 1987) as well as the insula (Oppenheimer et al., 1990; Cechetto, 1994), a region highly implicated in interoceptive processes.

The study of the functional connections between the heart and the brain has been called neurocardiology (Randell, 1984; Natelson, 1985; Brillman, 1993; Armour and Ardell, 1994; Cordero et al., 1995). It addresses not only the question of the brain and abnormal cardiac function, but also the central nervous system's control of normal cardiac function. Investigations in this field have focused almost exclusively on the effects of the brain on the heart. Undoubtedly, dependence on incoming information is necessary in order to optimize the usefulness of the outgoing information.

Closely linked to the issues of cardiovascular function and the brain is the concept of stress. What has been said above about interoception and emotion is directly relevant to stress. Both the fields of neurocardiology and stress are likely to benefit from assessment of the possible roles of interoceptive processes.

nine

THE ALIMENTARY TRACT

> Invisible light rays would make the body transparent, lay bare the vital organs, through the living flesh, of men and animals. . . . Not only will it be possible for a physician to actually see a living, throbbing heart inside the chest, but he will be able to magnify and photograph any part of it.
>
> John Elfreth Watkins, Jr., *Ladies Home Journal,* December, 1900

This chapter will focus on interoceptive processes related to the function of the alimentary, or gastrointestinal, tract. Because of the importance of brain function to interoceptive processes and the importance of imaging techniques in elucidating brain function, before describing the studies of interoception and alimentary function, a short methodological discussion of brain imaging techniques will be provided. Then, the data will be reviewed.

IMAGING

The fundamental function of interest in the brain is the action of neurons. There are, of course, essential actions at simpler levels—the actions of molecules and cell components such as synaptic vesicles that make up the neuronal cells; at more complex levels—for

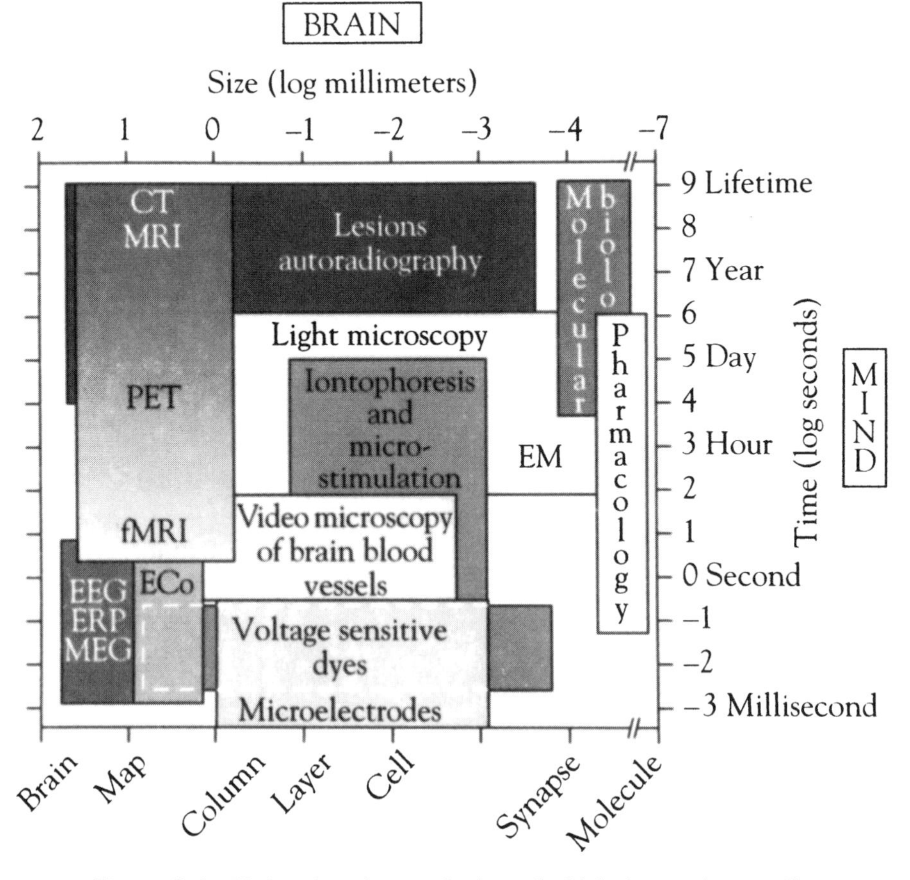

Figure 9–1. Various imaging methods, and which time and space dimensions for which they are most useful. The most commonly used methods relevant to interoception research are PET, fMRI, and EEG-related methods, ranging temporally from hours down to milliseconds and having spatial resolutions in the millimeter to centimeter range. [From Posner and Raichle, 1994, reprinted with permission of the holder of the copyright.]

example, all of the cells in aggregate that make up the amygdala, thalamus, or frontal cortex, and at comparable levels—all the cells in the brain that are not neurons such as glial cells and vascular components; but the neurons are "where the action is." For that reason, one would like to study brain function at the level of resolution in time and space of the actions of neurons. At the same time, however, it must not be forgotten that the purpose of the brain as a manager of information for the organism functions at a much more complex level than at the level of single neurons or even a single anatomical structure.

Different brain imaging techniques function at different levels of resolution in time and space, as illustrated in Figure 9–1. Some of the available methods will very briefly be reviewed here. The following books and article provide more information on the topic of imaging: Phelps et al. (1986), Andreasen (1989), Posner and Raichle (1994), Cohen and Bookheimer (1994), Lewis and Higgins (1996), Toga and Mazziotta (1996, 2000), Frackowiak et al. (1997), and Mazziota et al. (2000).

Imaging and Electrical Methods

Interoceptive processes are more about function than about structure. For that reason, little application of structural brain imaging has been applied to the understanding of interoception. Functional imaging, on the other hand, is likely to be very important. Probably the most commonly used functional imaging techniques in human research studies are PET and functional magnetic resonance imaging (fMRI), although single-photon emission computed tomography (SPECT) is also commonly used.

PET technology can be used to measure metabolic processes such as cerebral blood flow and cerebral glucose metabolism in the brain (linked to local neuronal activity) or other organs such as the heart (and also ligand binding to quantify receptor binding). Measuring blood flow, multiple scans can be performed with only about 15 minutes between scans. The limit of anatomical resolving power is 2–3 mm. Temporal resolution is seconds to minutes. Magnetic resonance imaging typically is used to image anatomical structures. An application of fMRI can be used to measure cerebral blood flow. SPECT can also be used to measure functional activity. While it is likely that the organ of most interest to understanding interoception is the brain, it is the function of the visceral organs that are being sensed. Studies that link interceptive processes to both brain function and heart function (e.g., Goldstein et al., 1990) would be of particular interest.

Some advantages that PET has over SPECT include the capacity for absolute quantification (instead of just relative—one region to another), better spatial resolution, shorter half-life isotopes (thus, more opportunity for within-subject studies), and more opportunity to use biologically relevant substances. The advantages fMRI has over PET include better temporal and spatial resolution. The major disadvantage is that fMRI now is mainly used to measure only cerebral blood flow.

Measurement of brain electrical activity also can be used as an imaging technique. Two commonly used methods are the electroencephalogram (EEG) and evoked potentials. Both methods measure amplified electrical signals from the surface of the head. With the EEG an array of electrode sensors measures activity at various points that reflect function in different regions of the cortex. This pattern of activity across the cortex can be viewed as a form of imaging. It can identify differential electrical activity reflective of differential neuronal activity in various areas of the cortex. It is often referred to as EEG topographical mapping.

Evoked potentials are measurements of electrical activity linked to a specific event. In order to be able to improve the signal-to-noise ratio, the EEG signal is usually averaged over multiple presentations of the evoking sensory event. Often these evoked potentials are recorded from a single brain region, such as the vertex, but multiple signals from multiple brain regions can be recorded. Measurement of evoked potentials from multiple brain regions can be viewed as a form of imaging as well.

Imaging Data: Emotion

Functional imaging techniques have been applied to a variety of brain processes relevant to interoception. Somatosensory processes have been studied (e.g., Phelps et al., 1986; Ginsberg et al., 1987; Andreasen, 1989; Frackowiak et al., 1997). So far, the major application to interoception has been in the alimentary (gastrointestinal) system, although EEG, evoked potential, and a few imaging studies with cardiac interoception have also been done. Several of these alimentary studies are described below. Before describing imaging studies of alimentary interoception, a brief summary of imaging studies addressing brain regions relevant to emotion and, therefore, to interoception will be presented.

Imaging techniques have been applied to the understanding of emotion for many years (see Davidson and Irwin, 1999). For example, LeDoux et al. (1983) demonstrated in rats that a neutral sound selectively increased blood flow in the auditory pathways. When the tone had been conditioned in a fear conditioning procedure subsequent tone presentation also activate the amygdala and hypothalamus. Imaging studies of classical fear conditioning have been recently reported in humans. Both subcortical regions and cortical regions were activated (Fredrikson et al., 1995a, 1995b). Although the amygdala did not show activation, it did correlate significantly with electrodermal changes indicative of fear (Furmark et al., 1997). In another study also using a shock unconditioned stimulus (UCS), activation occurred mainly in the right hemisphere (Hugdahl, 1998). Conditioned fear reactions were generally associated with changes in the amygdaloid and medial prefrontal cortex areas (e.g., Zubieta et al., 1999).

Other means of provoking emotion-related experiences have been used, including imagining, recall, and presentation of stimuli with emotional content. Imagining non-emotional circumstances led to increased blood flow in the dorsolateral prefrontal and posterior temporal cortex, while the emotional situation was associated with flow increases in the medial prefrontal and anterior temporal cortical areas (Partiot et al., 1995). Emotional recall activated the anterior insula, while visual emotional stimuli activated a more complex area, including the amygdala (Reiman et al., 1997; Lane et al., 1997). Recall of sadness was associated with insular activation, while happiness activated the frontal area. Across several studies experimentally induced mood change led to activation in regions including the orbitofrontal cortex, amygdala, cingulate cortex, prefrontal cortex, thalamus, anterior insula, and hypothalamus (sad mood), and orbitofrontal cortex, midbrain, amygdala, prefrontal cortex, thalamus, and hypothalamus (elation) (Pardo et al., 1993; Gemar et al., 1996; Baker et al., 1997; Schneider et al., 1997; Reiman et al., 1997; Lane et al., 1997; Mayberg et al., 1999). Some regions demonstrated deactivation during emotion as well. Using glucose measurement, activation of the amygdala was associated with recall of emotional visual material (Cahill et al., 1996).

Presentation of faces with emotional expressions activated regions including the amygdala, inferior temporal cortex, and medial thalamus (negative affect faces), and prefrontal cortex, anterior cingulate, fusiform gyrus, anterior temporal lobe, and entorhinal cortex (positive affect faces) (George et al., 1993; Dolan et al., 1996; Morris et al., 1996; Paradiso et al., 1997). Other studies have reported involvement of regions relevant to interoception (George et al., 1996; Kosslyn et al., 1996; Liotti et al., 2000), especially the frontal cortex, cingulate gyrus, insula, and amygdala. Functional overlap occurs between viewing faces with emotional content and the subjective experience of various mood states, but they are not synonymous, and overlap occurs among the regions activated, but they are not identical, either. In all these situations the regions activated are regions for which evidence exists for association with emotion from other sources, as reviewed in Chapter 4.

In addition to affective induction by viewing faces or recalling situations, pharmacologic methods have been used to induce emotional experiences, including our studies with isoproterenol already described. Procaine, a substance that produces a range of emotional, visceral–somatic, and psychosensory experiences (including fear, panic attacks, and hallucinations in some subjects), increased anterior paralimbic blood flow more than whole brain (Servan-Schreiber et al., 1998). Subjects who experienced the most fear had the greatest flow increase in the left amygdala (Ketter et al., 1996). In summary, regions that were typically involved were those that have been shown in other research (see Chapter 4) to be involved in emotion. And regions implicated in emotional processing are at least indirectly implicated in interoception.

Cerebral activation is associated with increases in brain glucose metabolism and blood flow due to increased energy needs. Zajonc (1985), following a theory originally proposed almost 100 years ago, hypothesized that changes in facial muscles produced by emotion modulate cerebral blood flow, which in turn influences emotional experience. This would be consistent with a James–Lange–like idea, that subjective experience follows somatic changes. Unfortunately, several reasons exist to doubt this idea. First, the actual pattern of cerebral arterial flow is unlikely to be affected significantly by changes in facial flow. Second, while increases in energy needs lead to increases in flow in specific regions associated with the brain activity under consideration, it is not clear that the opposite reliably occurs, that is, that increased flow leads to activation. Finally, large changes in cerebral blood flow induced pharmacologically by caffeine, for example, do not appear to lead to reports of major emotional changes (Cameron et al., 1990b).

Imaging Data: Alimentary Tract and Interoception

A broad view of the alimentary tract encompasses not just the esophagus to the rectum, but the taste receptors in the mouth, the olfactory receptors, and processes such as hunger and thirst. Imaging studies of sensory awareness in the alimentary system include studies of these processes. Kinomura et al. (1994) used PET and measurement of cerebral blood flow to determine which brain regions were activated by salty taste in normal right-handed humans. Pure water was used as the control substance. Activation was in more regions in the right than in the left hemisphere. In the right, regions significantly activated were the caudate nucleus, thalamus, hippocampus, and cingulate, middle temporal, lingual, and parahippocampal gyri. On the left were the superior and transverse temporal gyri and the insula. King et al. (1999), using fMRI, found in normal men that sweet taste activated the inferior insula, and hemispheric differences were not reported.

Denton et al. (1999) investigated regions of the brain activated by thirst and satiation of thirst with PET and measurement of cerebral blood flow. They used an infusion of hypertonic saline to induce thirst and water to satiate it. They found that the area of strongest activation during maximum thirst was the posterior cingulate. Other areas that they identified included the anterior cingulate, amygdala, parahippocampal gyrus, insula, thalamus, and mesencephalic region. They concluded that "... maximum thirst was uniquely associated with diffuse activation of the limbic and paralimbic cortex ..." (p. 5306).

The word *disgust* refers to the function of taste (and smell) and involves the upper alimentary tract (Phillips et al., 1998). Phillips et al. (1997) used fMRI and viewing of faces with different expressions (disgust, fear, neutral) to determine brain regions associated with the emotion of disgust. They reported replicating prior studies that indicated that viewing fearful faces activated the amygdala bilaterally. Viewing of faces with the expression of disgust activated the anterior insular cortex on the right and "... structures linked to a limbic cortico–striatal–thalamic circuit" (p. 495). Interestingly, a link has been observed among responses to viewing faces showing disgust, abnormal recognition of these faces by people with a disorder of function of the basal ganglia (Huntington's disease), and obsessive–compulsive disorder (Rozin, 1997). In summary, sensory activation of the rostral end of the alimentary tract is associated with activation of the insula, the cingulate gyrus, and various associated limbic–paralimbic structures.

Moving down the alimentary tract, studies have focused on the brain regions activated by esophageal stimulation. Using fMRI and a graded mechanical stimulus in normal subjects, Binkofski et al. (1998) found that the mildest stimulus that produced awareness (well localized albeit weak retrosternal sensation) produced activation in the parietal opercular cortex bilaterally, which they believed included the secondary somatosensory cortex. As intensity was increased, the primary somatosensory cortex and the right premotor cortical area also were activated. Greater increases in intensity activated the insula bilaterally, and then finally the anterior cingulate. They concluded that the secondary somatosensory cortex was associated with sensation, and, as stimulation became painful, limbic areas were recruited.

In another study of esophageal sensation (Aziz et al., 1997; Anderson, 1998) involving PET measurement of cerebral blood flow and normal subjects, the regions activated by non-painful and painful stimuli again were studied. Including all subjects, non-painful stimuli activated the central sulcus, the insular cortex, and the frontal–parietal operculum bilaterally and produced deactivation in the right prefrontal area. Increasing the stimulus so that it produced pain continued to activate these regions and also activated the anterior insular region on the right and the anterior cingulate gyrus. Combining the data from this study along with Binkofski et al. (1998), somatosensory areas were most consistently activated by non-painful esophageal stimulation, the insular region also was activated apparently by somewhat more intense stimulation, the right premotor area showed involvement, and, finally, when the sensation became painful anterior cingulate activation was observed. Thus, there was a gradient of regional recruitment as stimulus intensity increased. Data showing the activation pattern of one subject from this study are illustrated in Figure 9–2.

A third study of esophageal sensory processing in the brain used non-painful electrical stimulation and measurement of activation in cortical regions with magnetoencephalography in normal subjects (Schnitzler et al., 1999). Stimulation of the lip and of somatosensory afferents in the median nerve were used as control conditions. Stim-

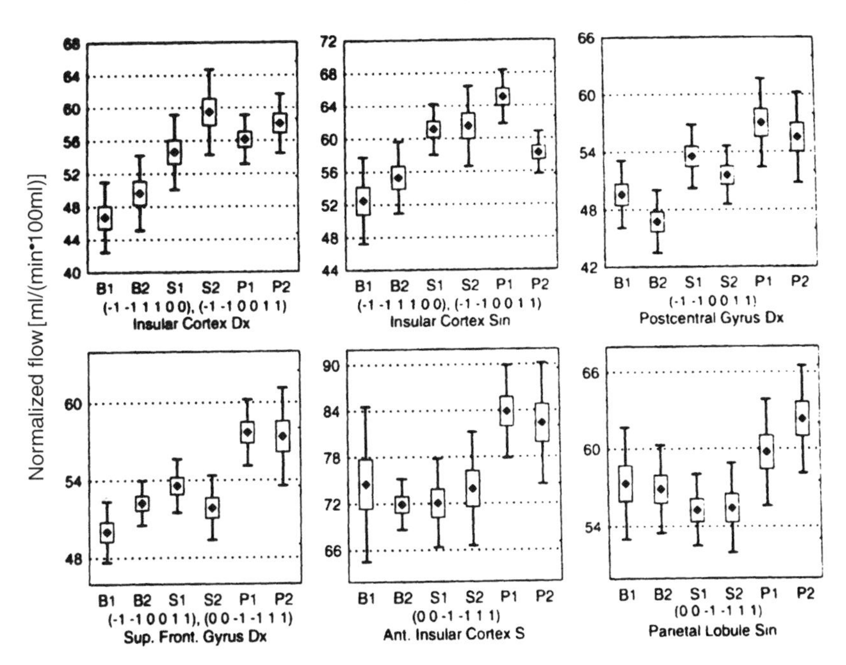

Figure 9–2. Means, standard errors, and standard deviations of regional cerebral blood flow from one subject who received esophageal stimulation. B: baseline; S: definite sensation without pain; P: painful sensation. (Dx: right; Sin: left). Different patterns are apparent, but it is clear that in some regions both painful and non-painful stimulation produced increases in cerebral blood flow. [From Andersson, 1998, reprinted with permission of the holder of the copyright.]

ulation of the control areas produced activation in both the primary and the secondary somatosensory regions. Stimulation in the distal esophagus did not produce activation in the primary somatosensory area, but the secondary somatosensory area was clearly activated bilaterally. No other regions showed significant activation with this imaging method. A stimulus frequency effect that might be related to habituation was demonstrated for esophageal, but not for median nerve, stimulation. These results agree with the earlier data indicating involvement of the secondary somatosensory cortex in esophageal sensory processes.

Kamath et al. (1998) studied evoked potentials associated with esophageal stimulation. Using healthy subjects, they electrically stimulated the distal esophagus at varying intensities and measured the evoked potential produced by these stimulations. They observed a change in the cardiac heart rate power spectrum consistent with an increase in vagal tone, present at all stimulus intensities. The evoked potential change was related to stimulus and intensity and with awareness of stimulation. Thus, stimulation even below the threshold of awareness affected cardiac dynamics, while occurrence of evoked potential was more closely associated with awareness.

The same research group (Hollerbach et al., 1998) determined evoked potentials produced by both electrical and mechanical (balloon distention) stimulation of both the proximal and distal esophagus. They determined that both types of stimulation at both sites produced evoked potentials. Based on different conduction velocities, they concluded that the afferent signals from the two different types of stimulation were carried through different nerve fibers and, thus, different pathways. Symptoms were not systematically reported in this study. In contrast, Hobson et al. (2000) concluded that both pathways—vagal and spinal—mediated cortical evoked potentials produced by esophageal stimulation.

In another study Hollerbach et al. (1999) focused more on the awareness issue. Using healthy subjects, they determined the evoked potential pattern from both esophageal stimulation and anticipated stimulation. The actual stimulus produced evoked potentials with peak intensity in the centro-parietal cortical area, while the evoked response to the anticipated (but not delivered) stimulus occurred over the frontocentral cortex. Of relevance to interoception, these studies clearly demonstrated that visceral afferent impulses from the esophagus can affect cortical function and that aspects of awareness can change the cortical location of the observed potentials.

A few studies in the tradition of the Russian and Eastern European theme of investigation have been published in journals that would be considered Western journals. For example, Slucki et al. (1965) studied monkeys that had loops implanted in the jejunal region of the alimentary tract. They also had implanted brain electrodes. Rhythmic inflations and deflations of the jejunal loops were used to stimulate mechanoreceptors. These rhythmic stimulations of the jejunum were used successfully as a discriminative stimulus. The pattern of discrimination training and discriminability of the rhythmic jejunal stimulus are illustrated in Figure 9–3. Onset (but not offset) of the stimulus was associated with desynchronization–activation pattern of the EEG in all three regions in which the electrodes were placed—the reticular formation, hypothalamus, and cor-

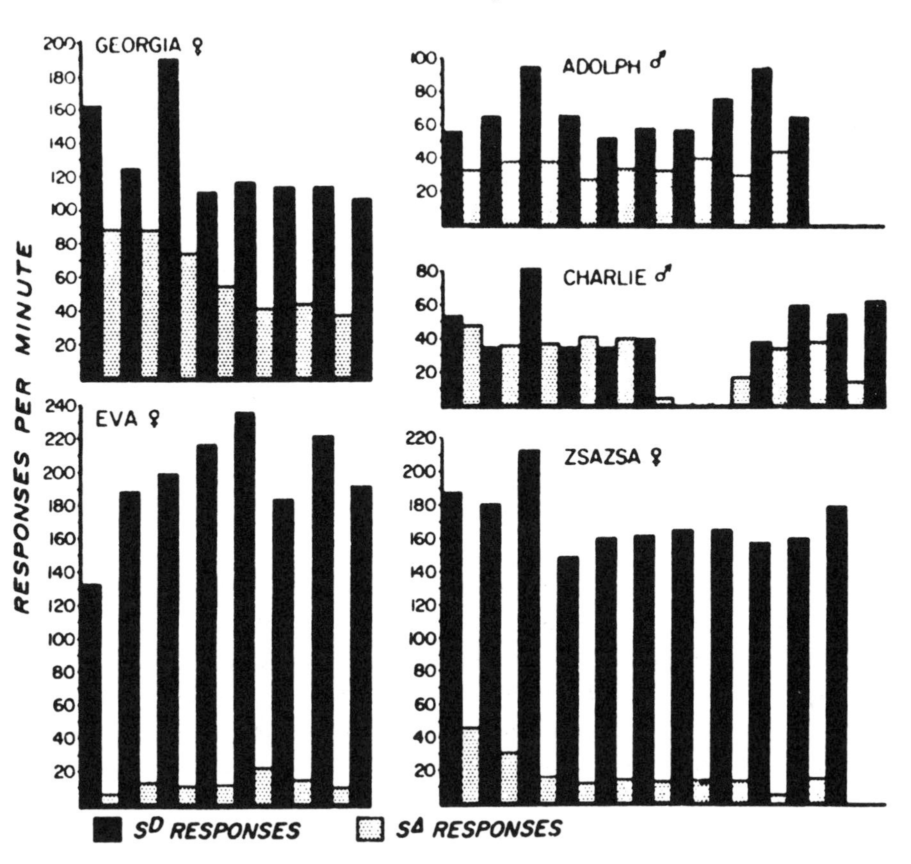

Figure 9–3. Results of training in five monkeys who were reinforced for responding only during jejunal distention as the discriminative stimulus (S^D). Four of five clearly made the discrimination, with two showing near-perfect discrimination. [From Slucki et al., 1963, reprinted with permission of the holder of the copyright.]

tex. This study is notable because it showed synchronous brain activation at several different levels of the central nervous system.

More recently, Belyaeva (1991; also Beliaeva, 1990) demonstrated EEG activation resulting from stimulation of mechanoreceptors in the stomach. The pattern of activation was correlated with the intensity of the visceral stimulus. Following the Pavlovian conditioning paradigm, the ability to extinguish this activation pattern with stimulation—that is, habituation of the response—was observed.

Access to the upper (e.g., esophagus) and lower (e.g., rectum) gastrointestinal tract is technically somewhat easier than is access to the intervening areas. Thus, in addition to studying brain regions activated by taste and esophageal stimulation, the other region that has been studied is the rectum. Rothstein et al. (1996) studied the effects of rectal balloon distention on cerebral blood flow measured with PET and on cerebral evoked potentials. Activation was measured at different distentions, but pain was never produced. Distention produced an evoked-potential response, the magnitude of which correlated with the amount of distention. The most robust changes in cerebral blood flow occurred in the precentral gyrus, the postcentral gyrus, and the thalamus. Using SPECT in normal subjects, Bouras et al. (1999) showed cerebral blood flow increases in the anterior cingulate with rectal distention. This activation did not correlate with magnitude of distention.

Silverman et al. (1997) performed a PET cerebral blood flow study with normal subjects and individuals with irritable bowel disease. They studied subjects both when painful rectal stimulation was occurring and when it was not occurring but anticipated. For normal subjects the anterior cingulate cortex was the region most activated by painful stimulation, and the intensity of activation correlated positively with the pain intensity ratings. No such correlation was observed for those with bowel disease. Anticipation of pain also activated the anterior cingulate in normal subjects but not those with disease. In contrast, the dorsolateral prefrontal cortex was associated with anticipation in those with bowel disease. Here again, as in the studies of esophageal stimulation, activation of the anterior cingulate was related only to stimuli intense enough to produce pain (in normal subjects).

NON-FUNCTIONAL IMAGING STUDIES OF ALIMENTARY INTEROCEPTION

So far in this chapter, the methods of imaging have been briefly noted, and the application of functional imaging techniques to emotion and interoception in the alimentary tract have been described. In the remainder of the chapter, other non-imaging interoception studies of the alimentary tract will be reviewed, including a review of nervous control of the gastrointestinal tract, normal interoceptive processes in this system, and abnormalities of sensory processing in individuals with gastrointestinal disorders.

Although taste and smell are usually considered senses directed towards stimuli in the external environment, they are also clearly part of the alimentary tract. Detailed review of the senses of taste and smell are beyond the scope of this volume. Such reviews can be found, for example, in standard textbooks of neuroscience (e.g., Haines, 1997; Zigmond et al., 1999; Kandel et al., 2000). Discussion of these senses in relation to interoception will be brief, and only a few examples of relevant studies will be provided.

Schapiro et al. (1971) reported that elimination of olfactory input in dogs over several months led to a decrease in gastric secretory function. (The same effect was observed by Shapiro et al., 1970a, 1970b, with visual and auditory and vestibular deprivation.) In rats, lesions of the central nucleus of the amygdala produced a severe impairment in aversive heart rate conditioning in which an olfactory stimulus served as the CS. It was not due to a decrement in either the unconditioned response or the orienting response (Sananes and Campbell, 1989). Aguero et al. (1997) reported that lesions of the parabrachial nucleus, an important part of the visceral sensory pathway in the brain (see Chapter 7), produced an impairment in the ability of rats to learn a gustatory–olfactory discrimination. The results of these studies demonstrate that a clear relationship exists among olfactory sensory functioning and other functions involved in interoception.

Like smell, taste is related to visceral function. For example, studies in rats (Lasiter, 1983; Mackey et al., 1986; Bermudez-Rattoni and McGaugh, 1991) demonstrated that lesions in the insular cortex disrupt conditioned taste aversion learning. Lenz et al. (1997) demonstrated that microampere electrical stimulation of the ventralis caudalis parvocellularis internis nucleus of the thalamus in humans during awake neurosurgical procedures produced a complex sensory experience including taste and smell as well as somatic and visceral sensations. This region has been called the gustatory thalamus, although the specifics of its exact function are under discussion (see Reilly, 1998). Particular tastes and appetites, for example for salt (sodium), involve visceral and central nervous system functions and regions implicated in interoception (see Johnson and Thunhorst, 1997; McCaughey and Scott, 1998). Being the proximal part of the alimentary tract, it is not surprising that the functions of taste and smell are related to other visceral functions and processes.

Interoceptors, mainly mechanoreceptors but also thermoreceptors and chemoreceptors, are present in the alimentary tract. The great interest during the 1990s in visceral sensory processes associated with the alimentary tract was motivated in part by the theory that the major functional bowel disorders (non-cardiac chest pain, functional dyspepsia, and irritable bowel disease—IBS) are due to abnormalities in visceral sensation. These disorders, like many disorders in other physiological systems, are called functional because evidence exists for disorders of function without identification (so far) of disorders of structure. Much research involving the study of visceral sensory processes in the alimentary tract is motivated by attempts to understand these putative disorders of sensation and perception. General discussions of alimentary function and dysfunction can be found in Emel'ianenko and Raitses (1976), Mayer and Raybould (1990), Lynn and Friedman (1993), Mayer and Gebhart (1994), Hu and Talley (1996), Rao (1996), Bueno et al. (1997), Schmulson and Mayer (1998), Naliboff et al. (1998), and Drossman (1998).

Visceral Afferent Mechanisms

A general description of the nervous system basis of visceral sensation was provided in Chapter 7. Here information about visceral afferent mechanisms directly applicable to the alimentary system is provided. Some overlap occurs with the material presented earlier.

The major visceral sensory neuronal receptor is the free nerve ending. The enteric nervous system, along with the afferents and efferents connected to it, control a variety of processes associated with the bowel, including smooth muscle function and bowel contraction, secretion of acid and other substances, absorption through the bowel wall, and blood flow (especially vasodilation). The sensory information is not a simple all-or-none. For example, the sensory response to mechanical stimulation (and also the motor response) has dynamic and static phases. Fibers that carry impulses produced by an increase in distention (A-delta fibers) differ from those that carry information concerning the persistence of distention (C fibers). Rate of change of distention affects sensory impulses as well.

Despite the ability of the enteric system to function independently, it has important afferent and efferent connections to the central nervous system through the sympathetic and parasympathetic branches of the autonomic nervous system. On the parasympathetic side, the efferents pass through the parasympathetic ganglia, and the afferents pass through the vagus nerve to the nodose ganglia. Both sensory mechanoreceptors and chemoreceptors send impulses through the vagus. On the sympathetic side, the efferents pass through the sympathetic ganglia, and the afferents pass through the dorsal root ganglia. Splanchnic sympathetic afferents appear to mediate nociception and carry signals produced by stimuli that threaten or produce tissue damage. Output from the central nervous system affects gastrointestinal function associated with higher brain functions such as stress, emotional changes, and conditioned effects. As described in more detail below, changes in visceral perception have been implicated in disorders of the gastrointestinal tract (Tougas, 1999; Delvaux, 1999).

The main afferent pathways involve parasympathetic sensory fibers in the vagus nerve passing through the nodose ganglion to the solitary tract in the brainstem, other parasympathetic pelvic nerves, and sympathetic spinal afferents. The vagus nerve carries sensory (as well as motor) impulses originating in most of the gastrointestinal tract, from the esophagus to the proximal colon and the caecum. The pelvic nerves carry the remainder of the parasympathetic afferents (and efferents) originating in the lower end of the gastrointestinal tract. Mechanoreceptors, chemoreceptors, thermoreceptors, and osmoreceptors all send afferent information through these tracts. As an example, the cells of the nucleus of the solitary tract respond to intestinal distention, but that response does not occur if the vagus nerve is cut (vagotomy).

Until recently it was believed that only the spinal afferents, that is, the sympathetic side of the visceral afferent sensory system, carried pain impulses. This view has been superseded. It is now believed that the vagus nerve is also involved. The vagus may carry impulses but may also modulate the pain impulse–carrying function of the spinal afferents. The mechanoreceptors respond to distention such that the impulses are proportional to the intensity of the stimulus (i.e., how much distention occurs). Spinal sensory nerve endings also are found in the mesentery and peritoneal ligament. Furthermore, like the vagal afferents, the spinal afferents respond not only to mechanical stimulation; chemical sensitivity also has been demonstrated. The neurotransmitters in the two afferent systems may overlap, but they are not exactly the same. For example, the neuropeptides found in the two systems are different.

Motor control of the alimentary tract involves afferent as well as efferent tracts from the central nervous system. For example, the gastrocolic reflex and intestino–intestinal reflexes are influenced by central excitatory and inhibitory inputs. Afferent sensory responses to various stimuli are dependent on gastrointestinal tone, that is, by the momentary functional status of the portion of the gut in question. The setting of that tone is an efferent function. The gastrointestinal tract is an excellent example of the hypothesis that effective efferent control of any organ requires a highly developed and well functioning afferent system as well.

Visceral Sensation

Information about the intensity and characteristics of a visceral stimulus is coded in part in the pattern of firing of the afferent nerve fibers. One theory holds that the encoding includes information about the noxious versus non-noxious aspects of the stimulus. Another holds that there are separate nociceptors. Evidence exists for both theories. Different mechanisms could occur in different pathways and/or with different stimuli. Another theory is that silent nociceptors exist, nerve endings that under normal circumstances do not encode noxious stimuli. Under some circumstances, however, such as chemical irritation, these receptors become nociceptors, presumably only temporarily (but perhaps persistently under other circumstances).

At least two types of nerve fibers have been identified as being involved in conducting visceral sensory impulses, C fibers and A-delta fibers. The C fibers are in the muscle layer of the gut, the serosa, and the mesentery. They adapt slowly and thus are mainly responsive to tonic (consistent ongoing) stimulation. In contrast, A-delta fibers, which are more rapidly conducting, are in the mucosa, respond to phasic (fluctuating) stimulation, and are rapidly adapting. The A-delta mucosal fibers are polymodal (respond to mechanical and other—e.g., chemical—stimuli), respond to stroking of the bowel, and respond less strongly to distention or contraction. The C fibers respond to distention and related stimuli, that is, they are mainly mechanoreceptors.

In general, it is said that non-painful sensations are "vague and emotionally colored." Anatomical localization often is diffuse and vague, but such sensations referable to the alimentary tract as nausea, severe hunger, and overfed satiety often are quite intense and specific in sensory characteristics. Painful stimuli also are often vague when it comes to localization, but the pain itself—from such stimuli as chemical irritation, distention, torsion, traction, forceful contraction (especially against a blockage), inflammation, and ischemic changes and necrosis—can be excruciating and anything but vague.

Distention is one of the most robust sources of gastrointestinal sensation. Links exist among relaxation responses in different parts of the gastrointestinal tract. For example, duodenal distention can produce relaxation in the stomach. Holzer and Raybould (1992) demonstrated that both vagal and spinal afferents were involved in the control of this coordinated response. On the efferent side, even though important control mechanisms exist in the enteric system, central nervous system mechanisms are quite important, for example, in the development of pathological changes. Henke et al. (1991) reviewed studies that demonstrated the involvement of the amygdala and other limbic system structures with emotional changes in the development of stress ulcers and individual differences in stress ulcer pathology. A variety of neurotransmitters, especially glutamate, are involved in these processes.

Pain associated with these gastrointestinal processes can be relieved by sympathectomy, and possibly relieved, reduced, or modified by vagotomy. Visceral sensory processes, probably including pain and intensity of sensory experience, can be affected by the functioning of the autonomic nervous system. Furthermore, for distention, for example, awareness is greater for rapid distention than for slow (and for this reason, in research studies investigators believe that slow changes are more susceptible to response bias). Pain in the alimentary tract is further discussed below.

An important concept that has already been illustrated is convergence. Sensation from non-visceral sources can mask visceral sensation. Convergence is the theory that this masking occurs because of nervous system gating. That is, the sensory impulses are carried in the same neuronal pathways—that is, converge—and the non-visceral impulses dominate, as if they close a gate, not allowing the visceral impulses to pass. With respect to pain, the concept of sensory gating is many years old, and studies described below with normal subjects indicate it happens when simultaneous visceral and nonvisceral (such as impulses from the skin) sensory stimuli occur. It should be noted that, logically, gating and apparent convergence could occur even if the pathways carrying the visceral and nonvisceral impulses were different but under some type of common control.

Above the spinal cord, information is processed in more detail, leading to such perceptual aspects as awareness of discriminative components of the stimulus (e.g., location and strength), which are less precise with visceral stimuli, but not nonexistent. Interestingly, studies have shown that normal subjects and patients with functional bowel disorders demonstrate differences in localization of the experimentally induced stimulation (distention).

Descending efferent control can affect sensation, just as sensation can affect motor control. For example, descending pathways exist that can modulate the gastrointestinal (and other visceral) pain experiences. Anatomically, the periaqueductal gray, nucleus raphe magnus, and lateral reticular formation have been implicated. Neurochemically, and not surprisingly, opioid peptides are involved. The function of these descending pathways appear also to be influenced by processes of convergence.

Pain in the Alimentary Tract

Although pain is not the most common sensory impulse originating in the gut, either under normal circumstances or in association with functional bowel disorders, it is the most intense and probably the most troubling for people with the various disorders. Pain has been discussed in various places in earlier chapters and earlier in this chapter as well. Some details of the physiology of pain will be discussed here. The topic of pain as a visceral sensory event also will be discussed in the next chapter.

Pathological processes in the gastrointestinal tract can produce hyperalgesia and/or allodynia. For example, chemicals released from injured tissues or inflammatory cells, (see Mayer and Gebhart, 1994, their Figure 3, p. 279) can convert a stimulus that otherwise would not have been painful into one that is. Sensitization can occur at the level of the dorsal horn cell (see Bueno et al., 1997). The quality of the pain can differ from that seen under normal circum-

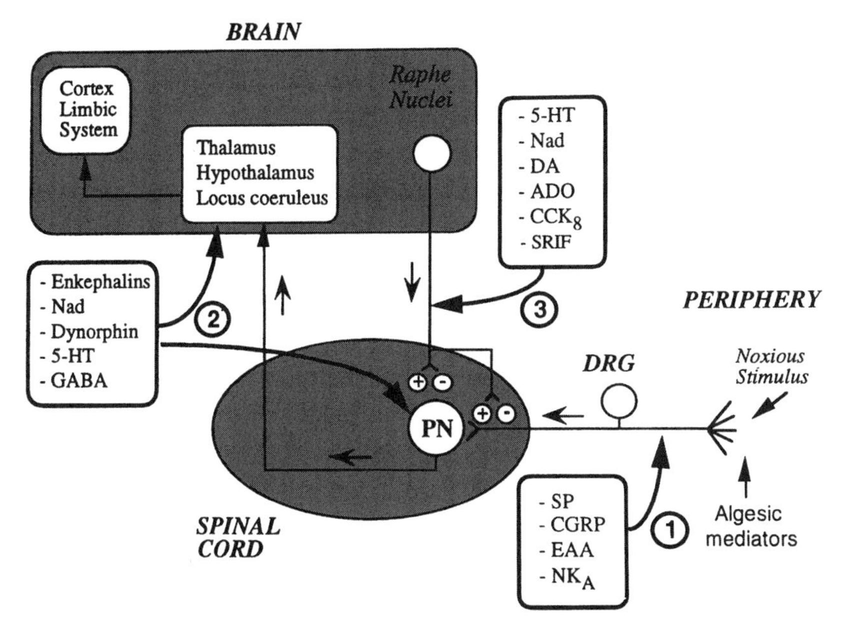

Figure 9–4. Diagram of sites where the indicated neuromediators were implicated in nociceptive pathways from the gastrointestinal tract to the spinal cord and brain. (See original publication for unfamiliar abbreviations.) [From Bueno et al., 1997, reprinted with permission of the holder of the copyright.]

stances. A noxious stimulus can increase the gain such that further stimulation will have an even greater than expected pain-producing effect. Descending modulation by the central nervous system is known to affect sensory processes. Some are transitory, but others can be longer-lasting, even virtually life-long. It has been speculated (see Mayer and Gebhart, 1994) that very early changes such as neonatal tissue irritation may be especially susceptible to permanent functional alterations. These pain mediators are illustrated in Figure 9–4.

What is the locus of the disease process? If sensory abnormalities are part of the etiology, is the abnormal process in the gut, in the peripheral nervous system, and/or in the central nervous system? Which pathways are most likely involved? What role does motor dysfunction play? Are there abnormalities in the interaction between ascending and descending functions? Are the abnormalities in the receptors or the biochemical mechanisms? What role do long-standing (e.g., gene expression) changes play? Which abnormalities are state and which trait? Does the enteric nervous system have a memory in the same sense that the central nervous system does? There are some clues. For example, certain injuries, such as peritoneal inflammation and surgical injury, lead to decreases in bowel motility rather than the increases often seen with functional bowel disorders, suggesting that these processes are not etiologically involved.

Interoception Data

Concerning awareness of gastrointestinal stimulation itself, a number of studies have been completed in both normal subjects and in individuals with various gastrointestinal disorders such as IBS. Some of the normal subject studies have been reviewed by Malagelada (1994) and in several of the other studies described below. Most were done in the 1990s.

Whitehead and Drescher (1980) addressed questions of *(1)* the relationship between awareness of gastric motility and ability to control gastric action with biofeedback, *(2)* the effects of training, and *(3)* the relationship between gastric and cardiac awareness. Feedback training was found to improve performance, and gastric and cardiac awareness were significantly positively correlated, but perception and control of gastric motility were not related.

In healthy subjects distention of the proximal duodenum produces a dose-related (i.e., amount of distention) relaxation of the stomach (Malagelada, 1994). Distention of the distal duodenum does not produce this effect. Of relevance to alimentary sensation, this effect can occur at a level of distention that is not sensed (i.e., is below the sensory threshold). Similar dose-dependent responses occur in the jejunum, with associated relaxation observed. In the jejunum, however, the relationship between amount of distention and sensation appears more tightly correlated than in the duodenum. The investigator also reported that the stomach and proximal duodenum were sensitive to hot and cold stimuli and to electrical stimuli.

Holzl et al. (1996) studied the relationship among gastrointestinal stimulus characteristics, behavioral effects of the stimulation, and awareness. They indicated that stimulus effects can occur at different levels, from simple detection—Has a (relevant) event happened?—to identification—What (how) is it (like)? (sensory quality, object/visceral event identity) (p. 201). Using this schema, they provide data supporting the hypothesis that detection can occur without awareness (i.e., reportability). Below awareness, one threshold seemed to exist, but above awareness a correlation was found between stimulus intensity and intensity of subjective perception. Localization of stimulation required subjective sensory awareness.

It has been reported that the application of somatic stimulation can affect awareness of visceral stimulation. For example, Coffin et al. (1994) studied the effects of the application of electrical nerve stimulation to the hand at two different sub-pain levels on awareness of different intensities of gastric and duodenal distention in normal subjects. Application of the electrical stimulus decreased awareness of distention of both the stomach and the duodenum. The higher intensity hand stimulus decreased awareness more than the lower. They suggested that these data support the hypothesis of a convergence of somatic and visceral afferent impulses onto the same pathways, probably in the spinal cord but possibly higher in the central nervous system, and that a gate-like mechanism controls flow of afferent information. Other studies also have reported an interaction between visceral and somatic nociceptive processes in both rats (Ness and Gebhart, 1991) and humans (Bouhassira et al., 1994).

Holzl et al. (1999) also studied the masking effects of visceral and somatic stimulation on each other. Using abdominal rather than hand stimulation, they reported that abdominal stimulation impeded perception of colonic distention, but not the opposite. In contrast to the above, they concluded that their findings contradicted the spinal sensory convergence model.

As already described, the sympathetic part of the autonomic nervous system (as well as the parasympathetic) is intimately involved in interoceptive processes. One means of modifying sympathetic tone is to apply negative pressure to the lower body. Using this method, Iovino et al. (1995) studied the effects of changes in sympathetic tone on perception of duodenal distention in healthy subjects. They also studied the effects of transcutaneous hand stimulation, as did Coffin et al. (1994). They found that sympathetic activation increased perception of duodenal distention as well as the reflex relaxation responses but did not affect perception of the somatic (i.e., hand) stimulus. Thus, sympathetic activation heightened gastrointestinal interoceptive awareness.

Another method to modify gastrointestinal awareness is with

medication. Amitriptyline, a tricyclic antidepressant with a strong anticholinergic effect, has been reported to benefit the pain associated with chronic bowel disorders. Gorelick et al. (1998) studied the effects of treatment with amytriptyline for three weeks in normal volunteers on perception of both cutaneous electrical stimulation and esophageal and rectal distention. Amitriptyline treatment increased the threshold for cutaneous awareness and pain but had no effect on visceral awareness or pain. Thus, in contrast to the effect of change of sympathetic tone on gastrointestinal interoception, modification of cholinergic function had no effect. The dose used was low, however, and a higher dose might have had an effect. Furthermore, Peghini et al. (1998) found that imipramine, another tricyclic antidepressant, did change pain thresholds to esophageal distention in normal subjects, and octreotide, an analog of somatostatin, changed distention threshold without having an analgesic effect (Johnston et al., 1999). In summary, these studies clearly document the existence of interoceptive processes in the alimentary tract and begin to address the question of mechanism(s).

DISORDERS

As noted above, functional bowel disorders are separate from but overlap disorders that are defined in considerable part by the location in the bowel in which the symptom focus occurs. For many of these patients, more than just one region of the bowel appears to be affected. Multiple factors affect the disease process and its severity, including age, sex, and fasting versus post-prandial condition. Although most of the studies of visceral sensation in these people involve mechanoreceptors, other receptors such as chemoreceptors are likely to be involved as well. Abnormalities of chemical mediators may be involved. A biopsychosocial model is appropriate to an understanding of these disease processes, including changes in interoceptive processes (see Drossman, 1998). So far, studies of visceral sensory changes in these disorders have focused mainly on functional dyspepsia and IBS, although at least one study has addressed non-cardiac chest pain. The findings from some of these studies will now be summarized.

Approximately one-third of people with non-cardiac chest pain have demonstrable abnormalities of motility, and the presence of acid in the esophagus (reflux) plays a role in symptom production. Stimulation tests have reported pain symptom reproduction at rates as low as less than 10% or as high as 70% or more, but in general the correlation between motility abnormalities and occurrence of symptoms is fairly weak.

Paterson et al. (1995) studied people with chest pain of undetermined etiology. They reported that pain induced by esophageal balloon distention was related to balloon volume for all subjects and that pain scores were higher for the chest-pain subjects than for either normal controls or subjects with demonstrated esophageal dysmotility, indicating that motility abnormalities were not, per se, associated with the sensory abnormality. For the chest-pain group only, pain increased with repeated distentions, indicating a sensitization process that did not occur in the other subjects.

Functional dyspepsia is a disorder with symptoms mainly referable to the stomach and upper small intestine. Gastritis due to *Helicobactor pylori* and hyperacidity may play a role in some patients, but for the majority the presence of *Helicobactor* is asymptomatic, and neither of these pathophysiological processes alone adequately explains the symptoms for the majority of patients. There does seem to be some role for factors such as stress, sleep abnormalities, changes in function of the sympathetic part of the autonomic nervous system, and conditioning processes, but the details are poorly understood.

Mertz et al. (1998b) compared responses to gastric distention in people with functional dyspepsia to those with organic dyspepsia (associated with demonstrable tissue injury or irritation) and normal subjects. Almost two-thirds of the functional dyspeptics, but no organic dyspeptics, were more sensitive to fullness, discomfort, and pain. The patients with the functional problem also were more likely to refer the pain to an atypical location and/or to describe pain during sham distention. Neither psychological measures nor altered motility explained the results. Mearin et al. (1991) and Salet et al. (1998) demonstrated that, in comparison to normal subjects, people with functional dyspepsia had greater pain and nausea responses to gastric distention but similar motor responses as well as similar responses to hand stimulation and to the cold pressor test.

Physiological correlates of this altered pain sensitivity are of interest. It has been reported that increased lipid content in the duodenum affected awareness of jejunal distention, but not electrical stimulation in the jejunum. Kanazawa et al. (2000) reported that people with functional dyspepsia had a normal sensory threshold but a lowered pain threshold to mechanical or electrical stimulation as indicated by evoked potentials. Some motoric abnormalities have been reported, although their relationship to sensory function is not clear. For example, Hausken et al. (1993) reported that motility of the gastric antrum is reduced by acute mental stress in normal subjects, but that stress did not have this effect in people with functional dyspepsia. They interpreted this finding as evidence of reduced vagal tone in the disease group. Mertz et al. (1998a) reported that low-dose amitriptyline (50 mg/day) did not normalize the altered perception to gastric distention in this group.

One estimate indicates that 10% to 20% of adult "Westerners" have IBS (see Lynn and Friedman, 1993). These patients report symptoms broadly referable throughout the gastrointestinal tract, including a globus sensation in the throat and gastroesophageal reflux, to the lower gastrointestinal signs and symptoms usually associated with the disorder. The bowel is often tender to abdominal pressure, although the pain, motility abnormalities, and general hypersensitivity to distention do not necessarily correlate with clinical symptoms. Disorders of both motility and sensation have been documented, but not all patients have disorders of both. It has been reported that the threshold for pain from distention of the bile duct may also differ in people with functional bowel problems. It has been observed that these people seem to be highly susceptible to response bias and to lower intestinal sensitization in the clinical situation. This sensitization process may have a conditioning component. In contrast to the claim that the cold pressor test is not abnormal in people with functional dyspepsia, it has been observed that it may be in IBS (the blood pressure response).

Mertz et al. (1995) studied rectal distention thresholds, symptom reports, and anorectal manometry in IBS patients in comparison to normal controls. More than 90% of the patients showed changes in visceral sensory function in response to rapid, but not to slow, distention. Symptoms clustered patients into three different groups, and findings were consistent over time.

Whitehead et al. (1997) replicated the prior finding that IBS patients had lower thresholds for rectal distention than did normal subjects. They also found that the differences observed were correlated with measures of the psychological factors of anxiety and somatization (but not with presence or absence of a history of sexual abuse). And Rossel et al. (1999) reported that these patients had lowered pain thresholds to rectal and sigmoid, but not to cutaneous, electrical stimulation.

Lembo et al. (1999) focused on pain-predominant IBS patients. In a survey of almost 450 patients, they found that symptoms clustered into four different groups. Less than one-third of the subjects complained of pain as their worst symptom, but pain was correlated with rectal hypersensitivity to the distention trial.

Turning to physiological differences, it was noted that these patients may have abnormal ascending inhibitory and descending excitatory reflexes (see Bueno, 1997). Evidence for an abnormality of central nervous system noradrenergic function comes from Dinan et al. (1990). They found that the growth hormone response to the tricyclic antidepressant desipramine was blunted in people with IBS. It was also found (Masand et al, 1997) that IBS patients have an increased prevalence of dysthymia, a less severe but more chronic form of depression than major depressive disorder. Major depressive disorder also shows blunted growth hormone responses.

Trimble et al. (1995) tested the theory that the various functional alimentary disorders are associated with a generalized disorder of gut sensitivity. They studied the response to esophageal and rectal bowel distention in individuals with functional dyspepsia and with IBS. Patients with functional dyspepsia and patients with IBS both had increased sensitivity to distention in the esophagus and the rectum. Consistent with earlier studies, no somatic sensory threshold abnormalities were observed. The investigators concluded that these results supported the theory of generalized disorder of gut sensitivity. Indeed, clinical pain in IBS can be referred to the back, shoulder, or even the thigh. In another study (Zighelboim et al., 1995) it was reported that gastric perception to distention was altered in IBS, but that rectal was not. Although this study also found an abnormality in IBS patients higher in the alimentary tract, the lack of abnormal rectal perception in this study was atypical.

Treatment methods such as the use of anti-spasmodics, bulk-forming agents, and pro-kinetic drugs generally have not been satisfactory in these patients (see Goldberg and Davidson, 1997; De Ponti and Malagelada, 1998). Thus, these results do not provide reliable clues as to the source of the visceral sensory abnormalities in the disorders. Evidence exists of abnormalities of serotonin function in functional bowel disorders and also the D2 dopamine receptor. What is clear is that interoceptive processes, including dysfunctions of those processes, do occur (e.g., Hegel and Ahles, 1992). Interoceptive abnormalities in the alimentary tract exist in many people with functional bowel disorders. Future studies need to target even more peripheral and nervous system changes, including more careful ascertainment of symptoms, separating sensation from perception.

PART THREE

Related Topics and Summary

ten

PAIN, PROPRIOCEPTION, PHANTOM LIMB, AND BODY IMAGE

. . . our bodies themselves, are they simply ours, or are they *us?*

William James, *The Principles of Psychology*, Volume One, 1950, p. 291

Visceral awareness is a component of a larger awareness, a larger sense, an awareness of the "I," or the "me." The "I" is usually not thought of as essentially a bodily awareness. If one loses a finger in an accident, the sense of self is not lost and does not even seem to be in any fundamental way diminished. The same can be said for the visceral organs. Nobody seems to lose her or his sense of self after an operation to donate a kidney or have a length of bowel removed. The self seems to exist independent of the body. But this apparent independence can and should be doubted. The essential role of visceral awareness in emotion, and the essential role of emotion in the sense of self, serve as reminders of how important bodily sense is. As already discussed, Damasio (1994), taking the Jamesian position, has spoken of the importance of "somatic markers" of emotion and the sense of the body as "in the background" to the sense of self.

This issue will be visited from a neurophilosophical point of view later, in Chapter 13. In this chapter, four topics related to this theme will be discussed. First will be a discussion of pain, including visceral pain, continued from earlier chapters. Then, proprioception, phantom limb phenomena, and the bodily sense of self will be reviewed. It is the hypothesis of this chapter that visceral awareness is at least potentially directly related to each of these themes, even if not so far directly addressed by the authors and investigators who have worked in these areas.

PAIN

Pain, defined as an unpleasant sensory and emotional experience (see Chapman and Nakamura, 1999), requires a central nervous system. Tissue damage can occur anywhere in the body, including the viscera, and pain can occur with or without ongoing tissue damage; but without a brain there can be no pain. There is an essential, perhaps even sine qua non, role of emotion and affective change in pain, even though it appears that pain sensations come before affect (see Price, 2000), and analgesic effects often depend on psychological affective processes.

In the earlier discussions of visceral sensation and visceral pain, the question of whether internal organs have specific nociceptors has been raised. Specific nociceptors with high thresholds, others that are silent nociceptors, and/or others that are intensity encoding have all been described or hypothesized. This area of research still contains many controversies (see Cervero and Janig, 1993; McMahon, 1997; Cervero and Laird, 1999).

Ascending pain tracts are formed by three-neuron chains. The observation that two pain pathways exist, at least from a functional point of view, dates back almost 100 years (see Cross, 1994), but more recently it has been shown to be true anatomically as well (see Price, 2000). One pathway (epicritic) carries the impulses of first-pain, involving the discriminative component of pain. The other pathway (protopathic) carries second-pain, involving the affective–motivational–emotional component. The discriminative pathway is phylogenetically newer, anatomically more lateral in the central nervous system, and more concerned with events in the environment (e.g., cutaneous sensation), while the affective–motivational–emotional pathway is older, more medial anatomically, and more concerned with bodily (visceral) functional states (see Cross, 1994; Trimble et al., 1997).

A descending pathway also is involved in modulatory effects that operate at the spinal cord level. Different descending modulatory effects appear to originate in the cortex, thalamus, and brainstem. Throughout the central nervous system, at all levels, close relationships exist between the pathways carrying somatic impulses and those carrying visceral impulses. For example, convergence of tactile and visceral pain inputs is transmitted in the dorsal columns and projected to the ventral posterior lateral thalamus. The need for electrical stimulation in the gastrointestinal tract that is persistent in time in order to elicit pain, which is referred and increases with more stimulation, indicates central nervous system processes of convergence and integration. The existence of these descending influences demonstrates that the experience of pain is not a purely passive process. Pain is an active interaction between the pain stimulus and the sensory and emotive perceptual apparatuses of the organism.

One review (Willis and Westlund, 1997) describes the ascending nociceptive pathways as including the spinothalamic, spinomesencephalic, spinoreticular, spinolimbic, and spinocervical system, as well as the dorsal column tracts. The descending pathway related to analgesia includes the periaqueductal gray, locus caeruleus, parabrachial area, nucleus raphe magnus, reticular formation, anterior pretectal nucleus, thalamus, areas of the limbic system, and cerebral cortical areas. The anterior cingulate gyrus is implicated especially in the affective aspects of pain processing, as are attention processes that implicate the cognitive apparatus and the cortex (Eccleston and Crombez, 1999; Peyron et al., 1999; Price, 2000). Other brain regions that are strongly implicated in both the affective aspects of pain and in interoceptive processes include the insula, amygdala, and S2 (also called SII) somatosensory region (see Price, 2000).

In addition to the close relationships between these two types of pathways, some types of pain appear to have aspects of both pain types. Although muscles, joints, and connective tissue are typically thought of as part of the non-visceral parts of the body, so-called deep pain in these tissues often is related more in character (e.g., diffuse, poorly localized) to visceral pain than to cutaneous pain (Gebhart and Ness, 1991).

Involvement of the autonomic nervous system in visceral sensation generally has already been described. Data exist that are specifically related to autonomic involvement in visceral pain. One long-standing concept is the existence of sympathetic dependent pain. Schott (1994) argued that such sympathetic pain syndromes as causalgia and reflex sympathetic dystrophy may not actually be sympathetic, but rather "... the visceral afferents which, confusingly, also travel within the autonomic nervous system ... may subserve 'sympathetic dependent' pain" (p. 397). To support this argument, this author pointed out that *(1)* many disorders of the autonomic nervous system are painless, *(2)* pain treatments aimed specifically at modifying autonomic function often are not successful, and *(3)* a broadening of the definition of visceral afferent to include fibers that run with the autonomic fibers (e.g., on blood vessels) would clarify the source of these pain disorders. On the other hand, the successful use of tricyclic antidepressant drugs to treat chronic and atypical pain appears to support a role for monoamines in at least the affective aspects of pain, thus implicating the peripheral and/or central autonomic mechanisms and pathways. Interactions appear to exist among pain, blood pressure, and opioid receptors (Bruehl et al., 1999).

The role of vagal afferents in visceral pain appears to be primarily modulatory. They can increase or diminish pain. For example, evidence exists for involvement of vagal afferents in the effect of morphine on pain control (Randich and Gebhart, 1992). Reviews of the treatment and pharmacologic aspects of visceral pain include Cross (1994), Cervero and Laird (1999), and Millan (1999). The sensory characteristics of visceral pain have been outlined in table format (Table 10–1) by Cervero and Laird (1999).

Several studies of anatomical regions in the brain associated with pain have used functional imaging techniques. Approximately 20 years ago Buchsbaum et al. (1983) investigated the effects of unpleasant electrical stimulation to the right forearm on glucose metabolism in normal subjects. They found that metabolism increased most in the postcentral region of the cortex on the left, consistent with crossed activation of the somatosensory area.

Talbot et al. (1991) used PET measurement of cerebral blood flow to identify regions activated by painful and non-painful heat stimuli to the forearm in normal subjects. They reported that the painful stimuli activated the contralateral anterior cingulate gyrus and primary and secondary somatosensory gyri, while the non-

Table 10–1. Sensory Characteristics of Visceral Pain and Related Pain Mechanism

Psychophysics	Neurobiology
Not evoked from all viscera	Not all viscera are innervated by sensory receptors
Not linked to injury	Functional properties of visceral sensory afferents
Referred to body wall	Viscerosomatic convergence in central pain pathways
Diffuse and poorly localized Extensive divergence in central nervous system	Few "sensory" visceral afferents
Intense motor and autonomic amplification	Mainly a warning system, with a substantial capacity for reactions

[Adapted with permission of the holder of the copyright from Cervero and Laird, 1999.]

painful control condition activated only the primary somatosensory area. Jones et al. (1992), in a response to Talbot et al., reported that with a different statistical analysis method the contralateral thalamus also was activated. They also stated that they did not observe activation in the somatosensory area and hypothesized that the difference in results was due to methodological differences.

Coghill et al. (1994) applied a painful heat stimulus and a non-painful vibratory stimulus to the left forearm of normal subjects. They reported contralateral activation in the primary and secondary somatosensory areas, anterior cingulate, anterior insula, thalamus, and frontal cortical supplemental motor area. The non-painful stimulus activated somatosensory cortex and posterior insula, and the strongest difference between painful and non-painful stimulation was observed in the anterior insula. One of their conclusions was that non-painful stimulation produced activation that was essentially completely contralateral, while painful stimulation produced activation that was primarily contralateral but activated regions on the same side in cortical and thalamic regions as well.

Casey et al. (1994, 1999) studied painful and non-painful forearm heat stimuli in normal subjects. Activated regions included the contralateral thalamus, cingulate cortex, primary and secondary somatosensory areas, and insula, along with ipsilateral secondary somatosensory cortex and thalamus and medial dorsal midbrain and cerebellar vermis. Using a different mode of pain induction, Coghill et al. (1998) found that pain induced by capsaicin injection produced a substantial decrease in global cerebral blood flow.

Several more recent imaging studies using painful heat stimulation have been reported (e.g., Craig et al., 1996; Vogt et al., 1996; Derbyshire et al., 1997, 1998; Derbyshire and Jones, 1998). In aggregate, these studies reported activation mainly in the following regions: anterior cingulate cortex, midcingulate cortex, perigenual cortex, posterior cingulate gyrus, cingulofrontal transitional cortical area, anterior and posterior insula, and thalamus. Other regions were less consistently activated. Contralateral activation was more common than was ipsilateral, but both were observed. The results of these studies were, thus, generally consistent with the earlier study results. In a related study Craig et al. (2000) reported that non-painful temperature sensation (graded cooling to the hand) significantly activated the insula, and Hanamori et al. (1998) reported that tail pinch pain in the rat produced changes in neuronal electrical activity in more than 75% of the cells studied in the medial area of the insular cortex. Ploghaus et al. (1999) found that anticipation of thermal pain activated regions in the medial frontal and insular cortices and cerebellum, near to but not the same as the regions activated by the pain itself.

Imaging studies of individuals with pain disorders have been reported. Silverman et al. (1997), as discussed in Chapter 9, reported that rectal pain activated the anterior cingulate in normal subjects but the left prefrontal area in people with IBS, and this same pattern was seen during expected but not delivered pain. In another study, Derbyshire et al. (1994) found that painful thermal stimuli activated the thalamus, anterior cingulate, lentiform nucleus, insula, and prefrontal cortex in both normal subjects and people with atypical facial pain. Changes in anterior cingulate and prefrontal areas seemed to differentiate patients from normal subjects.

In summary, differences existed between patients and normal subjects, but the regions involved in pain responses were generally the same in patients as were reported in normal subjects. Comparing these imaging studies of non-visceral pain to the studies described in Chapter 9, the main difference appears to be the activation of the primary somatosensory area in non-visceral pain, which is much less likely to occur with visceral pain. All the other structures involved seem to overlap considerably, with the insular and cingulate regions especially prominent (see Laurent et al., 2000; Ladabaum et al., 2000).

Beyond imaging studies of pain in these disorders, pain more generally as it is associated with various disorders is, for obvious reasons, of great importance and a focus of much research. Just a few examples will be offered here, briefly described, as illustrations of the areas of interest.

Angina pectoris is a visceral sensory abnormality in the thoracic area (see Sylvan, 1993; Harford, 1994; Foreman, 1999). Angina is usually considered to be due to cardiac ischemia, but it is clear that the majority of cardiac ischemic attacks are not associated with pain and other causes of angina (or angina-like sensations) exist, such as changes referable to the esophagus. Afferent as well as efferent neurons are found in the heart, but the exact nature of their function, the types of information they carry, and their potential role in cardiac pain are not well understood. Neurotransmitters implicated in ischemic change and cardiac pain include especially bradykinin and adenosine, and people with different syndromes, such as "Syndrome X" and asymptomatic myocardial ischemia, show different sensitivities to these neurotransmitters.

Abnormalities of pain sensitivity have been reported in people with psychiatric disorders. Marchand et al. (1959) observed more than 40 years ago that people experiencing psychosis have altered pain sensitivity to both somatic (fracture of the femur) and visceral (ulcer perforation and acute appendicitis) causes of pain. Others also have observed this (e.g., Rosenthal et al., 1990). Abnormalities of pain sensitivity and/or reporting also have been observed in depression (Bailey and Wootten, 1976; Alder and Gattaz, 1993; Fishbein et al., 1997), borderline personality disorder (Russ et al., 1992, 1994), and in some cases of anxiety (Roy-Byrne et al., 1985; Rizzo et al., 1985; Janssen et al., 1998; Asmundson et al., 1999). Fibromyalgia, a disorder often thought of as at the border between medical and psychiatric illnesses, shows distinct pain dysfunction (e.g., Clauw and Chrousos, 1997; Bennett, 1999). A review of the pain perception literature in psychiatric disorders may be found in Lautenbacher and Krieg (1994).

Pain in gastrointestinal disease will not be further reviewed here. Whitehead and Palsson (1998) pointed out that psychological factors, including attentional processes and disease attribution, probably contribute along with biological factors to the abnormal pain response to rectal distention in people with IBS. Visceral pain from both normal and pathological processes is an essential component of the overall topic of interoception.

PROPRIOCEPTION

Proprioception, defined most simply, is the organism's body-image sense of itself in space (and time). Proprioception is typically considered to be a function of the nervous system related to the somatic aspect of the body rather than the visceral, that is, related to muscles, joints, tendons, and vestibular functions, among others (Matthews, 1988; Keshner and Cohen, 1989; Kalaska, 1994), a kind of sixth sense. Cerebellar function is involved as well (Bloedel and Bracha, 1995). To the extent, however, that both visceral and somatic aspects make up one body connected to one nervous system, and to the extent that concepts such as Damasio's "somatic marker" hypothesis of emotion may eventually be confirmed, an approach to sensory–perceptual and motivational–emotive functions that encompasses the whole body seems worth consideration and exploration. It is for this reason that proprioception, along with phantom limb phenomena and body image, are being reviewed here. The theory of the existence of a cognitive map of the whole body, including visceral and somatic sensory and motor components, suggests that it would be potentially productive to think of these apparently disparate functions and processes as linked.

Several aspects of central body image maps have been gleaned from studies of proprioceptive processes. Maps are labile, changing from moment to moment, and are not completely bound to the anatomical limits, movements, and positions the body can actually assume. Furthermore, maps appear to be, at least in part, unconscious. Also, these maps contain not just the body in some disembodied way, but also relevant aspects of the external world. For example, they can be influenced by the sense of touch. They are maps of the body in relation to both the internal structure and status of the body itself as well as the environment immediately outside the body. It is likely that the cerebellum and the vestibular system contribute to body image maps, as do attentional processes. It has been argued that various specific functions related to proprioception and body sense reside in the right cerebral hemisphere (Joseph, 1988).

The functions of the cerebellum are particularly relevant to the question of how proprioceptive functions may relate to interoceptive processes. The cerebellum is intimately involved in the control of various reflex functions and in goal-directed movements of the arms and legs. In addition, it has been shown that the cerebellum is involved in certain learning processes, particularly defensive Pavlovian conditioning (Bloedel and Bracha, 1995; Beggs et al., 1999). The conditioning processes, in turn, are involved in emotional functions and are thus closely linked to visceral sensory and motor functions. While this linkage takes several steps, the hypothesis seems quite reasonable that these functions, from proprioception to interoception, are likely to be linked, at least indirectly.

PHANTOM LIMB PHENOMENA

The central hypothesis of interoceptive processing is that functions of at least some of the visceral organs can be sensed. For those organs that can be sensed, it follows that a change in that awareness should occur if the organ is removed. The most logical hypothesis might be that all sensations referable to that organ should disappear. However, the existence of the phantom limb phenomenon demonstrates that such sensations may not disappear. This section of this chapter will discuss the phenomenon of phantom body parts.

The most common phantom body part phenomena relate to limbs, especially the upper limbs. A phantom limb sensation is the subjective experience that a limb or part of a limb is still part of the body after it has become detached either by removal (e.g., surgically or by trauma) or by de-afferentation (disconnected from the rest of the nervous system, e.g., by brachial plexus avulsion). Although originally claimed not to occur, phantom phenomena also can occur with congenitally absent limbs. In addition to non-painful phantom sensation, phantom pain can and frequently does occur. Phantom phenomena were first described in the mid-seventeenth century, and the term was first used after the U.S. Civil War. Several helpful reviews of phantom phenomena are available (e.g., Postone, 1987; Katz and Melzack, 1990; Melzack, 1992; Ramachandran and Hirstein, 1998; Hill, 1999).

Although phantom limbs are the most common phantom phenomena, other phantom organs have been reported, including breasts, teeth, and corneas. Phantom sensory experiences have been reported in the visual and auditory systems as well, such as well formed hallucinations in individuals who are not otherwise experiencing any psychotic symptoms. These typically occur in people who have reductions in the normal functioning of the sensory system in question, such as people with cataracts.

Of course, of most interest to the theme of this volume is the question of whether or not phantom visceral organs occur. Quotes from two sources, below, indicate that they do, but it would also be important if, in contrast to phantoms limbs, phantom organs did not occur. (Would that mean that the mechanisms that account for the existence of phantom limbs—e.g., the hypothesized cortical map of the body—do not also contain the structures of the viscera?) Apparently, no systemic research on this subject has been done. The remainder of this section will describe research with other phantom body parts, under the assumption that this research will be relevant and generalizable, at least partly, to phantom visceral organs when such investigations are undertaken. It seems likely that the most relevant material will be that related to the central nervous system mechanisms.

> There is also a literature on the persistence of painful and nonpainful sensations associated with the removal or deafferentation of body structures other than the limbs, including . . . internal . . . organs. Ulcer pain has been reported to persist after vagotomy or subtotal gastrectomy with removal of the ulcer, labor pain and menstrual cramps after total hysterectomy, the sharp, burning pain of cystitis despite complete removal of the bladder. . . . Some patients report the sensation of a full bladder and a feeling that they are urinating even though the bladder has been completely removed. Sensations of passing gas and feces continue to be felt after the rectum has been removed (Katz and Melzack, 1990, p. 321).

> . . . [phantoms] have been reported after amputation of . . . even internal viscera . . . the acute pain of appendicitis following removal of the inflamed appendix. (Ramachandran and Hirstein, 1998, p. 1605).

Probably the most common phantom phenomenon is a nonpainful awareness of an upper limb just after its sudden removal.

This occurs in between 90% and 99% of such individuals, depending on the strictness of the definition. The most common symptom is a paresthesia, or a tingling sensation. Other common symptoms are numbness and (usually painful) cramping or burning. Additional sensory aspects, such as a cut present on the limb at the time of limb loss, are sometimes incorporated into the phantom. The occurrence of experiences of sweating or temperature change in the phantom implicates changes in autonomic nervous system function in phantom phenomena. Multiple sensory changes—touch, temperature, proprioceptive changes—may all occur together as parts of a total phantom experience.

The shape of the phantom is typically similar to the removed portion of the limb at the time of removal, but it can change over time. It has been speculated that the portion of the limb—for example, hand and fingers or foot—that is most extensively represented in the brain is most likely to remain. Posture and movement are often aspects of the phantom. Not only can some individuals volitionally move the phantom, but it often moves spontaneously, as in the associated postural–balance swinging motion of an upper extremity during walking.

The duration that a phantom remains is extremely variable. Some fade within days, while others last a lifetime. A phantom can disappear and reappear, even years later. They can even occur when a limb fails to form congenitally, although that is less common; 15%–20% of congenitally absent limbs are associated with phantom experiences. Indeed, it is not necessary that the limb actually be lost at all. Phantoms have been described in people with leprosy, multiple sclerosis, and paraplegia, although the percentage is considerably lower. The fundamental event appears to be a loss of afferent impulses from the relevant part of the body, regardless of whether it is still present.

Pain occurs commonly as part of the phantom experience. It is estimated that as many as 85% of all phantoms have pain, but there is great variation in the estimates, due mainly to variations and ambiguities in the definition of pain as separate from such feelings as tingling and burning. The location of phantom pain often will be the same as any pain present in the limb before its loss. Quality is somewhat more variable. Although it tends to persist if present for several months, it can change over time. Pain is made worse by psychological stress, cold, or contact with any irritating substances.

Factors in pain occurrence that are important include *(1)* the presence of pre-amputation pain, *(2)* the duration of pain pre-amputation and whether there was a pain-free period before limb loss, *(3)* the type, intensity, and location of any pain prior to limb loss, *(4)* emotional factors, and *(5)* input from other sensory modalities contributing to the experience (Katz and Melzack, 1990).

A number of theories address the etiology of phantom sensations. The earliest theories posited a peripheral source for the sensory impulses. However, the phantom experience did not necessarily follow the nerve distribution that would be expected, and local ischemia or anesthesia could modify the phantom sensation beyond the time when the local effects would be expected to dissipate. As noted, autonomic nervous system changes are implicated in at least some painful phantom experiences. The associations of decreased blood flow with thermal pain and EMG changes with cramping pain indicate a level of specificity to the putative autonomic changes that is related to the characteristics of the pain observed. Central nervous system changes also are hypothesized to be important. One theory implicates changes in the dorsal horn cells of the spinal cord. Loss of normal impulse input from the periphery is hypothesized to create abnormal firing patterns that lead to the abnormal sensory experiences (painful or non-painful). Arguing against this idea is the observation that paraplegic injuries are less, not more, likely to lead to phantom phenomena.

Theories that posit more central changes, above the spinal cord, have been offered. The gate-theory hypothesizes that nervous system functions map the parts of the body, their functions, and their interrelationships, a so-called body image, or body schema. The principal cortical and subcortical brain structures that have been implicated (not mutually exclusively) are the thalamus (especially ventral posterior) and the cortex (especially postcentral primary and secondary somatosensory). In the monkey, this type of information appears to be passed serially from the ventral posterior nucleus of the thalamus to the primary to the secondary somatosensory cortex (Pons et al., 1987). These maps are modifiable, which assumes neuronal plasticity and thus indicates that the body image itself is modifiable. Ramachandran and Hirstein (1998) noted that changes in phantom phenomena, and presumably body image, can occur in a month. They theorized that this process is actually one of uncovering, or unmasking, synapses and pathways that already exist. Cross-modality cortical reorganization has even been demonstrated in humans (Kujala et al., 2000).

Pain phenomena appear to be related to cortical and subcortical reorganization (or unmasking) events (Birbaumer et al., 1997; Montoya et al., 1998; Nicolelis et al., 1998). Cortical reorganization seems more likely to occur in people who experience painful phantoms than in those whose phantom is not painful. These observations imply that cortical reorganization processes might be important to the genesis of pain in the phantom—is phantom pain faulty rewiring? The occurrence of deep pain–like experiences suggests the involvement of processes akin to visceral pain processes.

Speculations have been raised about the role of psychological factors in phantom phenomena. Emotional factors such as anxiety do play a role. It is doubtful that any specific personality dysfunction is closely linked to this process, although some have suggested that a "rigid" personality style may make one more prone to phantom pain.

Melzack (Katz and Melzack, 1990; Melzack, 1992) offered a theory of phantom phenomena called a neuromatrix theory. He hypothesized that a nervous system map in the brain is largely genetically determined (see also Gallagher et al., 1998), but that it can be modified by experience involving "somatosensory memories." This theory posits a broadly distributed map residing in the classical thalamic–somatosensory cortex pathway and also in the reticular activating system, the limbic system (accounting for the emotional aspects, including pain—and perhaps thusly implicating involvement of the viscera), and the parietal cortex (linked to body awareness and such phenomena as body-part neglect and sense of self). Melzack (1992) provocatively hypothesized that "We do not need a body to feel a body" (p. 126). This theory has been criticized by Canavero (1994) and Hill (1999).

It seems that a multi-factorial theory would best explain phantom phenomena, including peripheral (e.g., autonomic nervous system) and central nervous system (e.g., dorsal horn cell changes) factors. The addition of superadded sensations (visual, tactile, motor) supports involvement of higher, integrative central nervous system centers as well as the fact that the central nervous system representation is not immutable, but no theory so far proposed is fully adequate.

How does the phantom phenomenon relate to interoception?

First, under a broad definition of interoception, phantom phenomena from anywhere within the body would be of interest. Second, the clinical observations that visceral organs can have phantom sensory experiences are important. Third, to the extent that visceral afferent information is an essential part of emotion and emotional factors are involved in phantom phenomena, visceral and phantom sensory events may be related. It seems that phantom organ experiences are rare, but has anybody systematically studied this? Are visceral organs part of the central nervous system map, the "somatic markers" of emotion, the body image?

BODY IMAGE

William James (1950) wrote of the perception of space and movement, as well as consciousness of the self, including the persistence of a coherent sense of a unique and unified self over time. He even observed the role of the internal organs in the perception of body motion. Proprioception and phantom body part phenomena are closely related to the body image. The relevance of body image to interoception is that the viscera make up much of the body and give the body much of its "heft", its "heart," and its "guts." James asked if our bodies belong to us or define us (see epigraph, above). Despite this, literature on body image in the neural and behavioral sciences is sparse. A brief discussion of this scant literature will be presented here, and the role of the body in consciousness of self will be discussed in Chapter 13.

In the psychiatric literature the focus of interest in body image mainly relates to attitudes toward the body and body image disorders (McCrea et al., 1982; Lacey and Birtchnell, 1986). Issues such as internal versus external focus (the social dimension—one's own sense versus the expectation one has about how one is perceived by others), the effects environmental factors have on one's own sense, and short-term versus long-term bodily sense are raised, as are concerns about whether this self–other dichotomy is a form of incorrect dualistic thinking.

Psychiatric disorders can affect body image, including gender distortions, subjective feelings of disintegration, depersonalization, and loss of bodily boundaries. Disorders of body image have been ascribed to a variety of factors—neurologic, drugs, phantom limbs, responses to an actual deformity, responses to a subjective sense of deformity without an actual deformity, and especially commonly, distortions of body size perception and the often related eating disorders.

Observations about body image and related disorders within the clinical neurologic literature date back to the beginning of the nineteenth century. Abnormalities of body schema and posture, right–left confusion, denial and neglect syndromes, denial of incapacitation, and false attribution of nonexistent body parts or functions are examples. Many of these dysfunctions are associated with abnormalities in the parietal lobe, especially in the non-dominant hemisphere. Cerebrovascular accidents ("strokes") and epileptic foci associated with body image abnormalities most commonly occur in this area of the brain. It has been suggested that the body part denial syndromes may be a disorder of attention in the right–left dimension. Denial comes from an attentional imbalance, which is why the most lateral parts of the body (e.g., arm) are most denied and neglected.

Each of the four topics in this chapter—pain, proprioception, phantom body parts, and body sense—relates to the subjective experience of the embodied self, even if that part of the body never existed except in a genetically determined map in the brain. James and Damasio, along with others along the way, argued that awareness of the viscera is fundamental to the existence of emotions. Perhaps such awareness is even more fundamental, not just essential to emotional experience but to the very conscious (and unconscious) experience of a unitary self. Such speculation takes one into the realm of neurophilosophy, which will be discussed in Chapter 13. For now, it is hoped that review of these four topics adds further credence to the hypothesis that visceral sensory impulses are, generally, fundamental.

eleven

STATE-DEPENDENT LEARNING AND DRUG DISCRIMINATION

> ... the symptoms [nausea, secretion of saliva, vomiting and sleep] are now the effect, not of the morphine . . . , but of all the external stimuli which previously had preceded the injection . . .
>
> Ivan Pavlov *Conditioned Reflexes*, 1960, p. 35–36

State-dependent learning is both a theory and a method. Stated simply, the theory is that a memory or a behavioral response produced in one state will show a decrement in likelihood of occurrence in a different state. The most common procedure (method) used to test if state-dependent learning has occurred is to train some subjects in a drug state and others in a non-drug state (or in a state produced by a different drug), and then to test half the subjects in each group in the same state as was present during training and the other half in the opposite state. If state-dependent learning has occurred, the subjects trained and tested in different states will demonstrate a decrement in comparison to those trained and tested in the same state, with possible confounding factors such as direct drug-induced decrements due to sedation controlled for. This model does not preclude the possibility that other drug effects on learning might or might not also be present, and there are procedures (such as administering the drug immediately after, rather than before, training) to test or control for these other possible effects.

One hypothesis concerning how state-dependent learning might occur is that the drug-induced state (or change of state induced by other means) functions as a discriminative stimulus (sometimes referred to in the literature as an interoceptive stimulus). In other words, the organism learns that performing the response is rewarded in the same drug state in which training occurred, and therefore during testing that response is more likely to be emitted in the same drug state (including non-drug, if that was the training state). A number of sources for reviews of this material (Thompson et al., 1970; Thompson and Pickens, 1971; Cameron and Appel, 1973; Overton and Winter, 1974; Connelly et al., 1975; Lal, 1977; Seiden and Dykstra, 1977; Colpaert and Rosecrans, 1978; Ho et al., 1978; Reus et al., 1979; Colpaert and Slangen, 1982; Glennon et al., 1983; Jarbe, 1986; Colpaert and Balster, 1988; Stolerman et al., 1989) and more recent articles on these topics (Nakagawa et al., 1993; Schramke and Bauer, 1997; Li and McMillan, 1998; Johanson and Preston, 1998) involving both human and infrahuman subjects are available. Findings in humans are generally comparable to those in non-humans (Kamien et al., 1993).

The relevance of these paradigms to interoception is that the stimuli to which subjects respond could include drug-induced changes in the functions of visceral organs and production of visceral afferent sensations (see quote at start of this chapter). It has been argued (e.g., Rosecrans and Chance, 1977; Seiden and Dykstra, 1977, pp. 423–424) that organisms respond to drug effects only in the central nervous system. This is not completely correct. While it is almost certainly true that organisms respond to the central nervous system effects whenever they are present, it is also true that organisms can respond to drug-induced effects in the periphery as well. Peripheral effects are neither necessary nor primary when central effects are also present, but they are at least sometimes sufficient and may play some role in control of responding, even when central effects also are produced (see below).

Despite the importance of this issue to understanding the mechanism(s) of drug discrimination, few studies have been done that directly addressed the question of whether drug-induced changes limited to the periphery can function effectively as discriminative stimuli. Studies have indirectly addressed this issue. For example, after a discrimination has been trained with a drug that does enter the brain, testing with a drug from the same class that, presumptively, produces the same peripheral effects but no central effects because it does not cross the blood–brain barrier produces responses by the subject as if the non-drug stimulus had been given (Ando, 1975; see Seiden and Dykstra, 1977, pp. 423–424). And administration of a peripherally acting antagonist of a drug that does enter the brain—morphine as the drug to be discriminated and naltrexone methobromide as the peripherally acting antagonist—did not prevent the subjects from responding as if morphine had been given (Locke and Holtzman, 1985). These studies do not demonstrate that strictly peripherally acting drugs cannot function effectively as drug stimuli. What they do demonstrate is that in situations in which training has occurred with drugs that do enter the brain, peripheral effects are, at maximum, relatively weak.

A few studies have addressed the role of peripheral cues directly. Colpaert et al. (1975) determined that isopropamide, a peripherally acting anticholinergic drug, could function as a discriminative stimulus versus saline in rats trained to bar press for food. After 45 training sessions, a larger number than often is required, four of the rats showed statistically significant discrimination, and the other two showed trends in the same direction. The stimulus effect of isopropamide was dose dependent and generalized to other anticholinergic drugs.

Epinephrine (adrenaline) does not cross the blood–brain barrier. Studies in both Rhesus monkeys and humans have demonstrated that the peripheral effects of epinephrine can function as discriminative stimuli. Studies with related drugs such as isoproterenol are directly related to this, because isoproterenol is also an adrenergically stimulating agent that does not cross the blood-brain barrier.

Schuster and Brady (1964) investigated the efficacy of intravenous epinephrine versus saline as a discriminative stimulus for food reinforcement. A clear discrimination was demonstrated, but similar to the results with isopropamide, the rate of conditioning was slow relative to drug discrimination studies that used effective drugs that entered the brain. Evidence from this study also indicated that the monkeys were aware of the infusions of the solution, although they were directly into the superior vena cava (one of the veins that returns blood from the body into the heart), indicating interoceptive awareness of the infusion flow itself as well as the effects of the epinephrine. Cook et al. (1960) showed that both epinephrine and norepinephrine also could function as conditioned stimuli for leg flexion, and Weinberger et al. (1984) demonstrated that epinephrine could function as a CS even if the subject was under general anesthesia during conditioning. (GSR, conditioned sweating, can occur under anesthesia as well—Gruber et al., 1968.)

Frankenhaeuser and Jarpe (1963) studied the physiological and subjective effects of epinephrine infusions at different doses in humans. Heart rate, blood pressure, and 12 subjectively rated variables were studied, along with concentration and memory performance variables. The study did not literally involve a discrimination paradigm, but the subjective variables can be considered a variation of that methodology. An increase in subjective ratings occurred that was dose dependent, with substantial individual variation in threshold dose. The highest infusion rate given was 0.20 mcg/kg/min, and the highest mean subjective effect was approximately a 40% increase in average symptom ratings over placebo baseline.

Cameron et al. (1990a) also investigated the effects of intravenous epinephrine. They assessed choice (a forced choice for the subject to choose whether an active substance—epinephrine—or an inactive substance—saline—was given), and several subjective ratings as well as heart rate, blood pressure, and catecholamine blood levels. The correlational relationships among these variables also were assessed. Different epinephrine doses were associated with different choice responses, physiological variables, and subjective ratings. For the purpose of this discussion it was clear that the pharmacologic changes of epinephrine in the periphery produced discriminable effects. Likelihood of choice of the active substance (i.e., the subject perceived that the active drug, epinephrine, had been given rather than saline) reached 90% (see Fig. 2–2). The subjective effects of the infusion correlated more strongly with choice than did the physiological effects, indicating that the choice response was most likely being mediated by awareness of the interoceptive signals produced by the drug, primarily changes in the action of the heart (r= +0.82, accounting for two-thirds of the variance).

These data demonstrated that drug-induced stimuli in the periphery are capable of exerting control over behavior, although they appear less robust than effects produced by drugs that enter the central nervous system. Several other state-dependent or drug discrimination phenomena have been demonstrated that are of potential relevance to interoception as well. These include effects of endogenous substances such as thyroid, hypothalamic-pituitary-adrenocortical (HPA) and other hormonal substances, and substances that interact with adrenergic or cholinergic receptor systems. Furthermore, some state-dependent effects occur that involve emotional changes or stimuli and discriminative/state-dependent effects from other stimuli that directly affect the brain, such as electrical stimulation. These topics will be briefly reviewed here.

Jones et al. (1978) found that thyrotropin releasing hormone (TRH) could function as a discriminative stimulus in rats versus saline. Generalization to other doses of TRH was observed, as well as to TRH administered into the brain ventricle. They did not, however, observe generalization to drugs such as amphetamine that were thought likely to have a stimulant effect similar to TRH. Stewart et al. (1967) found that rats could learn to discriminate progesterone, and Earley and Leonard (1979) demonstrated a state-dependent–like effect of androgen and estrogen on a learned taste aversion. Bluthe et al. (1985) reported that taste aversion conditioning could be produced with vasopressin, apparently based on its hypertensive effects. State-dependent effects of drugs or endogenous substances such as ACTH that affect the HPA axis (Pappas and Gray, 1971; Klein, 1972; Gray, 1975b) have been demonstrated. Schechter et al. (1983) obtained equivocal results in a study attempting to demonstrate drug discrimination of ACTH 4–10, while a later study (Riddle and Hernandez, 1989) reported a positive result.

So far data have been cited that demonstrate that drugs that enter the central nervous system can produce state-dependent and discriminable states, drugs that do not enter the brain can have similar effects, and hormonal substances that occur endogenously in the body can do the same if administered exogenously. What about substances that occur within the body and fluctuate naturally? The hormones already mentioned do so, although perhaps not typically at such high levels as those used by the experiments cited, and so do some others. One system that has been analyzed in some detail is changes in blood glucose levels, including the counter-regulatory hormonal changes that occur during hypoglycemia. These counter-regulatory changes include the catecholamines epinephrine and norepinephrine and several other hormones. The primary function of these counter-regulatory changes is to increase the circulating glucose level back toward normal. Both experimental and natural changes in blood glucose levels, including the counter-regulatory responses to hypoglycemia, have been studied with the goal of determining if these fluctuations produce discriminable changes.

Cameron et al. (1988) studied the change in awareness (symptoms) that occurred in normal subjects when hypoglycemia was produced by intravenous insulin administration. In addition to determining several symptoms, various counter-regulatory hormones were measured at three different times after administration of several different insulin doses. Results indicated that *(1)* subjects could distinguish insulin in a dose-dependent manner, *(2)* awareness occurred when blood glucose levels fell to approximately 50 mg/dl (normal fasting is about 70) and epinephrine rose from about 50 pg/dl to about 100–200, and *(3)* adrenergically produced symptoms such as tremor, sweating, and fast heart beat were most highly correlated with subjects' choice that an active substance (i.e., insulin) be administered. Thus, insulin produced interoceptive awareness mediated mainly by adrenergic activation but also by effects produced by the blood glucose reduction. Another investigative group published similar results (Schwartz et al., 1987; Mitrakou et al., 1991), noting that adrenergic symptoms occurred with less hypoglycemia (i.e., at a higher blood glucose level) than did neuroglycopenic symptoms. Discrimination of hypoglycemia also was demonstrated in rats (Duncan and Lichty, 1993). Cameron (1994) reviewed hypoglycemia and other situations involving adrenergic changes associated with symptoms.

Hypoglycemia unawareness is a serious clinical problem in many individuals with diabetes mellitus who use insulin to control their diabetes. This is due to the fact that the long-term presence of diabetes, especially in people who have experienced repeated and/or persistent blood glucose elevations, is associated with damage to nervous tissue (neuropathy), including the epinephrine-releasing adrenal medullary tissue and the autonomic nerves. This autonomic neuropathy has been shown to be associated with lack of awareness of acute low blood glucose levels (see Cameron, 1994). Furthermore, some individuals with diabetes claim they can discriminate high blood glucose levels from normal, as well as many of those without neuropathy claiming awareness of low levels. For these reasons, several investigative groups have tested if people with diabetes can, indeed, perform these interoception-like discriminations accurately.

One investigative group published a number of studies related to this question in the 1980s (Pennebaker et al., 1981; Cox et al., 1983, 1985; Gonder-Frederick et al., 1986, 1989; see also Pennebaker, 1982). They reported a number of findings with reference to this question: *(1)* Most individuals with diabetes could recognize blood glucose levels at a greater-than-chance rate. *(2)* The correlations of specific symptoms with actual glucose levels were moderately consistent over time but varied considerably across individuals. *(3)* These discriminations could be performed in different settings (hospital and home), with good accuracy for at least half the subjects. *(4)* Some gender difference in performance was observed. *(5)* Glucose changes were at least to some extent associated with subjects' moods. Cameron (unpublished) collected data that also indicates that subjects' estimates of their own blood glucose levels are correlated with actual levels (Fig. 11–1). Other investigative groups (Moses and Bradley, 1985; Diamond et al., 1989) have reported inconsistent results.

Several studies attempted to address the mechanisms associated with hypoglycemia awareness and unawareness. A number of groups (Hoeldtke et al., 1982; Heller et al., 1987; Berlin et al., 1987, 1988; Grimaldi et al., 1989; Moken et al., 1994; Meyer et al., 1998) documented the involvement of autonomic neuropathy and adrenergically related symptoms and mechanisms, although other mechanisms are important as well, and adrenergic changes are not necessary in order to observe effects. For example, beta-adrenergic blockade does not necessarily block hypoglycemic symptoms (Deacon et al., 1977; Blohme et al., 1981), and Mokan et al. (1994) reported that beta-adrenergic sensitivity was not diminished in these individuals. Howorka et al. (1998) observed diminished adrenaline responses to hypoglycemia in people with hypoglycemia unawareness but claimed that the autonomic changes were not sufficient to explain the results and hypothesized that central nervous system changes must contribute as well. Consistent with this hypothesis, Howorka et al. (1996, 2000) found that those subjects with unawareness had EEG changes not seen in subjects who did not experience unawareness, thus demonstrating central nervous system changes in these subjects specifically.

Pauli et al. (1991) reported a study directly related to intero-

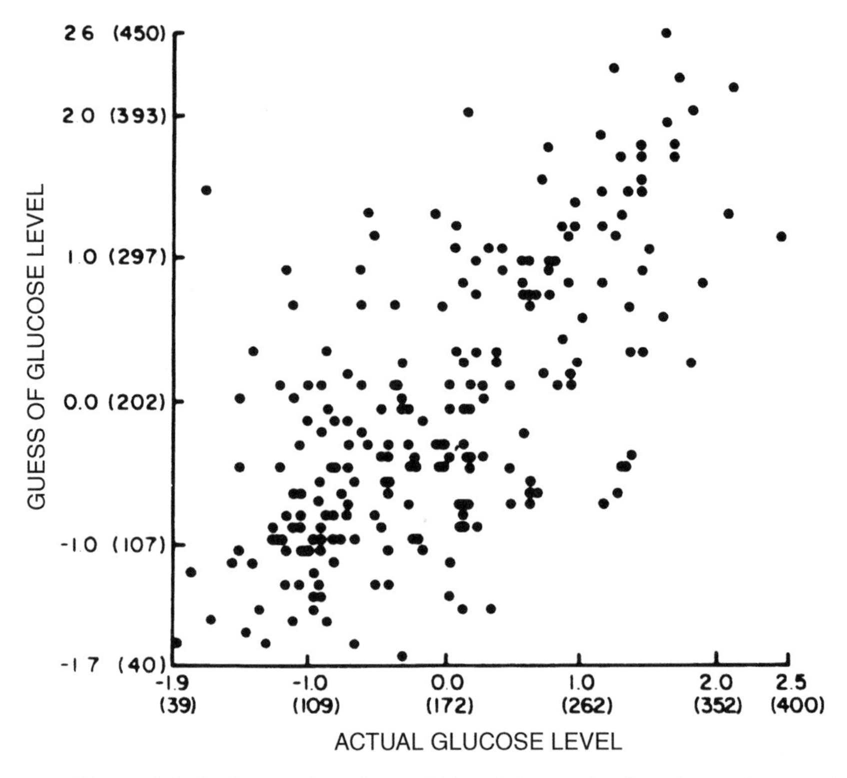

Figure 11–1. Scatterplot of actual blood glucose level against estimate of glucose level in hospitalized individuals with diabetes mellitus. Numbers in parentheses are "raw" data; outside parentheses are z-scores. A non-random tendency for guesses to approximate actual levels is apparent, which was also true when subjects were assessed individually (not shown). [Previously unpublished.]

ceptive processes in people with diabetic autonomic neuropathy. They assessed heart beat and dysrhythmia perception over 24 hours in people with diabetes with and without neuropathy versus normal subjects. Their results indicated that neuropathic nerve damage produced impaired heart beat perception and complete unawareness of dysrhythmias.

Cameron (1989) performed an experiment with normal subjects to assess the role of hypoglycemia-induced adrenergic counterregulatory changes in hypoglycemia awareness. This was done by combining insulin administration with administration of the beta-adrenergic blocking agent propranolol. On trials in which propranolol was not administered, adrenergic symptoms contributed to awareness of hypoglycemia. In contrast, on trials in which the high propranolol dose was given, other symptoms determined choice, but no decrement in awareness or accuracy was observed. Propranolol itself had no significant effect on choice (Fig. 11–2). These results suggest that hypoglycemia unawareness associated with diabetic autonomic neuropathy is due to more than just lack of adrenergic activation, since blocking adrenergic effects pharmacologically with a beta-adrenergic receptor antagonist did not block awareness in otherwise-normal subjects. Perhaps damage to afferent nerves is the cause. Notably, a very similar experiment in rats (Duncan and Hooker, 1997) also demonstrated with a drug-discrimination paradigm that propranolol did not block the hypoglycemia cue produced with insulin, as did a subsequent study in humans (Towler et al., 1993), an observation that dates back to the mid-1950s (French and Kilpatrick, 1955). In contrast to the lack of effect with beta-adrenergic blockade only, Towler et al. (1993) reported that pan-autonomic blockade (alpha-adrenergic, beta-adrenergic, and muscarinic cholinergic) did produce a decrement in awareness and concluded that cholinergic mechanisms also played a role in hypoglycemia awareness.

Reactive hypoglycemia is an excessive drop in blood glucose levels a few hours after food intake, presumably due to excessive insulin release in response to the ingestion of the meal. The status and validity of this syndrome is controversial. It has been attributed by some to psychological symptoms such as adrenergic-like symptoms incorrectly interpreted as low blood glucose levels (Permutt, 1976; Johnson et al., 1980; Hale et al., 1981; Taylor and Rachman, 1988; Palardy et al, 1989). Whatever the etiology, the individuals are perceiving symptoms relevant to interoceptive processes.

In summary, awareness of changes in concentrations of substances in the blood does occur, even without entry into the brain, including adrenergic and cholinergic drugs, various hormones, and glucose level. Damage to afferent autonomic nerves diminishes that awareness, but not completely. Both neuroglucopenic (due to low blood glucose) and adrenergic hyporesponsive mechanisms do sometimes play roles in awareness but do not appear to be the whole story.

As has been extensively demonstrated and emphasized throughout this volume, the two major divisions of the autonomic nervous system, the sympathetic, represented by adrenergic nerves, and the

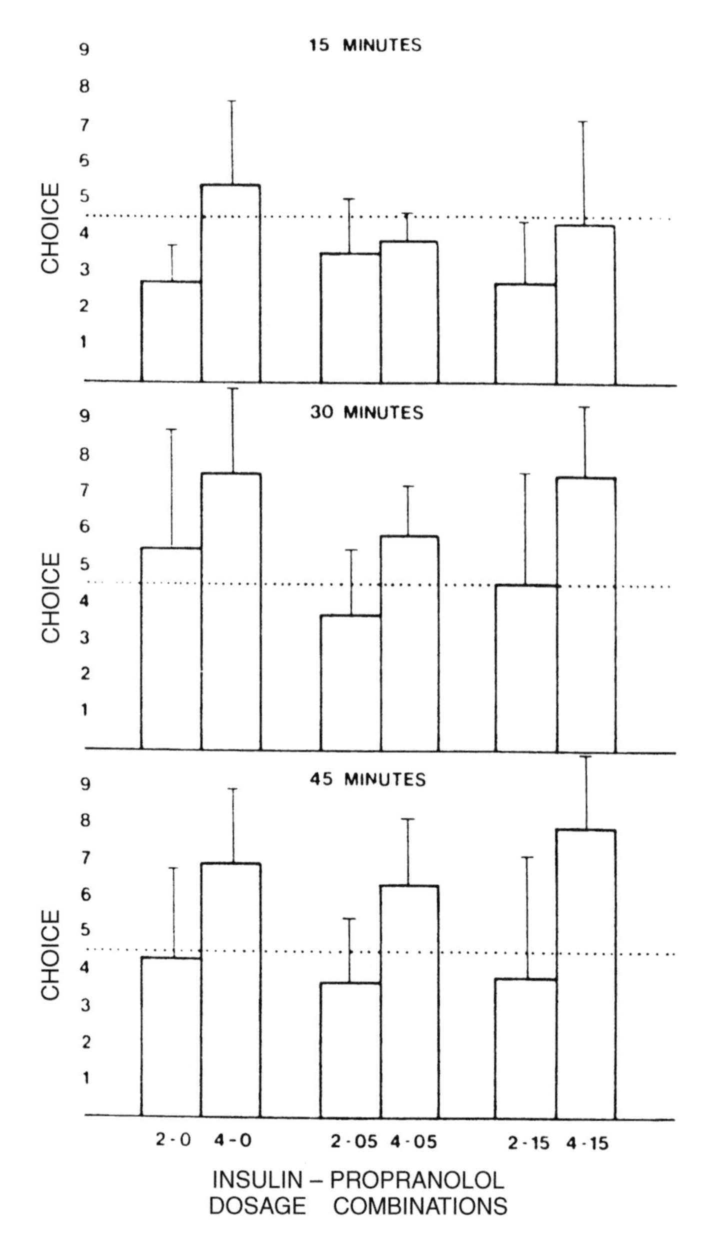

Figure 11–2. Choice score (1 = inactive substance, 9 = active substance, 4.5 = guess) at 15, 30, and 45 minutes after normal subjects were administered one of three intravenous doses of the beta-adrenergic blocking drug propranolol (0 = placebo, 05 = 0.05 mg/kg, 15 = 0.15 mg/kg) and one of two intravenous doses of regular insulin (2-2 units; 4-4 units). Results indicated that choice (subjective awareness) was affected by insulin dose and time after administration, but beta-adrenergic blockade had no effect, and there was no interaction. [From Cameron, 1989, reprinted with permission of the holder of the copyright.]

parasympathetic, mainly represented in the viscera by the vagus nerve, are highly involved in visceral sensory afferent impulse transduction. Other pathways, mainly chemical, also convey important afferent information. In the context of this chapter, so far the role of chemical information in controlling behavior has been emphasized, although some of the afferent signals generated by those substances that do not pass the blood–brain barrier are, in fact, undoubtedly being conveyed to the brain by nervous pathways originating in the periphery including the viscera. Even those drugs that do freely enter the brain may also produce peripheral impulses that reach the brain from the viscera and contribute to overall behavioral control. With that logic in mind, data will now be briefly reviewed involving drugs that have been shown to function effectively as discriminative stimuli that affect adrenergic and/or cholinergic receptors.

Clonidine is an agonist of the alpha2-adrenergic receptor. Bennett and Lal (1982) showed that clonidine functioned robustly as a discriminative stimulus in rats in a food reinforcement paradigm. This effect was antagonized only by yohimbine, an antagonist drug specifically for the alpha-adrenergic receptor, and other drugs that also are agonists of this receptor showed generalization of the discriminative effects. In a later study Jordan et al. (1993) replicated the stimulus effects of clonidine and showed that the cue effect was mediated through the alpha2-adrenergic receptor effects rather than through the imidazoline binding effect of clonidine. Lal and Yaden (1985) demonstrated an interaction between systemic autonomic physiological functioning and clonidine drug discrimination by showing that in spontaneously hypertensive rats a robust correlation existed between the stimulus and the anti-hypertensive effects. The same investigative groups (Spencer et al., 1988) also demonstrated that this discriminative blood pressure–reducing effect of clonidine could be classically conditioned to an odor CS, thus demonstrating combined operant and classical interoceptive effects. Browne (1981) showed that yohimbine, an alpha2-adrenergic receptor antagonist, also could function as a discriminative stimulus.

Stimulant drugs generally produce subjective effects similar to adrenergic activation. Thus, it could be expected that, similar to adrenergic agents, stimulants would function effectively as discriminative stimuli. Amphetamine is a stimulant that can function effectively either in a state-dependent learning paradigm (Roffman and

Lal, 1972; Schechter and Cook, 1975) or as a discriminative stimulus (Jones et al., 1974; Kilbey and Ellinwood, 1979; Porsolt et al., 1984). Caffeine is another effective stimulant drug (e.g., Holtzman, 1986), as are cocaine, (Jarbe, 1978; de al Garza and Johanson, 1985) ephedrine, and pseudoephedrine (Chait, 1994; Young and Glennon, 1998; Tongjaroenbuangam et al., 1998).

Cholinergic pharmacologic systems are mainly divided into nicotinic and muscarinic receptors. Evidence exists for efficacy of drugs affecting both types of receptors functioning as discriminative stimuli or producing state-dependent effects (Evans and Patton, 1970; Schechter and Rosecrans, 1972; Johansson and Jarbe, 1976; Pratt et al., 1983).

Epinephrine directly affects beta-adrenergic receptors, and clonidine and yohimbine affect alpha2-adrenergic receptors. Stimulant drugs function as either indirect effectors of adrenergic receptors or by producing activating effects that are subjectively similar to the effects of adrenergic drugs. Cholinergic drugs have their effects at either muscarinic or nicotinic receptors. In the studies cited involving drugs that enter the brain, it typically is not clear what role, if any, peripheral effects have, although studies designed to assess this have been reported. For example, Wagman and Maxey (1969) reported that a peripherally acting anticholinergic agent did not disrupt a somatic response-produced cue, while a centrally acting anticholinergic drug did. Bachman et al. (1979) used muscarinic and beta-adrenergic blocking drugs to demonstrate that the subjective effects of THC (an active ingredient of marijuana) did not depend on these peripheral autonomic effects, and Perkins et al. (1999) observed that blockade of the peripheral cues produced by nicotine did not attenuate discrimination of nicotine's effect. However, successful studies with substances that do not cross the blood–brain barrier make it likely that the peripheral (i.e., interoceptive) effects make at least some contribution to the effectiveness of these drugs in these paradigms.

Involvement of other receptors and other parts of the nervous system relevant to interoceptive processes is important as well. For example, Katner et al. (1999) found that a discriminative olfactory cue could reinstate alcohol-seeking behavior in rats, while an auditory cue did not, and that this discriminative effect could be antagonized by blockade of opiate receptors with naltrexone.

Interoceptive processes, of course, involve not just the organs and nerves of the viscera, but the central nervous system from the spinal cord to the cerebral cortex. For that reason, drugs or other stimuli that affect the central nervous system directly may also be involved in processes important to interoception. Some classes of drugs likely to have such potentially relevant effects—those affecting the adrenergic and cholinergic systems—have already been discussed. Two other areas of research in drug discrimination and state-dependent phenomena will be described because they also are likely to be at least indirectly relevant: state-dependent effects related to emotional stimuli and state-dependent or discriminative effects produced by electrical brain stimulation.

One group (Connelly et al., 1973, 1975, 1977) investigated the effects of an aversive conditioning procedure, tone with shock, on maintenance of a state-dependent aversive escape task involving chlordiazepoxide (a benzodiazepine drug) in rats. They found that presentation of the tone CS could disrupt the dissociation. They interpreted their results as indicating that the emotional–motivational characteristics of the tone prompted a memory retrieval process that mediated memory across otherwise dissociated states.

The question of the effects of emotional states on state-dependent learning also has been studied in humans. Macht et al. (1977) reported that the presence or absence of threat of electric shock could produce a state-dependent effect on memory for verbal material. Weingartner et al. (1977) described decrements in recall of verbal material by individuals with bipolar affective disorder (manic–depressive disease) based on switches between manic and non-manic normal mood states. Bower et al. (1978) also reported that mood states could affect word list recall but that the state-dependent effect was also highly dependent on the type of recall task used. Pearce et al. (1990) showed that pain might have state-dependent effects, and Jensen et al. (1989) demonstrated that effects of diazepam (an anxiolytic benzodiazepine drug) could produce state-dependent effects on the occurrence of conditioned emotion as measured by the GSR (skin conductance measure of sweating). Finally, Kunst-Wilson and Zajonc (1980) reported that the affective content of a stimulus could influence responses to stimuli even when the stimuli were so degraded that subjects could not overtly recognize the stimuli.

Like emotional states, other events within the brain may produce state-dependent or discriminable effects. A number of studies have shown that state-dependent effects can be produced by electrical brain stimulation. Electroconvulsive shock, which has effects across broad areas of the brain, is effective (DeVietti and Larson, 1971). Metrazol, a drug that produces seizures, also produces state-dependent effects (Kurtz and Palfai, 1973). Spreading cortical depression can produce similar effects (Pianka, 1976). Mayse and DeVietti (1971) compared the state-dependent effects of electroconvulsive shock to drug-induced state-dependent effects with pentobarbital.

Specific regions in the brain can be stimulated, or anesthetized, by direct application of electrical stimulation or local drug administration through an implanted electrode or cannula. Several regions have been assessed for the ability to produce state-dependent effects. McIntyre and Gunter (1979) reported that state-dependent effects could be obtained by electrical stimulation of the caudate nucleus, while effects from stimulation of the amygdala were less robust and might vary between the lateral and medial amygdaloid areas. Phillips and LePiane (1981) reported similar positive results with the caudate and several negative results with the amygdala.

The same research group (McIntyre et al., 1985; Stokes and McIntyre, 1981, 1985) found that unilateral kindling (repeated stimulation until a seizure occurs) produced evidence for state-dependent learning in the right hippocampus but less clearly in the left, which was true with or without separation of the two hemispheres from each other (split brain). Using a stimulation paradigm without kindling produced effects in both the right and left hippocampus. Using anesthetic injection into the septal region, Duncan and Copeland (1975) demonstrated a partial, or asymmetrical, state-dependent effect. The striatum has been implicated as well (Reus, 1986). Addressing the mechanism question, Modrow and Bliss (1979), using recording (not stimulating) electrodes, found that state-dependent learning with a barbiturate sedative drug produced changes in the activity of the medial forebrain bundle.

Drug discrimination tests of the efficacy of electrical brain stimulation and direct injection of drugs into the brain also have been done. For example, Hirschhorn et al. (1975) and Colpaert et al. (1977) demonstrated this effect with stimulation in the dorsal raphe nucleus and the lateral hypothalamus. Using infusions of morphine into parabrachial and ventral tegmental areas, Jaeger and van der Kooy (1996) were able to demonstrate that discrimination could be produced by direct injection into the parabrachial (but not ventral tegmental) area and that the motivational and discriminative effects of morphine could be differentiated.

In conclusion, only a few studies have directly tested the question of whether drugs that have pharmacologic effects strictly in the periphery can produce state-dependent learning or drug-state discrimination. The results, especially with epinephrine (adrenaline), do clearly demonstrate that such effects occur. They are less robust than effects of substances that enter the brain, which is perhaps consistent with the subtle sensory effects often seen in interoception generally. Sensory effects from a combination of central and peripheral events are likely to occur. Additional studies have documented the relevance of emotional factors and specific brain regions in these effects.

twelve

PSYCHIATRIC DISORDERS: ANXIETY, PANIC, DEPRESSION, AND SOMATIZATION

. . . the conscious ego: that it is first and foremost a body-ego.
Sigmund Freud, *The Ego and the Id*

A major question in the area of interoceptive processes is, Under what conditions do they appear to function improperly? Changes associated with medical or psychosomatic disorders—cardiac and gastrointestinal disorders—have already been discussed in Chapters 8 and 9. In this chapter, changes associated with psychiatric disorders will be addressed.

Psychiatric disorders are defined primarily by the presence of specific signs and symptoms. In several of these disorders, these symptoms include some that are referable to the body, and specifically to the viscera. Freud, writing early in the twentieth century, recognized the fundamental link between psychopathological processes and the body (see epigraph). Among the disorders in question, those that seem most relevant to interoception are the anxiety disorders and the somatoform disorders. Depressive disorders also are implicated through their commonly observed co-morbidity with panic attacks and other anxiety disorders. Among the anxiety disorders, those that most commonly involve somatic symptoms are panic disorder, with or without agoraphobia, and generalized anxiety disorder, with somatic symptoms sometimes seen in the other anxiety disorders as well, including those due to other conditions (including substance use). Among the somatoform disorders, somatization and somatoform pain disorder are most closely related to visceral sensation. Anxiety disorders will be discussed first.

ANXIETY AND ANXIETY DISORDERS

Anxiety Symptoms

In addressing the relationship between interoception and the various psychiatric disorders, what is generally considered are symptoms (i.e., subjective reports). These reports carry all the caveats discussed above concerning the relationship between subjective reports and awareness or detection of sensations (see Chapter 8). A few studies have focused more directly on detection issues, mainly with panic disorder and cardiac awareness (see below). Furthermore, the somatic symptoms associated with anxiety are not only visceral. For example, striate (so-called voluntary) muscle sensations and pains also are included, as are symptoms less specifically focused on one particular region or system in the body (e.g., dizziness, fatigue, agitation).

Somatic complaints are very common in anxious individuals (e.g., Kellner, 1988). Many such symptoms, for example, palpitations, breathlessness, trembling, and sweating, appear in two-thirds or more of individuals with anxiety neurosis. Kellner (1988) examined the interrelationships among anxiety, depression, and somatic symptoms. He reported that depression and anxiety were highly correlated, and that somatic symptoms were more strongly correlated with anxiety than with depression, findings consistent with other studies. He also noted that different symptoms are characteristic of anxiety versus depression.

Cameron et al. (1986) reported that factor analysis of anxiety symptoms in a sample of more than 300 anxious individuals with various anxiety disorders yielded five factors, including one that represented sympathetic nervous system changes and another that represented parasympathetic changes. More of the variance was accounted for when only individuals who had panic attacks were included. The association of anxiety (and stress) with symptoms referable to the different branches of the autonomic nervous system has been reported in many studies. Bandelow et al. (1996) also found evidence for a cardiorespiratory (sympathetic) factor.

Despite the close association between anxiety and somatic symptoms, this association is not strong in all anxious people. Some anxious individuals complain mainly of non-somatic psychological symptoms, while others describe many somatic sensations. Tyrer and Lader (1976) took measurements of central (e.g., auditory evoked potential) and peripheral (e.g., heart rate, skin conductance changes, finger tremor) physiological variables in normal subjects who were stressed in order to produce anxiety. They found significant but relatively weak correlations with the peripheral markers and generally stronger correlations with the central markers. Morrow and Labrum (1978) found that different measures of psychological anxiety showed stronger correlations than did physiological anxiety indexes in people hospitalized for myocardial infarction. Nonetheless, for both people with primary anxiety disorders and for anxious individuals in general (Bridges et al., 1968), those who complain of more somatic symptoms (such as cardiac—Hoehn-Saric et al., 1989a) have more physiological changes, indicating that they are aware of changes that can be demonstrated in the periphery.

The peripheral measures that traditionally have been used to demonstrate physiological changes in anxiety have included *(1)* changes directly connected to the autonomic nervous system—that is, sweat gland activity, heart rate, blood pressure, pupil size, salivary secretion; *(2)* other physiological measures, such as respiration, electromyogram, finger tremor; and *(3)* hormonal measures, such as, adrenal cortical and adrenal medullary functions, thyroid function, and changes in other hormones such as growth hormone (see Cameron and Nesse, 1988). Central nervous system measures have traditionally included EEG measures such as evoked potentials and contingent negative variation (see Lader and Marks, 1971; Lader, 1975). These physiological markers not only change in some but not all anxious people, but different measures change to a greater or lesser extent across individuals (i.e., substantial individual variation) and are associated with deterioration of task performance (Bond et al., 1974).

In addition to the markers listed above, more sophisticated peripheral and central markers associated with anxiety have been studied. Peripheral measures include measurement of receptors on peripheral tissues (e.g., Cameron et al., 1990c, 1996), assessment of receptor status by pharmacologic challenge studies (e.g., Nesse et al., 1984), comparison of hormone profiles and relation to symptoms across disorders (e.g., Starkman et al., 1990), and assessment of patterns of hormonal and physiological changes during symptom occurrence (Cameron et al., 1987). Central nervous system markers include use of pharmacologic challenge tests with substances that affect physiological function in the central nervous system (e.g., Abelson et al., 1991, 1992) and functional imaging techniques that will be reviewed below. These physiological markers may or may not correlate with interoceptive processes (that is not yet known), but they do relate to systems involved in visceral sensation.

More than 80 years ago Fraser and Wilson (1918) observed in people with irritable heart that "... minute doses [of adrenaline] produced a greater action in the patients than in the controls" (p. 29). This included objective discomfort and subjective throbbing in the head and chest in 90% of the patients but none of the controls. Thirty years later Wheeler et al. (1950) reported a high rate of symptoms referable to the chest and heart in people with neurocirculatory asthenia (anxiety neurosis), a finding supported by Levander-Lindgren and Ek (1962), while Kannel et al. (1958) reported that the ECG in these patients was normal. The evidence is now strong that a close relationship exists among changes in heart function, anxiety, and symptoms referable to the heart, including pain (Skerritt, 1983; Kawachi et al., 1994; Cannon, 1995). (Indeed, there is some evidence that heart function in young adults—at least in men—can predict aspects of mental health, possibly related to social anxiety, for decades—Phillips et al., 1987.) This link is supported by the efficacy of beta-adrenergic blocking drugs in reducing mild to moderate anxiety, primarily when peripheral cardiac symptoms are involved (Fonte and Stevenson, 1985; Tyrer, 1988).

Control of cardiac function involves both sympathetic and parasympathetic input. Another peripheral marker of anxiety, skin conductance (sweating), involves sympathetic control but with acetylcholine as the neurotransmitter. Anxious individuals show more conductance and habituate more slowly than do non-anxious people (Lader, 1967; Raskin, 1975; Watts, 1975; Knight and Borden, 1979; Chattopadhyay et al., 1980). This appears to be due in large part to those people who experience panic attacks (Birket-Smith et al., 1993). Thus, both cardiac function and skin conductance should provide peripheral physiological correlates of anxiety that may provide ways to relate interoceptive processes in anxiety to aspects of autonomic nervous system activity. Other markers of anxiety, not just related to autonomic function, include startle (Davis, 1983), the orienting reflex (Tan, 1964; Fredrickson, 1981), muscle tension (Sainsbury and Gibson, 1954), and skin temperature (Butscher and Miller, 1980).

Panic Disorder

Panic attacks can occur in a number of disorders, and their occurrence is the major defining characteristic of panic disorder. Four or more from a list of 13 symptoms are required to meet the criteria for the attacks. These 13 symptoms include several that are referable to visceral function, including "palpitations, pounding heart, or accelerated heart rate," "sensations of shortness of breath or smothering," "feeling of choking," "chest pain or discomfort," "nausea or abdominal distress," and "chills or hot flushes." Other symptoms are potentially at least indirectly related as well due to possible relations to autonomic nervous system functioning, including "sweating," "trembling or shaking," "feeling dizzy, unsteady, lightheaded, or faint," and "paresthesias (numbness or tingling sensation)." Thus, 10 of 13 symptoms are potentially at least indirectly related to visceral–autonomic sensory functioning.

Several studies have been published that have specifically addressed the issue of visceral sensory processes in people with panic disorder. They will be discussed later in this chapter. Before that, broader issues of symptoms and related physiological changes in panic disorder and generalized anxiety disorder, along with a discussion of respiratory changes and anxiety, will be presented. One theory of the pathophysiology of panic disorder (Gorman et al., 1989, 2000) implicated many of the brain regions relevant to visceral sensory processes—brainstem (nucleus of the solitary tract, parabrachial nucleus, periaqueductal gray, locus coeruleus), thalamus, amygdala, hypothalamus, limbic lobe, cingulate gyrus, and insular and medial prefrontal cortices.

Somatic complaints are an essential, defining part of panic. It does not appear to be the case, however, that panic patients claim to experience physical symptoms in a non-specific way (Goetz et al., 1989). Panic symptoms occurring in a given attack are specific and appear to occur in a specific order (Katerndahl, 1988). The earliest symptoms are those most directly related to the viscera—shortness of breath, palpitations, chest discomfort, and hot flashes, with choking, sweats, and dizziness coming soon after. Among patients presenting to primary care physicians, cardiac and gastrointestinal symptom clusters are very common (Katon, 1984). Studies that have monitored physiological changes during panic attacks that were not intentionally provoked experimentally demonstrated that *(1)* physiological changes do occur during panic attacks, *(2)* these changes vary in pattern and intensity and often are small in magnitude, and *(3)* they often are associated with a normal baseline (i.e., panic patients often do not have abnormalities of these physiological markers between attacks) (Lader and Mathews, 1970; Taylor et al., 1983; Freedman et al, 1985; Margraf et al., 1987; Cameron et al., 1987; Shear et al., 1992; Bystritsky et al., 1995; Leyton et al., 1996; Wilkinson et al., 1998). Some studies, nonetheless, have found evidence for heightened autonomic activation in these patients (Roth et al., 1986), which might be predicted in part by personality variables (King et al, 1988) and also in part by responses to stressful or nonstressful circumstance (Roth et al., 1998a; Wilkinson et al., 1998).

As already discussed in Chapter 9, non-cardiac chest pain is often caused by esophageal problems. Cardiac symptoms, as already noted, also are very common during panic attacks. The relationship among cardiac symptoms, cardiac disease, and panic disorder has been studied. One investigative group reported that 40% of patients with atypical angina have panic disorder (Basha et al., 1989), and one-third of angiography patients who had normal angiograms had panic disorder (Beitman et al., 1989). The presence of frightening cognitions in association with chest pain was highly associated with panic disorder and not primary heart disease (Fraenkel et al., 1996). Conversely, in patients with panic disorder, the presence of cardiac symptoms was not associated with a higher risk of underlying heart disease (Jolley et al., 1992).

Administration of pharmacologic substances known to be anxiogenic has been used as laboratory models with which to study the pathophysiology of panic. Prior to infusion, subjects sometimes report anxiety elevations in anticipation of administration. Anxiety before infusion predicts likelihood of experiencing an attack in response to substance administration (Yeragani et al., 1987b), while pre-infusion heart rate appears not to do so (Yeragani et al., 1987a). In contrast, during placebo infusions, individuals who experienced panic attacks did demonstrate higher heart rates and blood pressures (Balon et al., 1988).

Substances that produce adrenergic activation, both peripherally and centrally (e.g., yohimbine) or only in the periphery because they do not cross the blood–brain barrier (e.g., isoproterenol or epinephrine), including adrenergic symptoms, have been used to study panic. With isoproterenol, it has been reported that shortness of breath and trembling, along with cognitive symptoms, distinguished those who panicked from those who did not (Balon et al., 1990; Pohl et al., 1990). With epinephrine, both heart rate and breathing changes distinguished groups (van Zijderveld et al., 1997; Veltman et al., 1998). With yohimbine, those who panicked had higher heart rate, blood pressure, and MHPG—a catecholamine metabolite marker of adrenergic activation (Charney et al., 1987). With lactate, a highly panicogenic substance (which also produces less intense symptoms in healthy individuals) that lacks a well-specified pharmacologic mechanism of action, blood pressure changes (Ehlers et al., 1986) and breathing changes (Goetz et al., 1996) are the physiological and non-cognitive symptom markers most closely associated with the occurrence of panic.

It is important to note that elevation of peripherally acting adrenergically activating substances per se does not produce anxiety severe enough to qualify for a diagnosis of any anxiety disorder. It is not clear if peripherally acting substances directly produce anxiety at all or if they just produce anxiety-like somatic symptoms. They may produce anxiety indirectly through learned or unlearned cognitive reactions to the somatic symptoms in a kind of vicious cycle positive feedback system. This seems to be the most likely mechanism by which anxiety, including panic reactions, occurs when adrenergically activating substances are given to people with preexisting anxiety disorders. (It is not clear to what extent centrally acting adrenergic substances produce anxiety in the same manner.) Evidence for this scenario seems most clearly provided by studies of people with pheochromocytomas, tumors that produce very high levels of circulating norepinephrine and also very high levels of circulating epinephrine. It has been demonstrated that these individuals have somatic symptoms similar to panic, mainly those who have high epinephrine, but they do not reach the criteria for diagnosis of anxiety disorder because they do not have the intense psychic component (Starkman et al., 1985, 1990).

In summary, it appears that natural panic attacks are highly associated with subjective reports of symptoms referable to the viscera, especially cardiac and respiratory, but also gastrointestinal. Panic attacks induced experimentally are associated at least to some extent with cardiovascular sympathetic activation and respiratory changes. Those attacks induced by peripherally active adrenergic substances such as epinephrine and isoproterenol appear to do so through cognitive reactions to the awareness of the somatic symptoms produced by the substance administration. The importance of cognition in anxiety neurosis (moderate to severe anxiety of various kinds) is well known and described (e.g., Beck et al., 1974; Chambless et al., 2000).

Generalized Anxiety Disorder

The most recent version of the diagnostic criteria for psychiatric disorders (American Psychiatric Association, 1994) does not include somatic symptoms referable to the viscera or the autonomic nervous system more generally as part of the defining criteria of generalized anxiety disorder. Instead, somatic symptoms are included as associated features, including "cold, clammy hands" (due to localized sweating); "dry mouth; sweating; nausea or diarrhea; urinary frequency; trouble swallowing or a 'lump in the throat' " (p. 433). Earlier versions of this nomenclature system (DSM-III and DSM-III-R) did include such symptoms, called "autonomic hyperactivity," and included additional symptoms in the criteria such as "shortness of breath or smothering sensation," "palpitations or accelerated heart rate," "paresthesias," ". . . abdominal distress," and "flushes (hot flashes) or chills," along with the physical signs such as "high resting pulse" and "high respiration rate" (American Psychiatric Association, 1980, p. 233; American Psychiatric Association, 1987, p. 253). It would appear that the criteria were changed in order to focus on the cognitive component of anxiety as the defining feature of generalized anxiety disorder and more clearly to differentiate generalized anxiety disorder from panic disorder, but it is clear that generalized anxiety is believed often to involve somatic visceral–autonomic symptoms.

Very many studies of anxious individuals who experienced what now would be called generalized anxiety have been reported over the years, including many that studied somatic symptoms and visceral–autonomic physiological changes. Books published in the last 30 years are probably the most convenient sources of material on this subject (Lader and Marks, 1971; Lader, 1975; Tuma and Maser, 1985; Byrne and Rosenman, 1990), keeping in mind that many of them do not specifically discriminate among generalized anxiety disorder, panic disorder, and other anxiety disorders but often describe the individuals they studied in a way that implies that people with generalized anxiety constituted a considerable portion of those studied. A few results will be reviewed here as examples of the types of findings that have been reported.

Do people with generalized anxiety disorder actually have autonomic dysfunction? Results from one study suggest that generalized anxiety disorder patients may fall into one of two groups, those with and those without autonomic lability (with somatic symptoms), but the differentiation was not strong (Koehler et al., 1988). Lader (1967) and Raskin (1975) reported abnormal skin conductance, Sainsbury and Gibson (1954) described abnormal muscle tension, and Bond et al. (1974) indicated increased arousal occurred in patients who most likely would now be diagnosed with generalized anxiety. Abelson et al. (1991) described abnormal growth hormone responses to clonidine (an alpha2-adrenergically active drug), but normal heart rate, blood pressure, MHPG, and symptom reactions. Cameron and Nesse (1988) reviewed data indicating that catecholamine levels are elevated in generalized anxiety. Garvey et al. (1995) reported significant relationships between generalized anxiety symptoms and metabolites of serotonin and norepinephrine.

Not all investigators found differences, however. Hoehn-Saric et al. (1989b) measured autonomic and electromyographic (muscle) parameters in women with generalized anxiety versus normal subjects, both at baseline and during psychological stress tasks. They reported no increased activity in autonomic measures in patients under any condition. Birket-Smith et al. (1993) did not find evidence for abnormal skin conductance in generalized anxiety. And Davis et al. (1985) found evidence for a serotonin, but not a norepinephrine, abnormality in generalized anxiety. In summary, there does appear to be some evidence for abnormalities referable to visceral–autonomic function in generalized anxiety, but it is quite inconsistent, probably owing in large part to the heterogeneity of individuals studied.

Assuming that a visceral–autonomic abnormality does exist, are there any clues as to its nature? Thayer et al. (1996) found that people with generalized anxiety disorder showed evidence of reduced vagal control of the heart. Watkins et al. (1998) replicated that finding in healthy subjects not diagnosed with any anxiety disorder but who rated high on trait anxiety. Thus, vagal dysfunction is associated with generalized anxiety symptoms.

Respiratory Symptoms

Respiration provides an atypical situation with respect to interoception because respiratory function, although usually outside of acute consciousness and usually involuntary, within a certain range can be fully under voluntary control. Thus, respiration is functionally different from other visceral functions involving sensory awareness, and it is therefore difficult to be certain that respiratory sensations are fully analogous to interoceptive processes involving other visceral organs. Despite this, and to some extent because of it, sensory processes associated with respiratory function are of interest.

Interest in visceral physiological and psychological changes in response to respiration—for example, electrocardiographic changes in response to hyperventilation—dates back more than 50 years (Christensen, 1946; Coppen and Mezey, 1960). For at least 25 years it has been speculated in the literature that chronic hyperventilation is a cause for various psychological symptoms, including anxiety symptoms (Lum, 1975; Dalessio, 1978; Bass et al., 1983; Grossman and DeSwart, 1984; Hornsveld et al., 1990). Physiological abnormalities have been associated with hyperventilation (Missri and Alexander, 1978; Bechbache et al., 1979; Magarian, 1982; Grossman, 1983; Thyer et al., 1984; Freeman et al., 1986), and it has been observed that subjective breathlessness symptoms overlap with panic and agoraphobia (Garssen et al., 1983; Bass and Gardner, 1985; Bass et al., 1989).

Much of the interest in the pathophysiology of anxiety disorders, including somatic symptomatology, in the past 20 years and especially in the 1980s focused on panic disorder and included research involving respiratory changes. Several research reports concluded that people with panic disorder have a respiratory abnormality (e.g., Gorman et al., 1984; Hibbert, 1984; Cowley and Roy-Byrne, 1987) and that breathing increased concentrations of CO_2 can produce panic attacks very similar to natural panic attacks, including somatic symptoms (van der Hout and Griez, 1985; Fyer et al., 1987; Griez et al., 1987; Perna et al., 1994; Battaglia and Perna, 1995; Beck et al., 1997). Patients with chest pain but normal cardiac stress scintigrams were likely to have panic disorder and to have lower CO_2 levels (Maddock et al., 1998). Breath holding, which raises CO_2 levels, can also produce panic (Zandbergen et al., 1992), and evidence exists from ambulatory monitoring that panic patients have abnormal breathing patterns (Martinez et al., 1996). Significant co-morbidity has been reported between panic and a primary respiratory disorder, asthma (Perna et al., 1997). A drug that functions as a primary respiratory stimulant (doxapram) can induce panic attacks, primarily in people with preexisting panic disorder (Lee et al., 1993; Abelson et al., 1996). Nonetheless, while most investigative groups have identified abnormalities, some have reported less impressive results (van der Hout et al., 1987; Zandbergen et al., 1993).

Some research has addressed possible mechanisms by which respiratory changes might be related to panic. Family studies support a familial, and therefore possibly genetic, aspect not only to panic more generally but specifically to respiratory abnormalities. Horwath et al. (1997) found that smothering symptoms are a highly familial aspect of panic. Perna et al. (1995), Coryell (1997), and van Beek and Griez (2000) reported that individuals at high risk for panic (i.e., panic–free but with an affected first-degree relative) showed a high rate of panic induction by CO_2 inhalation. And Bellodi et al. (1998) reported that monozygotic twins of probands with panic were much more susceptible to CO_2-induced panic attacks than were dizygotic twins.

Studies have investigated changes in other physiological variables in association with respiratory changes in panic. One study (Asmundson and Stein, 1994) found no clear abnormalities. On the other hand, Bystritsky et al. (2000) observed autonomic abnormalities in some, but not all, panic patients during CO_2 inhalation. In subjects who were not diagnosed with panic, Wientjes and Grossman (1994) reported that both trait anxiety and hyperventilation effects contributed to psychosomatic symptoms. Klein (1993, 1994) offered a “suffocation false alarm” theory of panic attacks, which would implicate chemoreceptors, presumably in the brain. Not all studies, however, have provided clear support for respiratory abnormalities in people with panic disorder (e.g., Roth et al., 1998b).

Apparently, no studies have directly focused on the relationships among respiratory physiology, anxiety and panic, and interoception. Studies of interoception in panic have focused on the heart (see below). Studies of respiration and interoception in anxious individuals and normals would be methodologically complex to perform and interpret. Nonetheless, study of a physiological system in which voluntary and involuntary processes are so inextricably tied would be of substantial interest to understanding the full range of visceral sensory processes and functions.

INTEROCEPTION AND ANXIETY

Questions about the relationship between anxiety and somatic symptoms date back many years, but starting perhaps 20 to 30 years ago investigators began to focus on this question with more specific disorders. For example, Tyrer et al. (1980) assessed the relationship between self-report of heart rate and electrocardiographic measures of heart rate in three patient groups, anxious, phobic, and hypochondriacal, and reported that the phobic group was less accurate than were the other two groups. McLeod et al. (1986) reported that correlations among self-reports and four physiological measures at rest and during Stroop test–induced stress—sweating, palpitations, muscle tension, and trembling—were in the predicted direction but non-significant for people with generalized anxiety disorder. (The Stroop test is a procedure designed as an experimental stressor. After a presentation, subjects are asked to identify as rapidly as possible the color of the letters in which a word was written, and when the word itself was a different color—for example, the word "red" printed in green.) In contrast, Sturges and Goetsch (1996), studying only women, observed higher heart rates and more cardiac awareness during psychological stress or caffeine administration in high-anxious than in low-anxious subjects, but Sturges et al. (1998) reported that hyperventilation was not associated with more cardiac awareness on a heartbeat tracking task in high-anxious than in low-anxious women.

Focusing on panic disorder, Ehlers and Breuer (1992) reported several findings in studies with large patient samples. Patients with full panic disorder, less severe (infrequent) panic, and other anxiety disorders reported more cardiac awareness than did normal subjects. They also reported more gastrointestinal awareness, which bears on both the issue of cross-modality interoception as well as the possible interoceptive mechanism for gastrointestinal complaints (which are more common in anxiety patients). Panic disorder patients were more aware than were other patients of cardiac, but not gastrointestinal, function. In a second experiment using a heartbeat counting task, panic disorder patients were more accurate than were other patients, including those with simple (i.e., specific) phobias or infrequent panic attacks, and (consistent with earlier cardiac interoception data in normal subjects) men were more accurate than women. These data are illustrated in Figure 12–1. In a third study also using the counting task, panic disorder and generalized anxiety patients were more accurate than were depressed people.

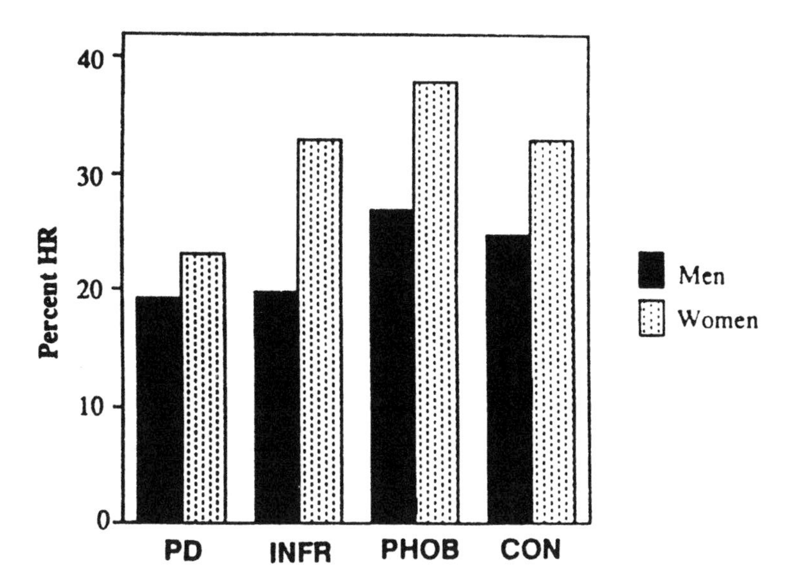

Figure 12–1. Mean heartbeat perception test error scores (percentage of heart rate) for PD (panic disorder), INFR (infrequent panic), PHOB (simple or specific phobias), and CON (controls). Men with panic attacks were most accurate. [From Ehlers and Breuer, 1992, reprinted with permission of the holder of the copyright.]

In contrast to these positive results, using the more difficult Whitehead procedure, Asmundson et al. (1993) observed that panic disorder patients did not demonstrate reliable cardiac awareness, either at rest or after hyperventilation. Using the heartbeat counting task, Antony et al. (1995) also found that panic patients were not more accurate than were people with social phobia or normal subjects either at rest or after exercise. Exercise did improve accuracy in all groups, although physical activity does not provoke panic attacks (O'Connor et al., 2000). Finally, Pauli et al. (1991) reported an ambulatory study involving cardiac perception over 24 hours in panic disorder patients and normals. Both groups perceived at least some of the heart rate accelerations recorded on Holter monitors, but patients were not more accurate than were normals.

Studies have addressed the effects of treatment on visceral awareness. For example, Antony et al. (1994) reported that accuracy of heartbeat counting was not changed on average by cognitive–behavioral treatment of panic disorder, although some patients did demonstrate changes.

Somatic perception in anxiety, especially panic disorder, has been reviewed by several groups (Barsky, 1992; McLeod and Hoehn-Saric, 1993; Ehlers et al., 1995; Ehlers and Breuer, 1996). These will now be reviewed from the point of view of the issues and factors raised.

Studies published in the 1980s suggested that high anxiety was associated with more accurate heartbeat awareness, but negative findings were also reported. The possibility that awareness was improved only during periods of arousal was considered, but using an experimental stress paradigm or pharmacologic induction of physiological activation did not clearly increase awareness, and as already described above, intercorrelations among various physiological markers themselves were not strong. Individual differences in patient awareness might be important, as it has been shown to be for normal subjects with cardiac awareness.

Symptom self-report has been used as the only measure of visceral (e.g., cardiac) awareness in almost all the studies in the literature thus far. Results from at least two studies indicated that panic disorder patients are not just more aware of their heart action, but also of their gastrointestinal function. In ambulatory research the actual rise in heart rate at times of reported panic and cardiac awareness is relatively small and not likely to be the only, or even the major, contributor to awareness. Other factors must contribute as well. Furthermore, although reports of palpitations often are associated with some change in heart action, as many as half the identified panic attacks during ambulatory monitoring were not associated with any clear increase in heart rate.

Ehlers and Breuer (1996) discussed in some detail data from laboratory objective assessments of heartbeat perception in people with panic disorder. Using methods that involved heartbeat detection (comparison of heartbeats to external signals, e.g., tones), they noted that several studies have failed to find differences between patients and normals, including data of their own and the study by Asmundson et al. (1993). Studying cardiac patients, many of whom probably actually had panic disorder, they found suggestive evidence that the patients did better than the controls. Ehlers and Breuer (1996), nonetheless, concluded that "In summary, none of the studies . . . showed that panic disorder patients perform better than controls . . ." in these types of studies (p. 169).

Using a heartbeat counting task, a task that might be easier than heartbeat detection but is also more prone to confounding, Ehlers and Breuer (1996) found support for panic disorder patients being more aware of heart action than controls. They identified four earlier studies that reported positive results, but two others that reported negative, and postural change (which did affect cardiac awareness in normals in earlier studies) did not necessarily help. In discussing possible reasons for the discrepant results, they noted that some patients in these studies were on medications that could have affected results and that presence of agoraphobia as part of the panic syndrome (in most cases signaling a more severely affected individual) heightened awareness. They also presented data indicating that instructions affect results (Ehlers et al., 1995) and that overall the mental tracking paradigm was valid in these subjects.

A major issue related to interoception in general is the question of the relationship of physiological changes to sensation–perception processes, and related to that is the issue of the relationships among the physiological variables themselves. For example, if it is hypothesized that interoception is related to heart action—in either normal people or people with some dysfunction such as panic disorder—then by definition the subjective and the physiological measures (or some aggregate measure of the physiological variables) should correlate above chance, even if other factors contribute as well. Furthermore, if it is also hypothesized that a more general unitary factor such as arousal is playing a role, then at least some of the individual physiological variables which define arousal should correlate above chance with each other. Does that occur?

It has already been noted that the intercorrelations among the physiological variables often used to assess arousal or thought to be activated in anxiety disorders often are low and not significant. The lack of significant correlations among the physiological variables or between the physiological and the subjective experiential variables is usually called desynchrony. McLeod and Hoehn-Saric (1993) discussed this issue and provided data from people with generalized anxiety disorder. They concluded that while the quantitative correlation coefficients are low, they are usually in the expected direction (McLeod et al., 1986). Zoellner and Craske (1999) reported that infrequent panickers performed more accurately on a heartbeat counting task than did control subjects, but that activation produced by caffeine ingestion did not change results. It is not clear what role other factors, especially prior knowledge, play in these categorical associations. Interestingly, Van der Does et al. (1997) found that among panic patients who were assessed for accuracy of heartbeat perception, there were two groups: those who were more accurate than controls and those who were not more accurate but had evidence of increased arousal. In other words, this study suggested that increased arousal can either disrupt or be mistaken for interoception of heart action.

All the investigators who have addressed interoception in panic have raised the issue of possible mechanisms. Arousal has already been mentioned. Another potential factor is variability. It has been observed that individuals can sense a change in function more reliably than an unchanging level, even if that level is somewhat abnormal. Panic disorder patients who go in and out of attacks might be especially prone to be aware of somatic changes. Related to that is reactivity. Panic disorder and other anxious patients might be more reactive to psychologically or physiologically provocative circumstances, making not only panic attacks but arousal fluctuations larger and more common than in normal subjects. Whether and how these people are generally more aroused or more reactive is unresolved and a matter of debate.

Other possible factors have been considered, as well. Attention to bodily processes, especially threat cues (i.e., the relationship between panic and fear), may heighten awareness. Consistent with the data (see Fig. 12–1) and Chapter 8 on cardiac interoception in normal subjects, men seem to be more attentive to internal sensations, while women are more attuned to external cues. Related to attention is expectation. Based on the prior experience of anxious individuals with previous anxious or panic events, they believe that somatic symptoms generally are harbingers of panic or increasing anxiety, and further, often believe that these panic or anxiety events are not only uncomfortable but also dangerous (e.g., "I am having a heart attack."). This has been called catastrophizing. It can set up a vicious cycle of more anxiety, more somatic symptoms, and so on. Kroeze et al. (1996) provided data indicating that reports of bodily sensations were more affected by attentional factors in panic disorder patients than in control subjects. It has been observed, for example, that panic disorder patients with more accurate cardiac perception seem to be at higher risk for relapse.

If panic disorder patients can sense their heartbeat and other somatic functions more intensely than normals, that may relate to why somatic symptoms are more prominently associated with anxiety than with other psychiatric disorders. It may also relate to illness maintenance, relapse, and treatment choice. For example, since individual differences in somatic awareness that are reliable over time appear to exist in panic disorder, those people with the most intense somatic symptoms should probably receive treatment aimed at the somatic symptoms, while others might need more psychologically based interventions. Specifically, these patients probably should receive some interoceptive deconditioning, and drugs such as tricyclic drugs (which often raise heart rate) may be less optimal in this subgroup of patients.

In summary, wide interest exists in the relationship between anxiety, especially panic disorder, and somatic symptoms, especially cardiorespiratory. Much of the research has involved self-report, making the issue of perception (as opposed to detection) very important in this area. A few studies have used more rigorous methodology to assess interoceptive ability in these patients, under the assumption that they are more aware than non-anxious individuals, but the results are contradictory and inconclusive. It is not yet completely clear what is the source of the increased somatic symptom complaints in these people. Van der Does et al. (2000) concluded that ". . . accurate HBP [heart beat perception] was uncommon, but more prevalent among panic disorder patients than among healthy controls, depressed patients, patients with palpitations and individuals with infrequent panic attacks" (p. 47).

Imaging Studies of Anxiety Disorders

A number of imaging studies using various techniques have been completed in people with anxiety disorders. These disorders are of potential relevance to understanding visceral sensory processes, especially panic disorder, because of the prominence of visceral–autonomic symptoms and because the brain changes observed were in regions implicated in visceral sensory–perceptual processes.

Early PET studies of panic disorder were reported by Reiman and colleagues. They initially reported that in patients who had panic attacks in response to lactate but not in patients who did not, they found an asymmetry of cerebral blood flow in the parahippocampal gyrus (Reiman et al., 1984). This same group, however, subsequently reported that this finding might have been an artifact. In other studies this group reported that panic produced by lactate increased cerebral blood flow bilaterally in the temporal poles, insular cortex, basal gangliar area, superior colliculus, and left cerebellar vermis (Reiman et al., 1989b). Anxiety produced by anticipation of shock increased cerebral blood flow in the temporal poles in normal subjects (Reiman et al., 1989a).

Other groups have investigated cerebral blood flow in panic disorder. Using single-photon emission computed tomography (SPECT), De Cristofaro et al. (1993) found a right–left asymmetry in the frontal cortex along with changes in the occipital and hippocampal areas, and Woods et al. (1988) found an abnormality in the frontal cortical region. Stewart et al. (1988) reported that lactate raised hemispheric blood flow in normal subjects and in patients who did not panic, but not in those who did. Activation of anterior limbic and paralimbic structures with procaine infusions, documented with PET, produced panic-like reactions in normal subjects (Servan-Schreiber et al., 1998). Consistent with the procaine data, induction of panic-like reactions in normal subjects with cholecystokinin-4 (CCK-4) (Benkelfat et al., 1996; Javanmard et al., 1999) produced cerebral blood flow (CBF) increases in the claustrum-insular region. (Cocaine also produces subjective effects that correlate with changes in CBF—Pearlson et al., 1993)

Several investigative groups have assessed glucose metabolic effects of panic. Reiman et al. (1986) observed an abnormally high whole brain metabolism. Nordahl et al. (1990) did not see such a whole brain increase but did observe some specific regional change that generally persisted after imipramine treatment, indicating that they were trait abnormalities (Nordahl et al., 1998). Bisaga et al. (1998) also reported glucose metabolic abnormalities in the hippocampal and parietal regions.

A number of investigators have studied potential EEG changes in people with panic disorder. Knott et al. (1996, 1997) reported greater EEG power in the delta, theta, and alpha frequency but less in the beta band in panic disorder patients. Dantendorfer et al. (1996) reported an elevated rate of non-epileptic EEG abnormalities in panic, associated with a high rate of structural abnormalities using MRI, especially in the septo-hippocampal area. Wiedemann et al. (1998) reported a laterality pattern that differed in panic disorder versus normal subjects, and an overall increase in cortical activation. Using a pharmacologic challenge to raise anxiety (caffeine), however, did not reveal any differences in response between patients and normal subjects (Newman et al., 1992).

Other studies have determined if specific structural or other brain abnormalities occur in panic disorder. In addition to Dantendorfer et al. (1996), Wurthmann et al. (1998) described increased ventricular size, mainly in the prefrontal region. Shioiri et al. (1996) did not find any abnormality of phosphorus metabolism in the frontal lobes of panic disorder patients measured with magnetic resonance spectroscopy. In a single case of intracranial stimulation in a patient being treated for an intractable tremor but who also had panic disorder, Lenz et al. (1995) reported that the patient described stimulation in the principal somatosensory nucleus of the thalamus as producing an experience very similar to a panic attack. In summary, imaging data from people with panic disorder, although not completely consistent, are indicative of changes in many of the regions that are involved in interoceptive processes.

Other anxiety disorders have been studied with functional imaging and EEG techniques in addition to panic disorder. These have included generalized anxiety disorder, obsessive–compulsive disorder, people with co-morbidity of an anxiety disorder with some other disorder (mainly depression), social and specific phobias, and post-traumatic stress disorder.

Wu et al. (1991) reported pretreatment cerebral glucose metabolism decreases in white matter and basal ganglia, with increases in left occipital, right temporal, and right frontal areas in generalized anxiety disorder and correlations of brain changes with symptoms. Imaging studies of obsessive-compulsive disorder (OCD) have implicated, especially, medial prefrontal and basal ganglia structures (Baxter et al., 1988; Saxena et al., 1998). Co-morbidity of depression and anxiety occurs quite commonly. Bruder et al. (1997), measuring the EEG in depressed people with and without anxiety, demonstrated a hemispheric asymmetry in which non-anxious depressed people showed evidence of more hemispheric activation on the left, with more activation on the right for the co-morbid individuals.

In social phobics, using SPECT, Stein and Leslie (1996) found no abnormalities of regional cerebral blood flow, including no significant correlations of scan findings with symptom ratings. In contrast, using a SPECT ligand designed to measure dopamine reuptake sites, Tiihonen et al. (1997) reported that reuptake sites in the striatum were decreased in social phobia. (Dopamine in the prefrontal cortex has been implicated in drug-produced anxiogenic interoceptive cues—Broersen et al., 2000.) Studying blood flow with fMRI, Birbaumer et al. (1998) found more activation of the amygdala in social phobics than in normal subjects to presentation of face stimuli, while another fMRI study (Schneider et al., 1999) that involved conditioning to aversive odor stimuli reported that normal subjects demonstrated decreases in amygdala and hippocampal areas but increases in people with social phobias.

Using PET, Mountz et al. (1989) reported that, after controlling for hyperventilation, people with specific phobias showed no significant changes in regional CBF during exposure to the phobic object. O'Carroll et al. (1993), using SPECT measurement of CBF, found reductions in CBF bilaterally in posterior brain regions and the right temporal-occipital area, with no clear relationship between brain function changes and symptoms. Rauch et al. (1995), using PET, reported that specific phobia symptom provocation increased CBF significantly in the cingulate, thalamus, and insular, temporal, and somatosensory cortices and concluded that their results were consistent with activation of paralimbic areas. One investigative group published several studies of specific phobias using PET (Wik et al., 1993, 1996, 1997; Fredrikson et al., 1993, 1995a, 1995b, 1997). Using phobia-related visual stimuli, they reported increases in regional CBF in the secondary visual cortex and decreases in the hippocampus, cingulate, and prefrontal, orbitofrontal, and temporal cortices. They also reported correlations between cortical and thalamic areas.

Rauch et al. (1996) used symptom provocation to invoke anxiety in people with post-traumatic stress disorder (PTSD) and reported PET-measured CBF increases in visual, limbic, and paralimbic areas on the right and decreases in frontal and temporal areas on the left. Shin et al. (1997) found CBF increases in the anterior cingulate and right amygdala. Zubieta et al. (1999) observed increases in CBF measured with SPECT in PTSD in the medial prefrontal area, along with some tendency for these changes to be associated with symptoms. Bremner et al. (1997) measured cerebral glucose metabolism in PTSD subjects in response to the adrenergically activating drug yohimbine. In contrast to the increases seen with CBF, glucose metabolism was lower in PTSD subjects in the prefrontal, orbitofrontal, temporal, and parietal cortices. Hamner et al. (1999) hypothesized an essential role for the anterior cingulate region in PTSD.

Three groups (Rauch et al., 1997; Reiman, 1997; Lucey et al., 1997) reported results of data analyses in which they pooled data across several different anxiety disorders, including correlations of regional brain changes with symptoms. Rauch et al. (1997) found that across diagnoses, regions most consistently involved included the right frontal and medial orbitofrontal cortices and the bilateral insula, lenticular nuclei, and brainstem, with a robust correlation between subjective anxiety and brainstem activation. Lucey et al. (1997) found differences across groups mainly in the bilateral frontal cortex and right caudate. Whole brain CBF and caudate structures correlated with both anxiety and depression scores. Finally, specifically relevant to interoceptive processes, in discussing the association of various brain regions with different aspects of emotion, Reiman (1997) concluded that ". . . anterior insular regions appear to participate in the evaluation procedure that invests potentially distressing cognitive and interoceptive sensory information with negative emotional significance . . ." (p. 4). Overall, these disorders also appear to be associated with activation of brain regions related to and consistent with the somatic symptoms typically part of interoceptive processes.

DEPRESSIVE DISORDERS

The definition of major depressive disorder contains no symptoms directly referable to visceral sensory processes. However, because 25%–40% of people with major depressive disorder have panic attacks (and the rate is even higher if generalized anxiety symptoms are included), visceral sensory processes are relevant to depression.

A recent epidemiologic study assessed more than 1600 patients of general care physicians who had no psychiatric diagnoses but had at least three symptoms considered to be anxious, depressive, or somatizing (Piccinelli et al., 1999). Six distinct groups were identified, including one heavily loaded with somatic symptoms including the visceral symptoms abdominal and chest pain, nausea, "gas," diarrhea, shortness of breath, and hard heartbeat. Somatic symptoms were much less strongly represented in the other five groups, suggesting there might be an identifiable subgroup of such patients with a specific disorder of somatic sensory change. In a similar study, Spinhoven and van der Does (1997) compared panic disorder and depressive disorder patients to control subjects on measures of somatization and somatosensory amplification. No differences on the amplification variable were seen, but panic disorder patients were highest and controls lowest on somatization. Further, it has been suggested that panic patients with associated major depression may be particularly prone to chest pain complaints (Beitman et al., 1987), despite the fact that non-visceral somatic pain thresholds may be even higher in depressed individuals (Lautenbacher et al., 1994).

Consistent with the hypothesis that abnormalities of visceral sensation in depression would implicate involvement of the autonomic nervous system, several groups have assessed autonomic function in these patients. Although, over the long history of studies of autonomic function in depression, results have been mixed and inconsistent, substantial evidence exists for adrenergic dysfunction in depression (e.g., Maes et al., 1993; Manji et al., 1994) that is not explained simply by the co-occurrence of anxiety (e.g., Grunhaus et al., 1990; Karege et al., 1993; Townsend et al., 1998). In contrast, vagal function, at least for cardiac control, appears to be normal in depressed individuals (Lehofer et al., 1997). Berrios et al. (1995) observed that autonomic symptoms are more common in people with Parkinson's disease than in unaffected individuals, and they hypothesized that the anxious and depressive symptoms often observed in people with Parkinson's disease are caused by autonomic changes.

It has already been noted (Chapter 8) that reports of visceral sensations can be highly influenced by perceptual factors other than simply visceral organ function (e.g., expectations). Investigators have studied the specific question of how somatic symptoms relate to mood, including symptoms related to anxiety and depression. Persson and Sjoberg (1987) reported that somatic discomforts were strongly correlated with various aspects of mood, especially more diffuse discomforts. It could not be unequivocally determined if moods affected somatic symptoms, symptoms affected mood, or some third factor influenced both. Roth et al. (1976) reported that heart rate fluctuations over 24 hours correlated with fluctuations in mood state and that the patterns of correlations observed were different in men than in women, indicating a physiological link between mood states and visceral organ function. However, Johnston and Anastasiades (1990) could not replicate this finding.

A number of functional imaging studies of depression have been reported, and these studies have been reviewed and summarized (Morris and Rapaport, 1990; Drevets and Raichle, 1992; Soares and Mann, 1997; Leuchter et al., 1997; Kennedy et al., 1997; Drevets, 1998). Of relevance to interoception, brain regions involved in interoceptive processes that have been found to be abnormal in depression include especially the prefrontal cortex, amygdala, thalamus, and limbic and paralimbic structures. Both anatomic and functional abnormalities have been observed.

A more recent finding potentially ties together a number of themes related to interoception—depression, functional imaging, and function of the vagus nerve. Stimulation of the vagus nerve is effective in the control of some cases of otherwise intractable epilepsy (McLachlan, 1997; Amar et al., 1998). Studies have indicated that this technique might also benefit people with treatment-resistant depression (Rush et al., 2000; George et al., 2000). Imaging by EEG during stimulation indicated that, overall, no changes occurred (Hammond et al., 1990; Salinsky and Burchiel, 1993), but a change in somatosensory evoked potentials was reported (Naritoku et al., 1992). Imaging studies of regional cerebral blood flow with PET and SPECT (Ko et al., 1996; Henry et al., 1998; Vonck et al., 2000) demonstrated regional effects. The most consistent change was observed in the thalamus, but other regions that showed activation in at least one of the studies included the medulla, postcentral gyrus, posterior temporal cortex, hypothalamus, insular cortex, putamen, and cerebellum.

SOMATIZATION

Somatization disorder is explicitly defined by the presence of pain from four or more body sites or functions, two or more gastrointestinal symptoms, at least one symptom related to sexual or reproductive function, and at least one pseudoneurological symptom (including difficulty swallowing, or lump in the throat, and urinary retention). Thus, people with somatization disorders are highly focused on dysfunctions of somatic and visceral sensation. It appears that no systematic research has been done to address the question of the extent to which these individuals may have different sensory (i.e., interoceptive) thresholds versus other explanations for their symptom reports and complaints. A few studies of general relevance to this topic will briefly be described.

In general medical practice somatic symptoms are one of the most common reasons individuals present to outpatient clinics, with some individual symptoms present in 10% or more of cases but with the majority not having a discernible cause after an initial assessment (Kroenke, 1992). People judged to be neurotic had a one-week prevalence of approximately 40% for cardiac symptoms and approximately the same for gastrointestinal symptoms, approximately four times the rate for a comparison group (Kellner and Sheffield, 1973).

Hypochondriasis is another somatoform psychiatric disorder involving not a large number of symptoms per se, but rather an inappropriate preoccupation with fear of having an illness based on a misinterpretation of one or more bodily symptoms. Furer et al. (1997) concluded that a close relationship exists between hypochondriasis and panic disorder. On the other hand, Barsky et al. (1994) concluded that these two disorders overlap but are distinct. MacLeod et al. (1998) also found that hypochondriasis and anxiety were distinct, based on the understandings or attributions that people with the different diagnoses made about their bodily symptoms. Kellner (1988, 1991) reviewed the topic of the relationship between psychiatric and psychosomatic disorders and somatic symptoms.

Somatic symptoms are present in some other psychiatric disorders as well. Psychosis is often broadly defined as a loss of normal contact with reality. One way in which that loss of contact with reality occurs is when a patient develops either a delusion—a false belief—or a hallucination—a false sensory experience. Delusions or hallucinations can involve visceral processes, for example, a delusional belief that one's insides are "rotting," or a sensory–perceptual feeling that such a process is occurring.

To summarize, a number of psychiatric disorders exist that are likely to involve visceral sensory processes, particularly anxiety disorders and especially panic disorder. Some research has been done addressing this question, including imaging studies, but results so far are inconclusive. The broader question of the role of interoceptive processes in medical and psychiatric disorders is just beginning to be addressed.

thirteen

NEUROPHILOSOPHY: CONSCIOUSNESS AND BODILY AWARENESS

> Philopause (noun), a point at which a researcher, weary of or frustrated by rigorous laboratory-based science, begins to look for nonscientific, philosophical explanations instead . . . (Natalie Angier in *The New York Times*). This term, which is perhaps loosely derived from 'menopause,' was originally used specifically for neuroscientists, some of whom have tended, as they've grown older, to abandon science-based attempts to map the workings of the brain and have turned instead to more general speculations about the nature of consciousness. Owing largely to advances in artificial intelligence, molecular genetics, psychopharmacology, and brain imaging, in recent years these distinct approaches have been joined in an endeavor known as neurophilosophy—mitigating, perhaps, the impulse for neuroscientists to experience a philopause . . .
>
> Word Watch, *Atlantic Magazine*, August 1999, p. 96

Apropos of the epigraph above, and regardless of how one views the current interest among neuroscientists in philosophical questions (see Blomfield, 1992; Smythies, 1992), interoception clearly is logically related to bodily awareness and the sense of self, and these in turn are related to the issue of consciousness more broadly. A detailed review and discussion of these philosophical issues is beyond the purview of this book and the competence of this author. Nonetheless, because of their importance, issues of potential direct relevance will be noted here.

CONSCIOUSNESS

As with emotion, William James's writings have been very influential in theorizing about consciousness. He described every thought as ". . . part of a personal consciousness," and this "personal consciousness is sensibly continuous." (James, 1950, Vol. 1, p. 225). This has been called the stream of consciousness, which serves to highlight its continuity and its tendency to contain only one predominant thought at a time. These two aspects of consciousness, in turn, serve to reinforce the subjective sense that each conscious being consists of one (and, except under rare and unusual circumstances, only one) unique and persistent self. Understanding this essential link, James addressed the concept of consciousness of self in the very next chapter of his book.

Interest in trying to understand consciousness from a neurobiological perspective increased during the 1980s and 1990s, culminating in books and articles appearing mainly in the 1990s (e.g., Edelman, 1989; Crick, 1994; Churchland, 1995; Chambers, 1997; Delacour, 1997; Feinberg, 1997; Logothetis, 1999; Schacter, 2000). Such issues as the neural correlates of consciousness (e.g., Block, 1996), the neurotransmitters involved (e.g., Perry et al., 1999), and the function of consciousness (e.g., Baars, 1998; Cameron, 1998) have been discussed. The question of the phylogenetic level at which consciousness first appeared has been addressed. For example, Cabanac (1999) suggested that emotion, and therefore consciousness, is first present in reptiles. Such a conclusion is certain to be controversial but nonetheless is a reminder of the intimate link (at least hypothetically) between consciousness and emotion (and thus with interoception).

In order for a concept to have a meaning it is usually necessary that both its existence and its non-existence (i.e., presence and absence) occur. Thus, for the concept of consciousness to have a meaning, unconsciousness must also exist. The question of import to this exposition is, To what extent does such higher mental function material that is outside of consciousness influence thought, emotion, and behavior? Most specifically, Do interoceptive processes exist that are not clearly conscious but have effects on these higher mental functions nevertheless?

Freud is often credited with the discovery of the psychological unconscious, but its existence was hypothesized much earlier, for example, by a number of prominent philosophers. Freud's writings related to neuroscience, primarily the *Project for a Scientific Psychology*, continue to exert influence (e.g., see Bilder and LeFever, 1998). Evidence for the existence of non-conscious processes is strong, including involvement of functions likely to be related to interoception (see Miller, 1992). And Freud, writing about the hypothesized unconscious psychological mechanisms the ego and the id, recognized the intimate relationship between these unconscious psychological processes and the body (see quote at the beginning of Chapter 12). A few examples will now be provided.

It is common to assume that humans and, presumably, other species that demonstrate consciousness, such as primates, are aware of the information entering the nervous system through the senses. It appears, however, that a great deal of processing of the sensory information is necessary before awareness occurs, and before that level of processing, neural activity occurs outside of consciousness. This early processing can occur in the anatomical and/or the temporal domains. For example, it appears that humans and other primates are not aware of activation of the primary visual cortex (Crick and Koch, 1995), and visual information in brain regions and systems associated with coordination of movement occurs outside conscious awareness (Milner, 1998). In the temporal domain, it appears that a lag of between 0.35 and 0.50 seconds occurs before a sensory event can enter consciousness (Libet, 1993).

Mental functions and processing, of course, occur outside consciousness at other levels than at the level of sensory awareness. Studies in cognitive psychology have demonstrated effects not only at the level of experience but also at the levels of thought and action (e.g., Kihlstrom, 1987). As an example, it has been shown that subliminal presentation (i.e., visual masking) of words can influence the interpreted meaning of subsequent immediately presented target words (Greenwald, 1996).

Processes that are potentially directly associated with interoception include classical conditioning and emotional learning. Concerning classical conditioning, studies have indicated that delay conditioning (the CS and UCS overlap) is not disrupted by hippocampal lesions, while trace conditioning (the CS ends before the UCS starts, typically with a separation of 0.5–1.0 seconds) is disrupted by such lesions (Clark and Squire, 1998; Schacter, 1998). In humans it appears that successful trace conditioning is dependent not only on intact hippocampal function but also on awareness of the conditioning contingencies, while delay conditioning can occur outside awareness (Clark and Squire, 1998).

Emotional learning is another type of higher mental function likely to be very relevant to interoception. For example, using the masking technique and functional imaging (PET), Morris et al. (1998) demonstrated that without masking, presentation of an image of an angry face produced activation in the left amygdala only, while with masking by an image of a face with a neutral expression (which produces lack of conscious awareness of the angry face) activation occurred only in the right amygdala. Thus, awareness or lack of awareness of the presentation of the stimulus with emotional content affected the side of the brain that was activated. In a related series of case reports, Bechara et al. (1995) noted that in humans hippocampal damage disrupts acquisition of declarative facts, while damage to the amygdala disrupts autonomic conditioning. One author has argued that consciousness originated quite recently in human history in the breakdown of the separation of functioning of the two cerebral hemispheres (Jaynes, 1976).

Historically, part of the relatively low interest in studying visceral sensory processes seems to be due to the apparent lack of conscious awareness of these functions most of the time in most people. It has been assumed that this lack of conscious awareness implied a lack of importance, a lack of influence on cognitive processes and behavior. Hopefully, this discussion will convince the reader that lack of conscious awareness is not evidence for lack of importance. While some so-called higher mental functions (such as explicit memory tasks) do require awareness, others (such as implicit memory tasks) do not (Verfaellie and Keane, 1997). Putting it slightly differently, as Damasio (1994) noted, bodily awareness can be fundamental while still being "in the background."

BODILY AWARENESS AND CONSCIOUSNESS OF SELF

Some forms of conditioning and learning and some other cognitive functions, especially those related to emotion, connect consciousness as well as unconscious (and subconscious) higher mental functions to interoception. The most direct connections are between interoception and consciousness of self, that is, the body as a main component of the self and interoceptive processes as essential to awareness of the body. The issues of body image, related to pain, phantom limb, and so on have been discussed in Chapter 10. Here, they will be discussed briefly in the context of their relationship to consciousness. The role of visceral sensation in body image is a fundamental issue in the study of consciousness.

Cabanac (1996) discussed the origin and development of consciousness from an evolutionary perspective and argued that consciousness arose as a consequence of sensory processes. He further argued that, as an outgrowth of the structure of sensation, consciousness has a quadri-dimensional structure, including duration, quality, intensity, and affectivity. Cabanac noted that development of the understanding of sensation has led not only to identification of more than the five classical senses directed to the external world, but also to the recognition that visceral sensations contribute to the hypothesized development of consciousness (including the four visceral sensors described earlier—osmoreceptors, mechanoreceptors, chemoreceptors, and thermoreceptors).

Continuing the theme of the role of evolution, Chiel and Beer (1997) argued that the development of adaptive behavior (including consciousness as one evolved mechanism involved in behavioral adaptation, although not the focus of their discussion) includes not just an interaction between the brain and the environment external to the organism, but also the ongoing involvement of the body in this process in both motor and sensory aspects. That is, there are three, not two, continuously interacting arenas, realms, or worlds. The body is the intermediary by which the nervous system interacts with the external world, and a continuous give and take occurs that provides both constraints and opportunities. Based in large part on incoming sensory information, the development of body image as a physical object in the world is also an ongoing process created in significant part by sensory information—although cognitive maps of the body and the external world also are apparently, at least in part, inborn and genetically based and therefore more stable in conscious awareness (Berlucchi and Aglioti, 1997).

Damasio (1994, 1999) wrote about the relationships between feeling states and the body, including consciousness. An important point that he emphasized, also mentioned earlier, is that input from the body to the central nervous system need not reach consciousness to have significant influences in higher mental functions and behavior. He refers to the body as a ground reference and describes background feelings. Unlike James, he attributes importance to interoceptive processes not only with reference to emotion, but more broadly as a general factor in ongoing organismic functioning. He also highlights the fact that bodily input provides stability, perhaps contributing to the sense of the self as consistent and persistent over time. Finally, he points out that neurological patients with the so-called locked-in syndrome (in which input from the body to the central nervous system is greatly disrupted) appear to lose not the cognitive but, rather, the affective tone of their reactions to emotional situations and circumstances.

Vanderwolf (1998) wrote cogently on the relationship between, on the one hand, the brain and behavior and, on the other, the brain and what is called mind, pointing out that mind is typically thought to be that faculty or entity that contains such functions as sensation, perception, emotion, attention, and so on. For the purpose of this discussion, several specific points that Vanderwolf makes will be emphasized as issues of importance to interoception and neurophilosophy.

First, the functions of the mind do not themselves produce any awareness. Introspection does not lead to awareness of how thought occurs. As Libet (1993) demonstrated, neural activity, or neural processing, occurs in the cortex at least 0.35 second before individuals are even aware that mental activity is occurring.

Second, consistent with this observation that the mind is not accessible to conscious thought, the only two sensory realms that are accessible are the external world and one's own body. Putting it slightly differently, one is only conscious of what is outside the nervous system, not of the workings of the nervous system itself.

Third, consistent with the idea that the sensory and motor systems are, in fact, one system, cognitive functions apparently are related to motor processes, as supported by the observations that total muscle relaxation appears to lead to loss of conscious imagery and that phantom limb phenomena depend on the persistence of sensory feedback produced by residual muscular activity from the affected limb. (Perhaps the less frequent or robust occurrence of phantom phenomena when visceral organs are removed relates to the relative vagueness and diffuseness of visceral sensory experience.) Carried to its conclusion, the required role of the motor response in sensory experience would imply that to see or hear (i.e., become conscious of) something requires that, in some way, the organism responds to it. For example, to feel an emotion it is necessary to respond to an emotion-provoking situation. Indeed, normal organismic functioning might be structured such that one can also end an emotion only by behaving in certain predetermined, hard-wired, fixed-action pattern ways. If this analysis is correct, to a great extent consciousness is behavior, and (although much of the body can be removed without losing consciousness of self) the body plays an essential role.

The issue of the importance of the body to consciousness and sense of self has been written about by philosophers as well as neuroscientists. For example, Lakoff and Johnson (1999) wrote extensively about how the body and bodily awareness strongly influences philosophical meanings and metaphors. They refer to the embodied mind and describe how the influence of the body and bodily awareness can affect such concerns as the sense of self, understood from a philosophical point of view.

To summarize, consciousness of self is a primary type in the larger realm of subjective consciousness, and bodily awareness is essential to the concept of self (Bermudez et al., 1995). Further, a great deal of psychobiological functioning occurs outside consciousness, including many so-called higher mental functions. Visceral sensory processes contribute importantly, at both the conscious and unconscious levels, to bodily consciousness, the higher cognitive functions, and behavior, especially but not exclusively to emotional, or affective, behavior. In other words, interoception plays a basic role in consciousness and related issues of interest to neurophilosophy.

fourteen

SUMMARY: PAST AND FUTURE

> The heart has its reasons which reason knows nothing of.
>
> Blaise Pascal

WHENCE?

What has been said so far? It has been proposed—and it is considered one of the issues of fundamental importance to this exposition—that interoception should be defined as more than simply visceral sensations that reach awareness, but less than any and all sensory impulses that arise from within the body. It is proposed that interoception should be defined to have occurred if, and only if, such visceral sensory impulses can be shown to have affected the cognition or behavior of the whole organism, with or without awareness. In other words, it is broader than the realm of awareness but narrower than the realm of physiology.

This definition is intended to lie within the realms of both sensation and perception, but probably closer to perception, as discussed and distinguished in Chapter 8. In other words, it seems highly likely that visceral afferent impulses that have demonstrable effects on behavior do not have pure effects but rather have effects influenced by multiple other interpretive and modifying factors, such as, perception.

The roots of interest in the existence and functioning of interoceptive processes can be traced back many centuries but arose most clearly in the twentieth century mainly from three sources: *(1)* the James–Lange theory of emotion, *(2)* the demonstrations of conditioning of visceral functions with Pavlovian techniques, and *(3)* as an offshoot of interest in the one- versus two-process learning theory controversy and the study of operant conditioning methods to control visceral functions (biofeedback). In this context three points should be re-emphasized about James's version of the theory.

First, not all emotions were necessarily included. James spoke of the "coarser" emotions and was referring primarily to what Cannon later called "great emotion," especially fear and anger. The role of visceral afferent impulses in such emotional states as joy remains largely unexamined.

Second, James did not argue that fear per se or fear behaviors required visceral input. He argued that the visceral input contributed to, or even created, the feeling of fear, but neither the cognition nor the behavioral response.

Third, as a corollary, he did not argue that the emotion as a cognitive–behavioral–experiential feeling arose from the visceral event. A careful reading of what James wrote leads to the following understanding of what his theory proposed, in paraphrased form. "I see the bear" (perceptual–cognitive event), "I run" (behavioral response, including production of sensory input from visceral–somatic sources into the brain), "I am afraid" (experiential feeling). Thus, the visceral sensory input is related to the feeling and is a result of non-feeling aspects of the full emotional (e.g., fearful) event. Further, this theory of visceral involvement includes both the efferent and afferent sides of the innervation of the viscera. In other words, a closed loop in the nervous system is required. This is actually a visceral motor–sensory theory. It now seems incorrect that the behavioral component is required to experience the feeling of fear or any emotion, but the idea that the closed visceral motor–sensory nervous system loop contributes in a fundamental way to the feeling of at least some, if not all, emotions remains highly plausible now, just as it was more than 100 years ago.

The James–Lange theory concerns emotion. Emotion is very

closely related to motivation, indeed, they are almost synonymous, although they tend to focus on different aspects of the overall phenomenon. As a fundamental aspect of emotion, understanding of interoception is essential to a full understanding of motivational processes. Conditioning also is closely related to interoception because of the abundant evidence that changes in visceral functioning can be conditioned, at least with Pavlovian procedures and possibly with operant as well (although that remains unresolved).

The last three decades of the twentieth century saw substantial advances in the understanding of neuroanatomic, neurophysiologic, and neuropharmacologic structures and functions likely to mediate both learned and unlearned visceral sensory events. These structures include specific receptors, chemicals, and pathways running from the visceral organs themselves throughout the neuraxis up to various regions of the cortex. Results of these studies have demonstrated substantial consistency in the involvement of particular central nervous system areas in visceral sensory processes, such as the nucleus of the solitary tract in the brainstem, various so-called limbic system structures, and the somatosensory, medial frontal, cingulate, and (especially) insular cortices. The existence of a sensory–motor closed-loop seems an essential necessity to normal interoceptive function. Finally, brain hemispheric laterality appears to be highly involved—both the hypothesis that more medial brain structures may be more involved with control of internal bodily processes, as well as the observation that some control appears to be lateralized (e.g., the apparent predominance of cardiac control mechanisms in the right hemisphere).

By far the most extensive studies of interoception have involved the heart, with the gastrointestinal tract second. Methodological issues are very important. For example, procedures designed to determine if subjects can recognize the occurrence of individual heartbeats are probably more valid than those that involve the counting of multiple beats, and verbal reports without accompanying measurements of actual heart action are fraught with potential distortions. A number of factors related to subject (individual differences), methods (independent variables), and outcome (dependent variables) affect results. Cardiovascular dynamics appear to be a very important contributor to results, and there is some reason to believe that what is sensed during cardiac interoception is actually blood flow out of the heart into the large blood vessels—that is, the interoceptors are in the blood vessels and not in the heart itself. Somatic mechanoreceptors in the chest wall may also play an important role. Interoceptive awareness is typically increased by any event or situation that increases heart action, including, for example, exercise and emotion. A few studies using primarily EEG or other functional imaging methods have indicated that the brain regions predicted by other research to be involved in interoceptive processes are, in fact, activated in these situations.

The other system that has been the main focus of study is the alimentary, or gastrointestinal, system. Pain has been the focus of several of these studies, but sensations that produce awareness without pain have also been studied. Studies of both healthy people and individuals with bowel disorders such as irritable bowel disorder have been reported. The use of imaging methods to study emotion generally and brain activation patterns during gastrointestinal sensation specifically became extensive during the 1990s. The data indicate that interoception can occur throughout the alimentary tract, from esophagus to rectum.

Other visceral organs and other sensory phenomena potentially related to interoception have not been systematically investigated but are likely to be important. Visceral pain occurs not only in the heart and alimentary system. For example, mesenteric ischemia, or distention of the capsule of the liver, bladder, or ureters, all cause visceral pain that can be excruciating. Further, some visceral sensations from these organs occur that are not painful, for example, less extensive bladder distention. Bladder distention can function as an interoceptive stimulus (Soldoff and Slucki, 1974). A further understanding of how visceral sensory–perceptual processes can malfunction should improve considerably our understanding of disorders observed in the cardiovascular, gastrointestinal, psychiatric, and other psychophysiologic and psychosomatic systems.

Beyond pain and other sensations from the visceral organs themselves, a broader definitional perspective might best be adopted. Rather than considering interoceptive processes, perhaps defining an overall bodily sense (or more than one—bodily senses) might be more appropriate. In other words, does it truly make psychobiological sense to limit the range of sensory inputs of importance to the bodily area from the neck down to the tops of the thighs, omitting the upper extremities? Would it not be more appropriate to define (as has been done by others) a bodily sense, including interoception, proprioception, labyrinthine function (i.e., the experience of the body in space), and other afferent information from the body? In other words, might it not make more sense to think of all the body outside the nervous system but under the skin as a source of sensory input, just as the external world provides input through the so-called five senses? Such a perspective would incorporate such things as phantom limb phenomena and stimuli produced by drugs or hormonal substances in the body and might be an impetus to further study and understanding of body image and body schema phenomena.

Despite the belief in some quarters that understanding consciousness is a scientifically intractable problem, some neuroscientists are showing increasing interest in neurophilosophical questions. A fundamental—indeed, possibly *the* fundamental—problem of consciousness is separating self from the world. It would seem that interoception should play a pivotal role in consciousness of self, thus serving to further general understanding of consciousness.

WHITHER?

The first section of this chapter selectively summarized important findings from the earlier chapters. In this section, further issues will be raised, specifically with an eye toward future potential problems to be investigated and resolved within the field of interoception and the development and application of new methodologies (e.g., functional imaging). Certainly, it is not complete. It is only this author's (partial) list. It is highly likely that every reader will come to this point with her or his own most pressing or interesting questions for study.

When Does Interoception Occur?

Is interoception a continuous, ongoing process? Undoubtedly, afferent impulses arise from the viscera continuously to provide ongoing orderly physiological functioning. Do some ascend in the nervous system to a level required for normal physiological function (e.g., baroreceptor impulses into the brainstem), but no further? Or do all go higher into the brain? And if they do, what are their purpose and effects on total organismic functioning and behavior? Most do not reach consciousness. Which do and which do not, and how is that controlled or determined? For example, do those that reach consciousness arise from different populations of interoceptors than those that do not (e.g., in the case of cardiac interoception, does awareness arise from visceral receptors in the heart or in the great vessels, or from non-visceral somatic receptors in the chest wall)? Is it a correct general principle, as it seems, that awareness is more likely to occur when organ function is not in the resting state than when it is (e.g., the pain of an obstructed small intestine or a pounding heart during fearful emotion)? In general, is interoceptive function influenced by natural fluctuations in visceral organ function, such as those that occur as part of circadian or other biorhythmic variations?

Interoception and Emotion

Interest in the explicit relationship between visceral sensory processes and psychobiological functions has been referred to the James–Lange theory. How close is this relationship, and how can it be studied further? It has already been described how emotion is closely linked to motivation and stress reactivity, and to a lesser but still very important extent, to cognition. Indeed, Damasio's "somatic marker" hypothesis, as well as the ideas of James, imply that without afferent sensory input from the body, no complete emotional experience could exist because the feeling part would be missing. For example, the "brain in a bottle" scenario from science fiction, whatever else its subjective existence would be like, would have no feelings (it would be, perhaps, "cruel and unfeeling"). However, as Damasio suggests, the bodily input might be "in the background." The bodily input cannot be discerned and distilled from the total experience, but without this visceral–somatic input, the emotional experience would be bland and fundamentally different.

How can this theory be tested? One method would be to remove or diminish visceral–somatic input to the brain. This can occur in several different ways. First, it can be done experimentally. A small number of investigations have been reported in which pharmacologic means were used to produce total autonomic, or panautonomic, blockade (e.g., Esler et al., 1977; Towler et al., 1993). However, this technique is very difficult to use, and studies directed at answering the question of the relationship between autonomic input and emotion have not been reported. Further, autonomic blockade, even if complete, does not block all possible visceral–somatic input to the brain.

A second method is to study individuals in whom visceral–somatic input has been damaged. Two such situations are damage to the spinal cord and autonomic failure due to some disease process. It appears that no systematic studies of people with autonomic failure from the perspective of understanding interoceptive processes have been reported, and the few results that have been reported are not consistent with the hypothesis. For example, Berrios et al. (1995) found that in individuals with Parkinson's disease (PD) who had autonomic dysfunction, ". . . 'Autonomic' symptoms . . . were more frequent in PD patients than in healthy controls . . ." (p. 789). Further, the same caveat applies to this strategy as to the strategy of using pharmacological means to block the autonomic nervous system—other non-autonomic sources of potential visceral–somatic sensory input to the central nervous system exist.

Another approach is the investigation of people with spinal cord injuries. This method has the same logical flaw, however, in that some of the visceral input to the brain (especially the vagus nerve) bypasses the spinal cord, entering the brain at the level of the brainstem. Despite this, several studies have been reported addressing the question of interoceptive processes in these individuals. No consistent conclusion can be drawn because results have been mixed. Some results were consistent with the hypothesis that reduced bodily input would reduce emotional experience (Hohmann, 1966; Jasnos and Hakmiller, 1975; Winters and Padilla, 1986; Critchley et al., 2001), while others found no impairment of emotional experience (Linton and Hirt, 1979; Nestoros et al., 1982; Heidbreder et al., 1984; Chwalisz et al., 1988; Lundqvist et al., 1991).

Thus, considering all these studies, no final conclusion can be drawn, and for the methodological reasons cited, results of these studies are of interest to this question but do not provide a rigorous test. New, more rigorous methods are needed. In the meantime, it seems fair to say that to understand behavior one must understand emotion, and to understand emotion one must understand interoception. And, as a corollary to this dictum, the whole issue of the potential role of interoceptive processes in what can be called positive emotions must be addressed. Is there a role for visceral–somatic input into the overall experiences of joy, contentment, and pleasant surprise?

Further Elucidation of the Neural Apparatus

A number of structures in the central nervous system have been implicated, directly or indirectly, in interoception. At this point it seems that three areas of research in this realm are of most immediate interest. The first area is to determine which structures are involved directly in the processing of visceral sensory impulses, how these structures relate to efferent motor functions, and which are indirectly involved (i.e., because they are directly involved in associated functions such as processing emotional information—for example, the amygdala, possibly).

The second area is to determine which visceral structures actually produce afferent neural information that can participate in interoceptive functions. While the cardiovascular–respiratory and alimentary systems have been most clearly implicated, other systems might be involved as well, especially those producing input that could affect behavior but may rarely if ever reach consciousness. Felten and colleagues (e.g., Fuller et al., 1981; Felten et al., 1987; Ackerman et al., 1989) demonstrated noradrenergic innervation of the liver, spleen, and immune tissue. Do any of these nerves carry sensory information?

The third area involves determination of the relationships and interconnections among inputs from specific visceral structures and other somatic structures that might contribute to either the interoceptive or, more broadly, the general somatic sense. These include, among other structures and receptors, information from muscles, joints, the labyrinth of the ear, and other contributors to the proprioceptive sense and general bodily sense. Further, where and what are the receptors that sense (in however vague a way) changes in concentrations of drugs or systemic endogenous substances—hormones, cytokines, and so on? Are they sensed in the periphery?

How Does Interoception Develop?

How did interoception develop phylogenetically, and how does it develop ontogenetically? One author (Porges, 1997; see Chapter 4) offered a theory of the phylogenetic development of emotion that ties it closely to the differentiation of function of the autonomic nervous system, linking it to visceral afferent processes and thus at least implicitly to interoception. Cabanac (1996, 1999; see Chapter 13) argued that emotion arose first in reptiles and was related to sensory processes, including visceral sensory processes, again implicating interoception, at least as it is experienced in organisms, at that level of evolutionary development. To the extent that questions such as the phylogenetic development of emotion are important, so would seem to be the question of the phylogenetic development of visceral sensation.

Ontogenetically, how does interoception develop? What is the role of biological maturation versus learning? As is often the case, it appears that both are important. For example, an essential part of bowel and bladder training in children involves the development of their abilities not only to voluntarily hold or release the appropriate muscles, but also to sense fullness of the bladder and rectum (interoceptive functions). And these skills involve a combination of maturation of the nervous system and social learning, suggesting that interoception might involve such a combination as well. Such phenomena as behavioral responses to emotional stimuli, which appear to include a combination of learning and unlearned fixed action patterns, reinforce the hypothesis that interoception as it relates to emotion depends on both learned and unlearned mechanisms.

Disorders of Interoception

In Chapters 8, 9, and 12, medical and psychiatric—psychosomatic—disorders that are likely to involve dysfunctions of visceral sensory–perceptual processes were reviewed. Fairly strong evidence exists for hypersensitivity of sensory functions in the alimentary tract in several gastrointestinal disorders. The evidence is suggestive in the cardiovascular–respiratory system but not as strong, as well as in several psychiatric disorders, especially anxiety and somatoform disorders. This area of research is potentially especially fertile, particularly if the most up-to-date methods are used, including a replacement of sole dependence on verbal reports of symptoms with more sophisticated techniques of relating symptom reports to actual physiological events in the viscera and the central nervous system. Findings should lead to substantial improvements in treatment, especially as empirical findings gradually undermine common tendencies to view some disorders (e.g., irritable bowel syndrome) as functional, that is, not fully legitimate or real.

A new medical perspective is called for. While proceeding from a perspective of interest in individual symptoms to the broader, more unifying perspective of syndromes and diseases has substantially advanced understanding in all branches of medicine, the need to understand the patho-psychobiology of individual symptoms per se has not disappeared. Symptoms and syndromes do not co-vary perfectly. A given symptom can occur in many syndromes, and a given syndrome does not always have the same symptoms among people or even within an individual over time. A better understanding of the etiopathology of symptoms qua symptoms is needed.

Consciousness and Unconsciousness

The historical belief in the rational man (or woman) within Western thought has obscured for many centuries an appreciation of the fact that most nervous system activity, even the higher mental processes, occurs outside awareness. Whatever function consciousness plays in the behavior of humans and other organisms, however important it is and no matter how it evolved, it is only a part of all that is going on in the brain, a relatively thin veneer in the overall functions of cognition, motivation–emotion, and behavior. Just as other activities of the brain that are fundamental to the organism's higher mental processes usually occur outside awareness, does it not seem plausible and even probable that the same is true of visceral sensory–perceptual processes? Lack of awareness is not evidence of lack of importance, lack of awareness of important processes demands that other methods be used to distinguish between that that is occurring outside of consciousness from that that is not occurring at all. As already argued, it is time to broaden the definition of interoception to include these putative events occurring outside consciousness, and it is also time to devise the methods and means to study these events.

Molecular and Total Organismic Mechanisms

What are the potential roles of molecular biology and genetics in understanding interoception? This chapter was written within a fortnight after the formal announcement of the completion of the Human Genome Project. Despite that, specific discussions of molecular mechanisms and genetics (or at least familial transmission of traits) appear only in very limited ways in this volume. This reflects, to a considerable extent, the fact that the field has not yet matured to the point that such studies have been done, but it also reflects another issue, that is, the relationship between total organismic and molecular mechanisms.

Science is largely the success of reductionism and analysis—breaking natural phenomena down into smaller and smaller parts to understand them. However, the logic of reductionism does not inevitably lead back to synthesis—to the unified functioning of the whole. It is not the case that, once the reductionist agenda is complete, either the large structure will be clear or that effort can and will be put into making it clear. Furthermore, it seems unlikely that the reductionist agenda will be complete any time in the foreseeable future. For all the success of reductionist molecular analyses of biological phenomena, understanding of many phenomena will require not a "bottom-up" but rather a "top-down" approach. Interoception, as part of perception, is an integrated phenomenon that occurs only in whole organisms and must be studied at that level.

Why?

The final issues, the final questions, are, Why does interoception exist, what function does it serve, why did it evolve, and what selective reproductive advantage does it provide? The answer, of course, is not known, but one hypothesis can be offered. As James and Cannon understood, visceral afferent information is an essential part of emotion, of motivation, of the "what," but especially of the "why" of behavior. We act because we feel and because we are ready for fight-or-flight—for action. Whether aware or unaware of our visceral functions, they are nonetheless telling us effectively and efficiently when to act, often without our even taking (wasting) the time consciously and rationally to think about it. To paraphrase, as Blaise Pascal said more than 300 hundred years ago, "the heart"—and the rest of the viscera, the visceral autonomic nervous system, and the body in toto—"has its reasons which reason"—and logic and rational thought—"knows nothing of."

REFERENCES

Abboud, F. M. Ventricular syncope: Is the heart a sensory organ? N. Engl. J. Med. 320; 390–2, 1989.

Abelson, J. L., D. Glitz, O. G. Cameron, M. A. Lee, M. Bronzo, G. C. Curtis. Blunted growth hormone response to clonidine in patients with generalized anxiety disorder. Arch. Gen. Psychiatry 48; 157–162, 1991.

Abelson, J. L., D. Glitz, O. G. Cameron, M. A. Lee, M. Bronzo, G. C. Curtis. Endocrine, cardiovascular, and behavioral responses to clonidine in patients with panic disorder. Biol. Psychiatry 32; 18–25, 1992.

Abelson, J. L., R. M. Nesse, J. G. Weg, G. C. Curtis. Respiratory psychophysiology and anxiety: Cognitive intervention in the doxapram model of panic. Psychosom. Med. 58; 302–13, 1996.

Ackerman, K. D., S. Y. Felten, C. D. Dijkstra, S. Livnat, D. L. Felten Parallel development of noradrenergic innervation and cellular compartmentation in the rat spleen. Exper. Neurol. 103; 239–55, 1989.

Adam, G. Interoception and behavior: An experimental study. Budapest, Akademiai Kiado (Publishing House of the Hungarian Academy of Sciences), 1967.

Adam, G. Visceral perception: Understanding internal cognition. New York, Plenum Press, 1998.

Adam, G., W. R. Adey, R. W. Porter. Interoceptive conditional response in cortical neurones. Nature 209; 920–1, 1966.

Ader, R. Conditioned adrenocortical steroid elevations in the rat. J. Comp. Physiol. Psychol. 90; 1156–63, 1976.

Ader, R., N. Cohen. Behaviorally conditioned immunosuppression. Psychosom. Med. 37; 333–40, 1975.

Ader, R., N. Cohen. Behaviorally conditioned immunosuppression and murine systemic lupus erythematosus. Science 215; 1534–6, 1982.

Adler, G., W. F. Gattaz. Pain perception threshold in major depression. Biol. Psychiatry 34; 687–9, 1993.

Adolphs, R. The human amygdala and emotion. The Neuroscientist 5; 125–137, 1999.

Aggleton, J. P. The contribution of the amygdala to normal and abnormal emotional states. Trends Neurosci. 16; 328–33, 1993.

Aguero, A., M. Gallo, M. Arnedo, F. Molina, A. Puerto. The functional relevance of medial parabrachial nucleus in intragastric sodium chloride-induced short-term (concurrent) aversion. Neurobiol. Learn. Memory 67; 161–6, 1997.

Allen, G. V., C. B. Saper, K. M., Hurley, D. F. Cechetto. Organization of visceral and limbic connections in the insular cortex of the rat. J. Comp. Neurol. 311; 1–16, 1991.

Amann, J. F., G. M. Constantinescu. The anatomy of the visceral and autonomic nervous systems. Sem. Veterinary Med. Surg. 5; 4–11, 1990.

Amar, A. P., C.N. Heck, M. L. Levy, T. Smith, C. M. DeGiorgio, S. Oviedo, M. L. Apuzzo. An institutional experience with cervical vagus nerve truck stimulation for medically refractory epilepsy: Rational, technique, and outcome. Neurosurgery 43; 1265–76, 1998.

American Psychiatric Association. Diagnostic and statistical manual of mental disorders, third edition (DSM-III), 1980.

American Psychiatric Association. Diagnostic and statistical manual of mental disorders, third edition (DSM-III-R), 1987.

American Psychiatric Association. Diagnostic and statistical manual of mental disorders, fourth edition (DSM-IV), 1994.

Ando, K. The discriminative control of operant behavior by intravenous administration of drugs in rats. Psychopharmacologia 45; 47–50, 1975.

Andreasen, N. C. (Ed.). Brain imaging: Applications in psychiatry. Washington, D.C., American Psychiatric Press, 1989.

Andersson, J. L. R. Within-study repeated measurements to increase sensitivity for positron emission tomography activation studies. J. Cereb. Blood Flow Metab. 18; 319–31, 1998.

Antony, M. M., T. A. Brown, M. G. Craske, D. H. Barlow, W. B. Mitchell, E. A. Meadows. Accuracy of heartbeat perception in panic disorder, social phobia, and nonanxious subjects. J. Anx. Disord. 9; 355–71, 1995.

Antony, M. M., E. M. Meadows, T. A. Brown, D. H. Barlow. Cardiac awareness before and after cognitive-behavioral treatment for panic disorder. J. Anx. Disord. 8; 341–50, 1994.

Antrobus, J. S., J. S. Antrobus. Discrimination of two sleep stages by human subjects. Psychophysiology 4; 48–55, 1967.

Appenzeller, O. The autonomic nervous system, fourth edition: An introduction to basic and clinical concepts. Amsterdam, Elsevier, 1990.

Armony, J. L., D. Servan-Schrieber, J. D. Cohen, J. E. LeDoux. Computational modeling in emotion: Explorations through the anatomy and physiology of fear conditioning. Trends Cogn. Sci. 1; 28–34, 1997.

Armour, J. A., J. L. Ardell. Neurocardiology. New York, Oxford University Press, 1994.

Asmundson, G. I., N. J. Norton, G. R. Norton. Beyond pain: The role of fear and avoidance in chronicity. Clin. Psychol. Rev. 19; 97–119, 1999.

Asmundson, G. I., L. S. Sandler, K. G. Wilson, G. R. Norton. Panic attacks and interoceptive acuity for cardiac sensations. Behav. Res. Therap. 31; 193–7, 1993.

Asmundson, G. I., M. B. Stein. Vagal attenuation in panic disorder: An assessment of parasympathetic nervous system function and subjective reactivity to respiratory manipulation. Psychosom. Med. 56; 187–93, 1994.

Aston-Jones, G. S., R. Desimone, J. Driver, S. J. Luck, M. I. Posner. Attention. In M. J. Zigmond, F. E. Bloom, S. C. Landis, J. L. Roberts, L. R. Squire (Eds.). Fundamental neuroscience. San Diego, Academic Press, 1999, pp. 1385–409.

Asratyan, E. A. The initiation and localization of cortical inhibition in the conditioned reflex arc. Ann. N. Y. Acad. Sci. 92; 1141–59, 1961.

Augustine, J. R. The insular lobe in primates including humans. Neurol. Res. 7; 2–10, 1985.

Augustine, J. R. Circuitry and functional aspects of the insular lobe in primates including humans. Brain Res. Rev. 22; 229–44, 1996.

Aziz, Q., J. L. R. Anderson, S. Valind, A. Sundin, S. Hamdy, A. K. P. Jones, A. K. Foster, E. R. Langstrom, D. G Thompson. Identification of human brain loci processing esophageal sensation using positron emission tomography. Gastroenterology 113; 50–9, 1997.

Aziz, O., D. G. Thompson, V. W. Ng, S. Hamdy, S. Sarkar, M. J. Brammer, E. T. Bullmore, A. Hobson, I. Tracey, L. Gregory, A. Simmons, S. C. Williams. Cortical processing of human somatic and visceral sensation. J. Neurosci. 20; 2657–63, 2000.

Baars, B. J. Metaphors of consciousness and attention in the brain. Trends Neurosci. 21; 58–62, 1998.

Bachman, J. A., N. L. Benowitz, R. I. Herning, R. T. Jones. Dissociation of autonomic and cognitive effects of THC in man. Psychopharmacology 61; 171–5, 1979.

Bailey, R. I., L. W. Wootten. Pain, depression and phobia: A biochemical hypothesis. J. Int. Med. Res. 4 (suppl. 2); 73–80, 1976.

Baker, S. C., C. D. Frith, R. J. Dolan. The interaction between mood and cognitive function studied with PET. Psychol. Med. 27; 565–78, 1997.

Baklavadzhian, O. G., E. A. Avetisian, K. G. Bagdasarian, V. S. Eganova, L. B. Nersesian. The neuronal organization of the amygdala-visceral reflex arch. [Russian] Uspekhi Fiziologicheskikh Nauk. 27; 51–77, 1996.

Baldwin, H. A., G. F. Koob. Rapid induction of conditioned opiate withdrawal in the rat. Neuropsychopharmacology 8; 15–21, 1993.

Balon, R., A. Ortiz, R. Pohl, V. K. Yeragani. Heart rate and blood pressure during placebo-associated panic attacks. Psychosom. Med. 50; 434–8, 1988.

Balon, R., V. K. Yeragani, R. Pohl, J. Muench, R. Berchou. Somatic and psychological symptoms during isoproterenol-induced panic attacks. Psychiatry Res. 32; 103–12, 1990.

Bandelow, B., M. Amering, O. Benkart, I. Marks, A. E. Nardi, M. Osterheider, C. Tannock, J. Tremper, M. Versiani. Cardio-respiratory and other symptom clusters in panic disorder. Anxiety 2; 99–101, 1996.

Banuazizi, A. Discriminative shock avoidance learning of an autonomic response under curare. J. Comp. Physiol. Psychol. 69; 236–46, 1972.

Barnes, C. D., O. Pompeiano (Eds.). Neurobiology of the locus coeruleus: Progress in brain research, Vol. 88. Amsterdam, Elsevier, 1991.

Barsky, A. J. Palpitations, cardiac awareness, and panic disorder. Am. J. Med. 92 (suppl. 1A); 31S-4S, 1992.

Barsky, A. J., D. K. Ahern, J. Brener, O. S. Surman, C. Ring, G. W. Dec. Palpitations and cardiac awareness after heart transplantation. Psychosom. Med. 60; 557–562, 1998.

Barsky, A. J., M. C. Barnett, P. D. Cleary. Hypochondriasis and panic disorder: Boundary and overlap. Arch. Gen. Psychiatry 51; 918–25, 1994.

Barsky A. J., B. Hochstrasser, A. Coles, J. Zisfein, C. O'Donnell, K. A. Eagle. Silent myocardial ischemia: Is the person or the event silent? JAMA 264; 1132–35, 1990.

Basha, I., V. Mukerji, P. Langevin, M. Kushner, M. Alpert, B. D. Beitman. Atypical angina in patients with coronary artery disease suggests panic disorder. Internat. J. Psychiatry Med. 19; 341–6, 1989.

Basmajian, J. V. Control and training of individual motor units. Science, 141; 440–1, 1963.

Bass, C., R. Cawley, C. Wade, K. C. Ryan, W. N. Gardner, D. C. S. Hutchison, G. Jackson. Unexplained breathlessness and psychiatric morbidity in patients with normal and abnormal coronary arteries. Lancet 1; 605–9, 1983.

Bass, C., W. Gardner. Emotional influences on breathing and breathlessness. J. Psychosom. Res. 29; 599–609, 1985.

Bass, C., P. Lelliott, I. Marks. Fear talk versus voluntary hyperventilation in agoraphobics and normals: A controlled study. Psychol. Med. 19; 669–76, 1989.

Battaglia, M., G. Perna. The 35% CO_2 challenge in panic disorder: Optimization by receiver operating characteristic (ROC) analysis. J. Psychiatr. Res. 29; 111–9, 1995.

Baumann, L. J., H. Leventhal. I can tell when my blood pressure is up, can't I? Health Psychol. 4; 203–18, 1985.

Baxter, L. R., J. M. Schwartz, J. C. Mazziota, M. E. Phelps, J. J. Pahl, B. H. Guze, L. Fairbanks. Cerebral glucose metabolic rate in nondepressed patients with obsessive-compulsive disorder. Am. J. Psychiatry 145; 1560–3, 1988.

Beatty, J., C. Kornfeld. Relative independence of conditioned EEG changes from cardiac and respiratory activity. Physiol. Behav. 9; 733–6, 1973.

Bechara, A., D. Tranel, H. Damasio, R. Adolphs, C. Rockland, A. R. Damasio. Double dissociation of conditioning and declarative knowledge relative to the amygdala and hippocampus in humans. Science 269 ; 1115–8, 1995.

Bechbache, R. R., H. H. K. Chow, J. Duffin, E. C. Orsini. The effects of hypercapnia, hypoxia, exercise, and anxiety on the pattern of breathing in man. J. Physiol. 293; 285–300, 1979.

Beck, A. T., R. Laude, M. Bohnert. Ideational components of anxiety neurosis. Arch. Gen. Psychiatry 31; 319–5, 1974.

Beck, J. G., J. C. Shipherd, B. J. Zebb. How does interoceptive exposure for panic disorder work? An uncontrolled case study. J. Anx. Disord. 11; 541–56, 1997.

Beggs, J. M., T. H. Brown, J. H. Byrne, T. Crow, J. E. LeDoux, K. LeBar, R. F. Thompson. Learning and memory: Basic mechanisms. In M. J. Zigmond, F. E. Bloom, S. C. Landis, J. L. Roberts, L. R. Squire (Eds.). Fundamental neuroscience. San Diego, Academic Press, 1999, pp. 1411–54.

Beitman, B. D., I. Basha, G. Flaker, L. DeRosear, V. Mukerji, J. W. Lamberti. Major depression in cardiology chest pain patients without coronary artery disease and with panic disorder. J. Affect. Disord. 13; 51–9, 1987.

Beitman, B. D., V. Mukerji, J. W. Lamberti, L. Schmid, L. DeRosear, M. Kushner, G. Flaker, I. Basha. Panic disorder in patients with chest pain and angiographically normal coronary arteries. Am. J. Cardiol. 63; 1399–1403, 1989.

Beliaeva, G. S. The neurophysiological mechanisms of the perception of interoceptive signals. Fiziologicheskii Zhurnal SSSR Imeni I. M. Sechenova 76; 855–62, 1990.

Bellodi, L., G. Perna, D. Caldirola, C. Arancio, A. Bertani, D. DiBella. CO_2-induced panic attacks: A twin study. Am. J. Psychiatry 155; 1184–88, 1998.

Belyaeva, G. S. Some neurophysiological mechanisms of the perception of interoceptive signals. Neurosci. Behav. Physiol. 21; 441–7, 1991.

Benarroch, E. E. Central autonomic network: Functional organization and clinical correlations. Armonk, N.Y., Futura Publishing Company, Inc., 1997.

Benkelfat, C., J. Bradwejn, E. Meyer, M. Ellenbogen, S. Milot, A. Gjedde, A. Evans. Functional neuroanatomy of CCK4-induced anxiety in normal health volunteers. Am. J. Psychiatry 152; 1180–4, 1995.

Bennett, D. A., H. Lal. Discriminative stimuli produced by clonidine: An investigation of the possible relationships to adrenoceptor stimulation and hypotension. J. Pharmacol. Exper. Therap. 223; 642–8, 1982.

Bennett, R. M. Emerging concepts in the neurobiology of chronic pain: Evidence of abnormal sensory processing in fibromyalgia. Mayo Clin. Proc. 74; 385–98, 1999.

Bergman, J. S., H. J. Johnson. The effects of instructional set and autonomic perception on cardiac control. Psychophysiology 8; 180–90, 1971.

Bergman, J. S., H. J. Johnson. Sources of information which affect training and raising of heart rate. Psychophysiology 9; 10–39, 1972.

Berlin, I., A. Grimaldi, C. Payan, C. Sachon, F. Bosquet, F. Thervet, A. J. Puech. Hypoglycemic symptoms and decreased beta-adrenergic sensitivity in insulin-dependent diabetic patients. Diabetes Care 10; 742–7, 1987.

Berlin, I., A. Grimaldi, C. Landault, F. Zoghbi, F. Thervet, A. J. Puech, J. C. Legrand. Lack of hypoglycemic symptoms and decreased beta-sensitivity in insulin-dependent diabetic patients. J. Clin. Endocrinol. Metab. 66; 273–8, 1988.

Berlucchi, G., S. Aglioti. The body in the brain: Neural basis of corporeal awareness. Trends Neurosci. 20; 560–4, 1997.

Bermudez, J. L., A. Marcel, N. Eilan (Eds.). The body and the self. Cambridge, MA, The MIT Press, 1995.

Bermudez-Rattoni, F., J. L. McGaugh. Insular cortex and amygdala lesions differentially affect acquisition on inhibitory avoidance and conditioned taste aversion. Brain Res. 549; 165–70, 1991.

Bernstein, I. L. Learned taste aversions in children receiving chemotherapy. Science 200; 1302–3, 1978.

Bernstein, I. L. Taste aversion learning: A contemporary perspective. Nutrition 15; 229–34, 1999.

Berntson, G. G., S. Hart, M. Sarter. The cardiovascular startle response: Anxiety and the benzodiazepine receptor complex. Psychophysiology 34; 348–57, 1997.

Berrios, G. E., C. Campbell, B. E. Politynska. Autonomic failure, depression and anxiety in Parkinson's disease. Br. J. Psychiatry 166; 789–92, 1995.

Bilder, R. M., LeFever, F. F. (Eds.). Neuroscience of the mind on the centennial of Freud's project for a scientific psychology. Ann. N.Y. Acad. Sci., Vol. 843, 1998.

Binkofski, F., A. Schnitzler, P. Enck, T. Frieling, S. Posse, R. J. Seitz, H. J. Freund. Somatic and limbic cortex activation: A functional magnetic resonance imaging study. Ann. Neurol. 44; 811–5, 1998.

Birbaumer, N., W. Grodd, O. Diedrich, U. Klose, M. Erb, M. Lotze, F. Schneider, U. Weiss, H. Flor. fMRI reveals amygdala activation to human faces in social phobics. Neuroreport 9; 1223–6, 1998.

Birbaumer, N., W. Lutzenberger, P. Montoya, W. Larbig, K. Unertl, S. Topfner, W. Grodd, E. Taub, H. Flor. Effects of regional anesthesia on phantom limb pain are mirrored in changes in cortical reorganization. J. Neurosci. 17; 5503–8, 1997.

Birk, L., A. B. Crider, D. Shapiro, B. Tursky. Operant electrodermal conditioning under partial curarization. J. Comp. Physiol. Psychol. 62; 165–6, 1966.

Birket-Smith, M., N. Hasle, H. H. Jensen. Electrodermal activity in anxiety disorders. Acta Psychiatr. Scand. 88; 350–5, 1993.

Bisaga, A., J. L. Katz, A. Antonini, C. E. Wright, C. Margouleff, J. M. Gorman, D. Eidelberg. Cerebral glucose metabolism in women with panic disorder. Am. J. Psychiatry 155; 1178–83, 1998.

Black, A. H. Transfer following operant conditioning in the curarized dog. Science 155; 201–3, 1967.

Black, A. H. Autonomic aversive conditioning in infrahuman subjects. In Brush, F. R. (Ed.). Aversive conditioning and learning. New York, Academic Press, 1971, pp. 3–104.

Black, A. H. The operant conditioning of central nervous system electrical activity. In Shapiro, D., T. X. Barber, L. V. DiCara, J. Kamiya, N. E. Miller, J. Stoyva (Eds.) Biofeedback and self-control 1972. Chicago, Aldine Publishing Company, 1972, pp. 96–142.

Black, A., G. A. Young, C. Batenchuk. Avoidance training of hippocampal theta waves in Flaxedilized dogs and its relation to skeletal movement. J. Comp. Physiol. Psychol. 70; 15–24, 1970.

Bleecker, E. R., B. T. Engel. Learned control of ventricular rate in patients with atrial fibrillation. Psychosom. Med. 35; 161–70, 1973.

Blessing, W. W. Inadequate frameworks for understanding bodily homeostasis. Trends Neurosci. 20; 235– 9, 1997.

Block, N. How can we find the neural correlate of consciousness? Trends Neurosci. 19; 456–9, 1996.

Bloedel, J. R., V. Bracha. On the cerebellum, cuteomuscular reflexes, movement control and the elusive engrams of memory. Behav. Brain Res. 68; 1–44, 1995.

Blohme, G., I. Lager, P. Lonnroth, U. Smith. Hypoglycemic symptoms in insulin-dependent diabetics: A prospective study of the influence of beta-blockade. Diabete Metabolisme 7; 235–8, 1981.

Blomfield, O. H .D. Encounter with neurophilosophy. Austral. New Zealand J. Psychiatry 26; 277–83, 1992.

Bluthe, R. M., R. Dantzer, M. Le Moal. Peripheral injections of vasopressin control behavior by way of interoceptive signals for hypertension. Behav. Brain Res. 18; 31–9, 1985.

Bond, A. J., D. C. James, M. H. Lader. Physiological and psychological measures in anxious patients. Psychol. Med. 4; 364–73, 1974.

Boring, E.G. The sensation of the alimentary canal. Am. J. Psychol. 26; 1–57, 1915.

Bouhassira, D., R. Chollet, B. Coffin, M. Lemann, D. LeBars, J. C. Willer, R. Jian. Inhibition of a somatic nociceptive reflex by gastric distention in humans. Gastroenterology 107; 985–92, 1994.

Bouras, E. P., T. J. O'Brien, M. Camilleri, M. K. O'Connor, B. P. Mullan Cerebral tomography of rectal stimulation using single photon emission computed tomography. Am. J. Physiol. 277; G687–94, 1999.

Bower, G. H., K. P. Monteiro, S. G. Gilligan. Emotional mood as a context for learning and recall. J. Verbal Learn. Verbal Behav. 17; 573–85, 1978.

Brady, J. P., L. Luborsky, R. E. Kron. Blood pressure reduction in patients with essential hypertension through metronome-conditioned relaxation: A preliminary report. Behav. Therap. 5; 203–9, 1974.

Bremner, J. D., R. B. Innis, C. K. Ng, L. H. Staib, R. M. Salomon, R. A. Bronen, J. Duncan, S. M. Southwick, J. H. Krystal, D. Rich, G. Zubal, H. Dey, R. Soufer, D. S. Charney. Positron emission tomography measurement of cerebral metabolic correlates of yohimbine administration in combat-related posttraumatic stress disorder. Arch. Gen. Psychiatry 54; 246–54, 1997.

Bremner, J. D., J. H. Krystal, S. M. Southwick, D. S. Charney. Noradrenergic mechanisms in stress and anxiety: I. Preclinical studies. Synapse 23; 28–38, 1996a.

Bremner, J. D., J. H. Krystal, S. M. Southwick, D. S. Charney. Noradrenergic mechanisms in stress and anxiety: II. Clinical studies. Synapse 23; 39–51, 1996b.

Brener, J. A. A general model of voluntary control applied to the phenomena of learned cardiovascular change. In P. A. Obrist, A. H. Black, J. Brener, L. V. DiCara (Eds.). Cardiovascular psychophysiology: Current issues in response mechanisms, biofeedback, and methodology. Chicago, Aldine, 1974, pp. 365–91.

Brener, J. Control of internal activities. Br. Med. J. 37; 169–74, 1981.

Brener, J., C. Ring. Sensory and perceptual factors in heart beat detection. In D. Vaitl, R. Schandry (Eds.). From the heart to the brain: The psychophysiology of circulation-brain interaction. Frankfurt, Peter Lang, 1995, pp. 193–221.

Bridger, W. H., S. R. Schiff, S. S. Cooper, W. Paredes, G. A. Barr. Classical conditioning of cocaine's stimulatory effects. Psychopharmacol. Bull. 18; 210–4, 1982.

Bridges, P. K., M. T. James, D. Leak. A comparative study of four physiological concomitants of anxiety. Arch. Gen. Psychiatry 19; 141–5, 1968.

Brillman, J. Neurocardiology. Neurological Clinics of North America 11; 239–496, 1993.

Brodsky, M. A., D. A. Sato, L. T. Iseri, L. J. Wolff, B. J. Allen. Ventricular tachyarrhythmia associated with psychological stress: The role of the sympathetic nervous system. JAMA 257; 2064–7, 1987.

Brodsky, M., D. Wu, P. Denes, C. Kanakis, K. M. Rosen. Arrhythmias documented by 24 hour continuous electrocardiographic monitoring in 50 male medical students without apparent heart disease. Am. J. Cardiol. 39; 390–5, 1977.

Broersen, L. M., F. Abbate, M. G. Feenstra, J. P. de Bruin, R. P. Heinsbroek, B. Olivier Prefrontal dopamine is directly involved in the anxiogenic interoceptive cue of pentylenetetrazol but not in the interoceptive cue of chlordiazepoxide in the rat. Psychopharmacology 149; 366–76, 2000.

Broman, J. Neurotransmitters in subcortical somatosensory pathways. Anat. Embryol. 189; 181–214, 1994.

Brondolo, E., R. C. Rosen, J. B. Kostis, J. E. Schwartz. Relationship of physical symptoms and mood to perceived and actual blood pressure in hypertensive men: A repeated-measures design. Psychosom. Med. 61; 311–8, 1999.

Brooks, C. M., K. Koizumi, A. Sato (Eds.). Integrative functions of the autonomic nervous system. New York, Elsevier/North-Holland Biomedical Press, 1979.

Brown, B. B. Recognition of aspects of consciousness through association with EEG alpha activity represented by a light signal. Psychophysiology 6; 442–52, 1970.

Brown, C. C., R. A. Katz. Operant salivary conditioning in man. Psychophysiology 4; 156–60, 1967.

Browne, R. G. "Anxiolytics antagonize yohimbine's discriminative stimulus properties. Psychopharmacology 74; 245–9, 1981.

Bruder, G. E., R. Fong, C. E. Tenke, P. Leite, J. P. Towey, J. E. Stewart, P. J. McGrath, F. M. Quitkin. Regional brain asymmetries in major depression with or without an anxiety disorder: A quantitative electroencephalographic study. Biol. Psychiatry 41; 939–48, 1997.

Bruehl, S., J. A. McCubbin, R. N. Harden. Theoretical review: Altered pain regulatory systems in chronic pain. Neurosci. Biobehav. Rev. 23; 877–90, 1999.

Bruggemann, J., T. Shi, A. V. Apkarian. Viscero-somatic neurons in the primary somatosensory cortex (SI) of the squirrel monkey. Brain Res. 756; 297–300, 1997.

Bruggemann, J., T. Shi, A. V. Apkarian. Viscerosomatic interactions in the thalamic ventral posterolateral nucleus (VPL) of the squirrel monkey. Brain Res. 787; 269–76, 1998.

Buchsbaum, M. S., H. H. Holcomb, J. Johnson, A. C. King, R. Kessler. Cerebral metabolic consequences of electrical cutaneous stimulation in normal individuals. Human Neurobiol. 2; 35–8, 1983.

Bueno, L., J. Fioramonti, M. Delvaux, J. Frexinos. Mediators and pharmacology of visceral sensitivity: From basic to clinical investigations. Gastroenterology 112; 1714–43, 1997.

Burstein, R. Somatosensory and visceral input to the hypothalamus and limbic system. In G. Holstege, R. Bandler, C. S. Saper (Eds.). The emotional motor system: Progress in brain research, vol. 107. Amsterdam, Elsevier, 1996, pp. 257–67.

Butscher, D. S., G. E. Miller. The relationship of cognitively induced anxiety and hand temperature reduction. J. Psychosom. Res. 24; 131–6, 1980.

Butcher, K. S., D. F. Cechetto. Receptors in lateral hypothalamic area involved in insular cortex sympathetic responses. Am. J. Physiol. 275; H689–96, 1998.

Bykov, K. M. The cerebral cortex and the internal organs. New York, Chemical Publishing Company, 1957.

Bykov, K. M., I. T. Kurtsin. The corticovisceral theory of the pathogenesis of peptic ulcer. Oxford, Pergamon Press, 1966.

Byrne, D. G., R. H. Rosenman (Eds.). Anxiety and the heart. New York, Hemisphere Publishing Company, 1990.

Bystritsky, A., M. Craske, E. Maidenberg, T. Vapnik, D. Shapiro. Ambulatory monitoring of panic patients during regular activity: A preliminary report. Biol. Psychiatry 38; 684–9, 1995.

Bystritsky, A., M. Craske, E. Maidenberg, T. Vapnik, D. Shapiro. Autonomic reactivity of panic patients during a CO_2 inhalation procedure. Depr. Anx. 11; 15–26, 2000.

Cabanac, M. On the origin of consciousness, a postulate and its corollary. Neurosci. Biobehav. Rev. 20; 33–40, 1996.

Cabanac, M. Emotion and phylogeny. Jap. J. Physiol. 49; 1–10, 1999.

Cahill, L., R. J. Haier, J. Fallon, M. T. Alkire, C. Tang, D. Keator, J. Wu, J. L. McGaugh. Amygdala activity at encoding correlated with long-term, free recall of emotional information. Proc. Nat. Acad. Sci. 93; 8016–21, 1996.

Cameron, O. G. Beta-adrenergic blockade does not prevent hypoglycemia awareness in non-diabetic humans. Psychosom. Med. 51; 165–72, 1989.

Cameron, O. G. The symptoms of adrenergic activation. In O. G. Cameron (Ed.). Adrenergic dysfunction and psychobiology. Washington, D.C., American Psychiatric Press, Inc., 1994, pp. 237–55.

Cameron, O. G. The function of consciousness. Trends Neurosci. 21; 201, 1998.

Cameron, O. G., J. B. Appel. Conditioned suppression of bar-pressing behavior by stimuli associated with drugs. J. Exper. Anal. Behav. 17; 127–37, 1972.

Cameron, O. G., J. B. Appel. A behavioral and pharmacological analysis of some discriminable properties of d-LSD in rats. Psychopharmacologia 33; 117–34, 1973.

Cameron, O. G., J. B. Appel. Drug-induced conditioned suppression: Specificity due to drug employed as UCS. Pharmacol. Biochem. Behav. 4; 221–4, 1976.

Cameron, O. G., R. Buzan, D. S. McCann. Symptoms of insulin-induced hypoglycemia in normal subjects. J. Psychosom. Res. 32; 41–9, 1988.

Cameron, O. G., S. Gunsher, M. Hariharan. Venous plasma epinephrine levels and the symptoms of stress. Psychosom. Med. 52; 411–24, 1990a.

Cameron, O. G., M. A. Lee, G. C. Curtis, D. S. McCann. Endocrine and physiological changes during 'spontaneous' panic attacks. Psychoneuroendocrinology 12; 321–31, 1987.

Cameron, O. G., J. G. Modell, M. Hariharan. Caffeine and human cerebral blood flow: A positron emission tomography study. Life Sci. 47; 1141–6, 1990b.

Cameron, O. G., R. M. Nesse. Systemic hormonal and physiological abnormalities in anxiety disorders. Psychoneuroendocrinology 13; 287–307, 1988.

Cameron, O. G., C. B. Smith, M. A. Lee, P. J. Hollingsworth, E. M. Hill, G. C. Curtis. Adrenergic status in anxiety disorders: Platelet alpha2-adrenergic receptor binding, blood pressure, pulse, and plasma catecholamines in panic and generalized anxiety disorder patients and in normal subjects. Biol. Psychiatry 28; 3–20, 1990c.

Cameron, O. G., C. B. Smith, R. M. Nesse, E. M. Hill, P. J. Hollingsworth, J. A. Abelson, M. Hariharan, G. C. Curtis. Platelet alpha2-adrenoreceptors, catecholamines, hemodynamic variables, and anxiety in panic patients and their asymptomatic relatives. Psychosom. Med. 58; 289–310, 1996.

Cameron, O. G., B. A. Thyer, R. M. Nesse, G. C. Curtis. Symptom profiles of patients with DSM-III anxiety disorders. Am. J. Psychiatry 143; 1132–7, 1986.

Cameron, O. G., J. K. Zubieta, L. Grunhaus, S. Minoshima. Effects of yohimbine of cerebral blood flow, symptoms, and physiological functions in humans. Psychosom. Med. 62; 549–59, 2000.

Canavero, S. Dynamic reverberation: A unified mechanism for central and phantom pain. Med. Hypotheses 42; 203–7, 1994.

Canli, T. Hemispheric asymmetry in the experience of emotion: A perspective from functional imaging. The Neuroscientist 5; 201–7, 1999.

Cannon, R. O. The sensitive heart: A syndrome of abnormal cardiac pain perception. JAMA 273; 883–7, 1995.

Cannon, W. B. Bodily changes in pain, hunger, fear and rage. Boston, Charles T. Branford Company, 1953.

Cantril, H., W. A. Hunt. Emotional effects produced by injection of adrenalin. Amer. J. Psychol. 44; 300–7, 1932.

Card, J. P., L. W. Swanson, R. Y. Moore. The hypothalamus: An overview of regulatory systems. In M. J. Zigmond, F. E. Bloom, S. C. Landis, J. L. Roberts, L. R. Squire (Eds.). Fundamental neuroscience. San Diego, Academic Press, 1999, pp. 1013–26.

Carroll, D. Cardiac perception and cardiac control. Biofeedback Self Regul. 2; 349–69, 1977.

Casey, K. L. Forebrain mechanisms of nociception and pain: Analysis through imaging. Proc. Nat. Acad. Sci. 96; 7668–74, 1999.

Casey, K. L., S. Minoshima, K. L. Berger, R. A. Koeppe, T. J. Morrow, K. A. Frey. Positron emission tomographic analysis of cerebral structures activated specifically by repetitive noxious heat stimuli. J. Neurophysiol. 71; 802–7, 1994.

Cechetto, D. F. Central representation of visceral function. Fed. Proc. 46; 17–23, 1987.

Cechetto, D. F. Identification of a cortical site for stress-induced cardiovascular dysfunction. Integr. Physiol. Behav. Sci. 29; 362–73, 1994.

Cechetto, D. F., C. B. Saper. Role of the cerebral cortex in autonomic function. In A. D. Loewy, K. M. Spyer (Eds.). Central regulation of autonomic functions. New York, Oxford University Press, 1990, pp. 208–23.

Cechetto, D. F., J. X. Wilson, K. E. Smith, D. Wolski, M. D. Silver, V. C. Hachinski. Autonomic and myocardial changes in middle cerebral artery occlusion: Stroke models in the rat. Brain Res. 502; 296–305, 1989.

Cervero, F., R. D. Foreman. Sensory innervation of the viscera. In A. D. Loewy,

K. M. Spyer (Eds.). Central regulation of autonomic functions. New York, Oxford University Press, 1990, pp. 104–25.

Cervero, F., W. Janig. Visceral nociceptors: A new world order? Trends Neurosci. 15; 374–8, 1992.

Cervero, F., J. M. A. Laird. Visceral pain. The Lancet 353; 2145–8, 1999.

Cervero, F., J. F. B. Morrison (Eds.). Visceral sensation: Progress in brain research, vol. 67. Amsterdam, Elsevier, 1986.

Chait, L. D. Factors influencing the reinforcing and subjective effects of ephedrine in humans. Psychopharmacology 113; 381–7, 1994.

Chambers, D. J. Facing up to the problem of consciousness. In J. Shear (Ed.) Explaining consciousness—The 'hard problem.' Cambridge, Massachusetts, The MIT Press, 1997, pp. 9–30.

Chambless, D. L., A. T. Beck, E. J. Gracely, J. R. Grisham. Relationship of cognitions to fear of somatic symptoms: A test of the cognitive theory of panic. Depr. Anx. 11; 1–9, 2000.

Chapman, C. R., Y. Nakamura. A passion of the soul: An introduction to pain for consciousness researchers. Conscious. Cogn. 8; 391–422, 1999.

Charney, D. S., S. W. Woods, W. K. Goodman, G. R. Heninger. Neurobiological mechanisms of panic anxiety: Biochemical and behavioral correlates of yohimbine-induced panic attacks. Am. J. Psychiatry 144; 1030–6, 1987.

Chattopadhyay, P., E. Cooke, B. Toone, M. Lader. Habituation of physiological responses in anxiety. Biol. Psychiatry 15; 711–21, 1980.

Chernigovskiy, V. N. Interoceptors: Russian monographs on brain and behavior. Washington, D.C., American Psychological Association, 1967.

Chiel, H. J., R. D. Beer. The brain has a body: Adaptive behavior emerges from interactions of nervous system, body and environment. Trends Neurosci. 20; 553–7, 1997.

Christensen, B. C. Studies on hyperventilation. II. Electrocardiographic changes in normal man during voluntary hyperventilation. J. Clin. Invest. 25; 880–9, 1946.

Chronister, R. B., S. G. P. Hardy. The limbic system. In D. E. Haines (Ed.). Fundamental neuroscience. New York, Churchill Livingstone, 1997, pp. 443–54.

Churchland, P. M. The engine of reason, the seat of the soul: A philosophical journey into the brain. Cambridge, Massachusetts, The MIT Press, 1995.

Chwalisz, K., E. Diener, D. Gallagher. Autonomic arousal feedback and emotional experience: Evidence from the spinal cord injured. J. Personal. Soc. Psychol. 54; 820–828, 1988.

Clark, W. C. The psyche in the psychophysics of pain: An introduction to sensory decision theory. In J. Boivie, P. Hansson, U. Lindblom (Eds.). Touch, temperature, and pain in health and disease: mechanisms and assessment. Seattle, IASP Press, 1994.

Clark, C. R., G. M. Geffen, L. B. Geffen. Catecholamines and attention. II: Pharmacological studies in normal humans. Neurosci. Biobehav. Rev. 11; 353–64, 1987.

Clark, R. E., L. R. Squire. Classical conditioning and brain systems: The role of awareness. Science 280; 77–81, 1998.

Clauw, D. J., G. P. Chrousos. Chronic pain and fatigue syndromes: Overlapping clinical and neuroendocrine features and potential pathogenic mechanisms." Neuroimmunomodulation 4; 134–53, 1997.

Coffin, B., F. Azpiroz, J. R. Malagelada. Somatic stimulation reduces perception of gut distention in humans. Gastroenterology 107; 1636–42, 1994.

Coghill, R. C., C. N. Sang, K. F. Berman, G. J. Bennett, M. J. Iadarola. Global cerebral blood flow decreases during pain. J. Cereb. Blood Flow Metab. 18; 141–7, 1998.

Coghill, R. C., J. D. Talbot, A. C. Evans, E. Mayer, A. Gjedde, M. C. Bushnell, G. H. Duncan. Distributed processing of pain and vibration by the human brain. J. Neurosci. 14; 4095–108, 1994.

Cohen, M. S., S. Y. Bookheimer. Localization of brain function using magnetic resonance imaging. Trends Neurosci. 17; 268–77, 1994.

Cohn, J. N. Sympathetic nervous system activity and the heart. Am. J. Hypertens. 2; 353S–6S, 1989.

Colpaert, F. C., R. L. Balster. Transduction mechanisms of drug stimuli. Berlin, Springer-Verlag, 1988.

Colpaert, F. C., C. J. E. Niemegeers, P. A. J. Janssen. Differential response control by isopropamide: A peripherally induced discriminative cue. Europ. J. Pharmacol. 34; 381–4, 1975.

Colpaert, F. C., C. J. E. Niemegeers, P. A. J. Janssen. Haloperidol blocks the discriminative stimulus properties of lateral hypothalamic stimulation. Europ. J. Pharmacol. 42; 93–7, 1977.

Colpaert, F. C., J. L. Slangen (Eds.). Drug discrimination: Applications in CNS pharmacology. Amsterdam, Elsevier Biomedical Press, 1982.

Colpaert, F. C., J. A. Rosecrans (Eds.). Stimulus properties of drugs: Ten years of progress. Amsterdam, Elsevier/North-Holland Biomedical Press, 1978.

Connelly, J. F., J. M. Connelly, J. O. Epps. Disruption of dissociated learning in a discrimination paradigm by emotionally-important stimuli. Psychopharmacologia 30; 275–82, 1973.

Connelly, J. F., J. M. Connelly, R. Phifer. Disruption of state-dependent learning (memory retrieval) by emotionally-important stimuli. Psychopharmacologia 41; 139–43, 1975.

Connelly, J. F., J. M. Connelly, J. R. Nevitt. Effect of foot-shock intensity on amount of memory retrieval in rats by emotionally important stimuli in a drug-dependent learning escape design. Psychopharmacology 51; 153–57, 1977.

Cook, L., A. Davidson, D. J. Davis, R. T. Kelleher. Epinephrine, norepinephrine, and acetylcholine as conditioned stimuli for avoidance behavior. Science 131; 990–1, 1960.

Coover, G. D., B. R. Sutton, J. P. Heybach. Conditioning decreases in plasma corticosterone level in rats by pairing stimuli with daily feedings. J. Comp. Physiol. Psychol. 91; 716–26, 1977.

Coppen, A. J., A. G. Mezey. The influence of sodium amytal on the respiratory abnormalities of anxious psychiatric patients. J. Psychosom. Res. 5; 52–5, 1960.

Cordero, D. L, N. A. Cagin, B. H. Natelson. Neurocardiology update: Role of the nervous system in coronary vasomotion. Cardiovasc. Res. 29; 319–28, 1995.

Coryell, W. Hypersensitivity to carbon dioxide as a disease-specific trait marker. Biol. Psychiatry 41; 259–63, 1997.

Coull, J. T. Neural correlates of attention and arousal: Insights from electrophysiology, functional neuroimaging and psychopharmacology. Prog. Neurobiol. 55; 343–61, 1998.

Cowley, D. S., P. P. Roy-Byrne. Hyperventilation and panic disorder. Am. J. Med. 83; 929–37, 1987.

Cox, D. J., W. L. Clarke, L. Gonder-Frederick, S. Pohl, C. Hoover, A. Snyder, L. Zimbelman, W. R. Carter, S. Bobbitt, J. Pennebaker. Accuracy of perceiving blood glucose in IDDM. Diabetes Care 8; 529–36, 1985.

Cox, D. J., A. Freundlich, R. G. Meyer. Differential effectiveness of electromyographic feedback, verbal relaxation instructions and medication placebo. J. Consult. Clin. Psychol. 43; 892–8, 1975.

Cox, D. J., L. Gonder-Frederick, S. Pohl, J. W. Pennebaker. Reliability of symptom-blood glucose relationships among insulin-dependent adult diabetics. Psychosom. Med. 45; 357–60, 1983.

Craig, A. D. An ascending general homeostatic afferent pathway originating in lamina I. In G. Holstege, R. Bandler, C. S. Saper (Eds.). The emotional motor system: Progress in brain research, Vol. 107. Amsterdam, Elsevier, 1996, pp. 225–42.

Craig, A. D., E. M. Reiman, A. Evans, M. C. Bushnell. Functional imaging of an illusion of pain. Nature 384; 258–60, 1996.

Craig, A. D., K. Chen, D. Bandy, E. M. Reiman. Thermosensory activation of insular cortex. Nature Neurosci. 3; 184–90, 2000.

Crick, F. The astonishing hypothesis: The scientific search for the soul. New York, Simon and Schuster, 1994.

Crick, F., C. Koch. Are we aware of neural activity in primary visual cortex? Nature 375; 121–3, 1995.

Crider, A. B., G. E. Schwartz, S. Shnidman. On the criteria for instrumental autonomic conditioning: A reply to Katkin and Murray. Psychol. Bull. 71; 455–61, 1969.

Critchley, H. D., C. J. Mathais, R. J. Dolan. Neuroanatomical basis for first- and second-order representations of bodily states. Nature Neurosci. 4; 207–12, 2001.

Cross, S. A. Pathophysiology of pain. Mayo Clin. Proc. 69; 375–83, 1994.

Cutler, N. R., J. E. Hodes. Assessing the noradrenergic system in normal aging: A review of methodology. Exper. Aging Res. 9; 123–7, 1983.

Dahme, B., R. Richter, R. Mass. Interoception of respiratory resistance in asthmatic patients. Biol. Psychol. 42; 215–29, 1996.

Dales, R. E., W. O. Spitzer, M. T. Schechter, S. Suissa. The influence of psychological status on respiratory symptom reporting. Am. Rev. Respir. Dis. 139; 1459–63, 1989.

Dalessio, D. J. Hyperventilation. The vapors. Effort syndrome. Neurasthenia. Anxiety by any other name is just as disturbing. JAMA 239; 1401–2, 1978.

Damasio, A. R. Descartes' error: Emotion, reason, and the human brain. New York, Avon Books, 1994.

Damasio, A. R. The feeling of what happens: Body and emotion in the making of consciousness. New York, Harcourt, Brace and Company, 1999.

Dantendorfer, K., D. Prayer, J. Kramer, M. Amering, W. Baischer, P. Berger, M. Schoder, K. Steinberger, J. Windhaber, H. Imhof, H. Katschnig. High frequency of EEG and MRI brain abnormalities in panic disorder. Psychiatry Res. 68; 41–53, 1996.

Darwin, C. The expression of the emotions in man and animals. Chicago, The University of Chicago Press, 1965.

Davidson, R. J., M. E. Horowitz, G. E. Schwartz, D. M. Goodman. Lateral differences in the latency between finger tapping and the heart beat. Psychophysiology 18; 36–41, 1981.

Davidson, R. J., K. Hugdahl (Eds.). Brain asymmetry. Cambridge, Massachusetts, MIT Press, 1995.

Davidson, R. J., W. Irwin. The functional neuroanatomy of emotion and affective style. Trends Cogn. Sci. 3; 11–21, 1999.

Davidson, T. L. The nature and function of interoceptive signals to feed: Toward integration of physiological and learning perspectives. Psychol. Rev. 100; 640–57, 1993.

Davis, D. D., S. R. Dunlop, P. Shea, H. Brittain, H. C. Hendrie. Biological stress responses in high and low trait anxious students. Biol. Psychiatry 20; 843–51, 1985.

Davis, M. Potentiation of startle reflex behavior by anxiety: Neural localization and attenuation by diazepam. Psychopharmacol. Bull. 19; 457–65, 1983.

Davis, M. Neurobiology of fear responses: The role of the amygdala. In S. Salloway, P. Malloy, J. L. Cummings (Eds.). The neuropsychiatry of limbic and subcortical disorders. Washington, D.C., American Psychiatric Press, Inc., 1997, pp. 71–91.

Davis, M. Are different parts of the extended amygdala involved in fear versus anxiety? Biol. Psychiatry 44; 1239–47, 1998.

Deacon, S. P., A. Karunanayake, D. Barnett. Acebutalol, atenolol, and propranolol and metabolic responses to acute hypoglycaemia in diabetics. Br. Med. J. 2; 1255–7, 1977.

Deanfield, J. E., M. Shea, M. Kensett, P. Horlock, R. A. Wilson, C. M. DeLandsheere, A. P. Selwyn. Silent myocardial ischaemia due to mental stress. Lancet 2; 1001–5, 1984.

De Cristofaro, M. T., A. Sessarego, A. Pupi, F. Biondi, C. Faravelli. Brain perfusion abnormalities in drug-naive lactate-sensitive panic patients: A SPECT study. Biol. Psychiatry 33; 505–12, 1993.

Delacour, J. Neurobiology of consciousness: An overview. Behav. Brain Res. 85; 127–41, 1997.

de la Garza, R., C. E. Johanson. Discriminative stimulus properties of cocaine in pigeons. Psychopharmacology 85; 23–30, 1985.

Delvaux, M. M. Stress and visceral perception Canad. J. Gastroenterol. 13 (suppl. A); 32A–6A, 1999.

Denton, D., R. Shade, F. Zamarippa, G. Egan, J. Blair-West, M. McKinley, J. Lancaster, P. Fox. Neuroimaging of genesis and satiation of thirst and an interoceptor-driven theory of origins of primary consciousness. Proc. Nat. Acad, Sci. 96; 5304–9, 1999.

De Olmos, J. S., L. Heimer. The concepts of the ventral striatopallidal systems and extended amygdala. In J. F. McGinty (Ed.). Advancing from the ventral striatum to the extended amygdala. Ann. N.Y. Acad. Sci. 877; 1–32, 1999.

DePascalis, V., G. Palumbo, V. Ronchitelli. Heartbeat perception, instructions, and biofeedback in the control of heart rate. Internat. J. Psychophysiol. 11; 179–93, 1991.

DePonti, F., J. R. Malagelada. Functional gut disorders: From motility to sensitivity disorders. A review of current and investigational drugs for their management. Pharmacol. Therap. 80; 49–88, 1998.

Derbyshire, S. W., A. K. Jones. Cerebral responses to a continual tonic pain stimulus measured using positron emission tomography. Pain 76; 127–35, 1998.

Derbyshire, S. W., A. K. Jones, P. Devani, K. J. Friston, C. Feinmann, M. Harris, S. Pearce, J. D. Watson, R. S. Frackowiak. Cerebral responses to pain in patients with atypical facial pain measured by positron emission tomography. J. Neurol., Neurosurg., Psychiatry 57; 1166–72, 1994.

Derbyshire, S. W., A. K. Jones, F. Gyulai, S. Clark, D. Townsend, L. L. Firestone. Pain processing during three levels of noxious stimulation produces differential patterns of central activity. Pain 73; 431–5, 1997.

Derbyshire, S. W., B. A. Vogt, A. K. Jones. Pain and Stroop interference tasks activate separate processing modules in anterior cingulate cortex. Exper. Brain Res. 118; 52–60, 1998.

Derryberry, D., D. M. Tucker. Neural mechanisms of emotion. J. Consult. Clin. Psychol. 60; 329–38, 1992.

Deutsch, J. A., D. Deutsch. Physiological psychology. Homewood, Illinois, Dorsey Press, 1966, pp. 108–44.

deVietti, T. L., R. C. Larson. ECS effects: Evidence supporting state-dependent learning in rats. J. Comp. Physiol. Psychol. 74; 407–15, 1971.

Devinsky, O., M. J. Morrell, B. A. Vogt. Contributions of anterior cingulate cortex to behavior. Brain 118; 279–306, 1995.

Diamond, E. L., K. L. Massey, D. Covey. Symptom awareness and blood glucose estimation in diabetic adults. Health Psychol. 8; 15–26, 1989.

DiCara, L. V., J. J. Braun, B. A. Pappas. Classical conditioning and instrumental learning of cardiac and gastrointestinal responses following removal of neocortex in the rat. J. Comp. Physiol. Psychol. 73; 209–16, 1970.

DiCara, L. V., E. A. Stone. Effect of instrumental heart-rate training on rat cardiac and brain catecholamines. Psychosom. Med. 32; 359–68, 1970.

Dinan, T. G., S. Barry, S. Ahkion, A. Chua, P. W. N. Kelling. Assessment of central noradrenergic functioning in irritable bowel syndrome using a neuroendocrine challenge test. J. Psychosom. Res. 34; 575–80, 1990.

Dinsmoor, J. A., J. C. Bonbright, D. R. Lilie. A controlled comparison of drug effects on escape from conditioned aversive stimulation ("anxiety") and from continuous shock. Psychopharmacologia 22; 323–32, 1971.

Dolan, R. J., P. Fletcher, J. Morris, N. Kapur, J. F. W. Deakin, C. D. Frith. Neural activation during covert processing of positive emotional facial expressions." NeuroImage 4; 194–200, 1996.

Donovan, B. T. Humors, hormones, and the mind. London, MacMillan Press, 1988.

Drake, R. A. Left cerebral hemisphere contributions to tachycardia: Evidence and recommendations. Med. Hypotheses 19; 261–6, 1986.

Drevets, W. C. Functional neuroimaging studies of depression: The anatomy of melancholia. Ann. Rev. Med. 49; 341–61, 1998.

Drevets, W. C., M. E. Raichle. Neuroanatomical circuits in depression: Implications for treatment mechanisms. Psychopharmacol. Bull. 28; 261–74, 1992.

Drossman, D. A. Gastrointestinal illness and the biopsychosocial model. Psychosom. Med. 60; 258–67, 1998.

Droste, C. An experimental approach to symptomatic and asymptomatic myocardial ischemia. In D. Vaitl, R. Schandry (Eds.). From the heart to the brain: The psychophysiology of circulation-brain interaction. Frankfurt, Peter Lang, 1995, pp. 367–85.

Droste, C., H. Meyer-Blankenburg, M. W. Greenlee, H. Roskamm. Effect of physical exercise on pain thresholds and plasma beta-endorphins in patients with silent and symptomatic myocardial ischaemia. Europ. Heart J. 9 (suppl. N); 25–33, 1988.

Duncan, P. M., M. Copeland. Asymmetrical state dependency from temporary septal area dysfunction in rats. J. Comp. Physiol. Psychol. 89; 537–45, 1975.

Duncan, P. M., E. P. Hooker. The role of sympathetic arousal in discrimination of insulin-produced hypoglycemia. Behav. Pharmacol. 8; 389–95, 1997.

Duncan, P. M., W. Lichty. Discrimination of insulin-produced hypoglycemia in rats. Physiol. Behav. 54; 1099–102, 1993.

Dworkin, B. R., N. E. Miller. Failure to replicate visceral learning in the acute curarized rat preparation. Behav. Neurosci. 100; 299–314, 1986.

Earley, C. J., B. E. Leonard. Effects of prior exposure on conditioned taste aversion in the rat: Androgen-and estrogen-dependent events. J. Comp. Physiol. Psychol. 93; 793–805, 1979.

Ebstein, R. P., J. Stessman, R. Eliakim, J. Menczel. The effect of age on beta-adrenergic function in man: A review." Israel J. Med. Sci. 21; 302–11, 1985.

Eccleston, C., G. Crombez. Pain demands attention: A cognitive-affective model of the interruptive function of pain. Psychol. Bull. 125; 356–66, 1999.

Edelman, G. M. The remembered past: A biological theory of consciousness. New York, Basic Books, 1989.

Ehlers, A., P. Breuer. Increased cardiac awareness in panic disorder. J. Abnorm. Psychol. 101; 371–82, 1992.

Ehlers, A., P. Breuer. How good are patients with panic disorder at perceiving their heartbeats? Biol. Psychol. 42; 165–82, 1996.

Ehlers, A., P. Breuer, D. Dohn, W. Fiegenbaum. Heartbeat perception and panic disorder: Possible explanations for discrepant findings. Behav. Res. Therap. 33; 69–76, 1995.

Ehlers, A., J. Margraf, W. T. Roth, C. B. Taylor, R. J. Maddock, J. Sheikh, M. L. Kopell, K. L. McClenahan, D. Gossard, G. H. Blowers, W. S. Agras, B. S. Kopell. Lactate infusions and panic attacks: Do patients and controls respond differently? Psychiatry Res. 17; 295–308, 1986.

Eichler, S., E. S. Katkin. The relationship between cardiovascular reactivity and heartbeat detection. Psychophysiology 31; 229–34, 1994.

Elbert, T., H. Rau. What goes up (from the heart to the brain) must calm down (from the brain to the heart)!: Studies on the interaction between baroreceptor activity and cortical excitability. In D. Vaitl, R. Schandry (Eds.). From the heart to the brain: The psychophysiology of circulation-brain interaction. Frankfurt, Peter Lang, 1995, pp. 133–149.

Ellsworth, P. C. William James and emotion: Is a century of fame worth a century of misunderstanding?" Psychol. Rev. 101; 222–9, 1994.

Emel'ianenko, I. V., V. S. Raitses. Role of interoceptive signalization in modulating emotional-behavioral responses. Biulleten Eksperimentalnoi Biologii i Meditsiny 81; 264–6, 1976.

Engel, B. T. Electroencephalographic and blood pressure correlates of operantly conditioned heart rate in restrained monkeys. Pavlov. J. Biol. Sci. 9; 222–32, 1974.

Engel, B. T., P. Nikoomanesh, M. M. Schuster. Operant conditioning of rectosphincteric responses in the treatment of fecal incontinence. N. Engl. J. Med. 290; 646–9, 1974.

Epstein, R. The neural-cognitive basis of the Jamesian stream of thought. Conscious. Cogn. 9; 550–75, 2000.

Epstein, L. H., D. B. Stein. Feedback-influenced heart rate. J. Abnorm. Psychol. 83; 585–8, 1974.

Esler, M., S. Julius, A. Zweifler, O. Randall, E. Harburg, H. Gardiner, V. Dequattro. Mild high-renin essential hypertension: Neurogenic human hypertension? N. Engl. J. Med. 296; 405–11, 1977.

Essau, C. A., J. L. Jamieson. Heart rate perception in the Type A personality. Health Psychol. 6; 43–54, 1987.

Essman, W. B. Changes in the amnesic and aversive properties of avoidance conditioning with chlordiazepoxide. Pharmacol. Biochem. Behav. 1; 125–7, 1973.

Evans, H. L., R. A. Patton. Scopolamine on conditioned suppression: Influence of diurnal cycle and transitions between normal and drugged states. Psychopharmacologia 17; 1–13, 1970.

Ewing, D. J. Heart rate variability: An important new risk factor in patients following myocardial infarction. Clin. Cardiol. 14; 683–5, 1991.

Fahrenberg, J., M. Franck, U. Baas, E. Jost. Awareness of blood pressure: Interoception or contextual judgment? J. Psychosom. Res. 39; 11–8, 1995.

Fehm-Wolfsdorf, G., M. Gnadler, W. Kern, W. Klosterhalfen, W. Kerner. Classically conditioned changes of blood glucose level in humans. Physiol. Behav. 54; 155–60, 1993.

Fehr, F. S., J. A. Stern. Peripheral physiological variables and emotion: The James–Lange theory revisited. Psychol. Rev. 74; 411–24, 1970.

Feinberg, T. E. (Ed.) "Consciousness." Sem. Neurol. 17; 81–190, 1997.

Feldman, R. S., B. R. Kaada, T. Langfeldt. Effects of septal lesions and chlordiazepoxide (Librium) on avoidance behavior in rats. Pharmacol. Biochem. Behav. 1; 379–87, 1973.

Felten, D. L., S. Y. Felten, D. L. Bellinger, S. L. Carlson, K. D. Ackerman, K. S. Madden, J. A. Olschowski, S. Livnat. Noradrenergic sympathetic neural interactions with the immune system: Structure and function. Immunol. Rev. 100; 225–60, 1987.

Fendt, M., M. S. Fanselow. The neuroanatomical and neurochemical basis of conditioned fear. Neurosci. Biobehav. Rev. 23; 743–60, 1999.

Ferguson, M. L., E. S. Katkin. The relationship between visceral perception and emotion." In D. Vaitl, R. Schandry (Eds.). From the heart to the brain: The psychophysiology of circulation-brain interaction. Frankfurt, Peter Lang, 1995, pp. 269–81.

Ferguson, M. L., E. S. Katkin. Visceral perception, anhedonia, and emotion. Biol. Psychol. 42; 131–45, 1996.

Fields, C. Instrumental conditioning of the rat cardiac control system. Proc. Nat. Acad. Sci. 65; 293–9, 1970.

Fishbein, D. A., R. Cutler, H. L. Rosomoff, R. S. Rosomoff. Chronic pain-associated depression: Antecedent or consequence of chronic pain? Clin. J. Pain 13; 116–37, 1997.

Fisher, S. Organ awareness and organ activation. Psychosom. Med. 29; 643–7, 1967.

Fonte, R. J., J. M. Stevenson. The use of propranolol in the treatment of anxiety disorders. Hillside J. Clin. Psychiatry 7; 54–62, 1985.

Foreman, R. D. Mechanisms of cardiac pain. Annu. Rev. Physiol. 61; 143–67, 1999.

Frackowiak, R. S. J., K. J. Friston, C. D Frith, R. J. Dolan, J. C. Mazziotta (Eds.). Human brain mapping. New York, Academic Press, 1997.

Fraenkel, Y. M., S. Kindler, R. N. Melmed. Differences in cognitions during chest pain of patients with panic disorder and ischemic heart disease. Depr. Anx. 4; 217–22, 1996.

Frankenhaeuser, M., G. Jarpe. Psychophysiological changes during infusions of adrenaline in various doses. Psychopharmacologia 4; 424–32, 1963.

Fraser, F., R. M. Wilson. The sympathetic nervous system and the 'irritable heart of soldiers'. Br. Med. J. 2; 27–9, 1918.

Fredrikson, M. Orienting and defensive reactions to phobic and conditioned fear stimuli in phobics and normals. Psychophysiology 18; 456–65, 1981.

Fredrikson, M., H. Fischer, G. Wik. Cerebral blood flow during anxiety provocation. J. Clin. Psychiatry 58 (suppl. 16); 16–21, 1997.

Fredrikson, M., G. Wik, P. Anns, K. Ericson, S. Stone-Elander. Functional neuroanatomy of visually elicited simple phobic fear: Additional data and theoretical analysis. Psychophysiology 32; 43–8, 1995a.

Fredrikson, M., G. Wik, T. Greitz, L. Eriksson, S. Stone-Elander, K. Ericson, G. Sedvall. Regional cerebral blood flow during experimental phobic fear. Psychophysiology 30; 126–30, 1993.

Fredrikson, M., G. Wik, H. Fischer, J. Andersson. Affective and attentive neural networks in humans; A PET study of Pavlovian conditioning. NeuroReport 7; 97–101, 1995b.

Freedland, K. E., R. M. Carney, R. J. Krone, L. J. Smith, M. W. Rich, G. Eisenkramer, K. C. Fischer. Psychological factors in silent myocardial ischemia. Psychosom. Med. 53; 13–24, 1991.

Freedman, R. R., P. Ianni, E. Ettedgui, N. Puthezhath. Ambulatory monitoring of panic disorder. Arch. Gen. Psychiatry 42; 244–8, 1985.

Freeman, L. J., A. V. Conway, P. G. F. Nixon. Heart rate response, emotional disturbance and hyperventilation. J. Psychosom. Res. 30; 429–36, 1986.

Freire-Maia, L., A. D. Azevedo. The autonomic nervous system is not a purely efferent system. Med. Hypotheses 32; 91–9, 1990.

French, E. B., R. Kilpatrick. The role of adrenaline in hypoglycaemic reactions in man. Clin. Sci. 14; 639–51, 1955.

Frezza, D. A., J. G. Holland. Operant conditioning of the human salivary response. Psychophysiology 8; 581–7, 1971.

Fuller, R. W., S. Y. Felten, K. W. Perry, H. D. Snoddy, D. L. Felten. Sympathetic noradrenergic innervation of guinea-pig liver: Histofluorescence and pharmacological studies. J. Pharmacol. Exper. Therap. 218; 282–8, 1981.

Furedy, J. J., B. Damke, W. Boucsein. Revisiting the learning-without-awareness question in human Pavlovian autonomic conditioning: Focus on extinction in a dichotic listening paradigm. Integr. Physiol. Behav. Sci. 35; 17–34, 2000.

Furer, P., J. R. Walker, M. J. Chartier, M. B. Stein. Hypochondriacal concerns and somatization in panic disorder. Anxiety 6; 78–85, 1997.

Furman, S. Intestinal biofeedback in functional diarrhea: A preliminary report. J. Behav. Therap. Exper. Psychiatry 4; 317–21, 1973.

Furmark, T., H. Fischer, G. Wik, M. Larsson, M. Fredrikson. The amygdala and individual differences in human fear conditioning. NeuroReport 8; 3957–60, 1997.

Fyer, M. R., J. Uy, J. Martinez, R. Goetz, D. F. Klein, A. Fyer, M. R. Liebowitz, J.

Gorman. CO_2 challenge of patients with panic disorder. Am. J. Psychiatry 144; 1080–82, 1987.

Gallagher, S., G. E. Butterworth, A. Lew, J. Cole. Hand-mouth coordination, congenital absence of limb, and evidence for innate body schemas. Brain Cogn. 38; 53–65, 1998.

Gannon, L. The role of interoception in learned visceral control. Biofeedback Self Regul 2; 337–47, 1977.

Garcia, J., W. J. Hankins, K. W. Rusiniak. Behavioral regulation of the milieu interne in man and rat. Science 185; 824–31, 1974.

Garcia-Diaz, D. E., L. L. Jimenez-Montufar, R. Guevara-Aguilar, M. J. Wayner, L. Armstrong. Olfactory and visceral projections to the nucleus of the solitary tract. Physiol. Behav. 44; 619–24, 1988.

Gardner, R. M, J. A. Morrlee, D. N. Watson, S. L. Sandoval. Cardiac self perception in obese and normal persons. Percept. Motor Skills 70; 1179–86, 1990.

Garssen, B., W. van Veenendaal, R. Bloemink. Agoraphobia and the hyperventilation syndrome. Behav. Res. Therap. 21; 643–9, 1983.

Garvey, M. J., R. Noyes, C. Woodman, C. Laukes. Relationship of generalized anxiety symptoms to urinary 5-hydroxyindoleacetic acid and valillylmandelic acid. Psychiatry Res. 57; 1–5, 1995.

Gaston, K. E. Brain mechanisms of conditioned taste aversion learning: A review of the literature. Physiol. Psychol. 6; 340–53, 1978.

Gebhart, G. F., T. J. Ness. Central mechanisms of visceral pain. Can. J. Physiol. Pharmacol. 69; 627–34, 1991.

Gemar, M. C., S. Kapur, Z. V. Segal, G. M. Brown, S. Houle. Effects of self-generated sad mood on regional cerebral activity: A PET study in normal subjects. Depression 4; 81–8, 1996.

George, M. S., T. A. Ketter, D. S. Gill, J. V. Haxby, L. G. Ungerleider, P. Herscovitch, R. M. Post. Brain regions involved in recognizing facial emotion or identity: An oxygen-15 PET study. J. Neuropsychiatry Clin. Neurosci. 5; 384–94, 1993.

George, M. S., P. I. Parekh, N. Rosinsky, T. A. Ketter, T. A. Kimbrall K. M. Heilman, P. Herscovitch, R. M. Post. Understanding emotional prosody activates right hemisphere regions. Arch. Neurol. 53; 665–70, 1996.

George, M. S., H. A. Sackeim, A. J. Rush, L. B. Marangell, Z. Nahas, M. M. Husain, S. Lisanby, T. Burt, J. Goldman, J. C. Ballenger. Vagus nerve stimulation: A new tool for brain research and therapy. Biol. Psychiatry 47; 287–95, 2000.

Gershon, M. D. The second brain. New York, HarperCollins Publishers, 1998.

Gewirtz, J. C., M. Davis. Application of Pavlovian higher-order conditioning to the analysis of the neural substrates of fear conditioning. Neuropharmacology 37; 453–9, 1998.

Ghelarducci, B., L. Sebastiani. Classical heart hate conditioning and affective behavior: The role of the cerebellar vermis. Arch. Italien. Biol. 135; 369–84, 1997.

Ginsberg, M. D., F. Yoshii, S. Vibulsresth, J. Y. Chang, R. Duara, W. W. Barker,

T. E. Boothe. Human task-specific somatosensory activation. Neurology 34; 1301–8, 1987.

Glavin, G. B. Stress and brain noradrenaline: A review. Neurosci. Biobehav. Rev. 9; 233–43, 1985.

Glennon, R. A., J. A. Rosecrans, R. Young. Drug-induced discrimination: A description of the paradigm and a review of its specific application to the study of hallucinogenic agents. Medicinal Res. Rev. 3; 289–340, 1983.

Goetz, R. R., J. M. Gorman, D. J. Dillon, L. A. Papp, E. Hollander, A. J. Fyer, M. R. Liebowitz, D. F. Klein. Do panic disorder patients indiscriminately endorse somatic complaints? Psychiatry Res. 29; 207–13, 1989.

Goetz, R. R., D. F. Klein, J. M. Gorman. Symptoms essential to the experience of sodium lactate-induced panic. Neuropsychopharmacology 14; 355–66, 1996.

Goldberg, J., P. Davidson. A biopsychosocial understanding of the irritable bowel syndrome: A review. Can. J. Psychiatry 42; 835–40, 1997.

Goldberg, S. R., C. R. Schuster. Conditioned suppression by a stimulus associated with nalorphine in morphine addicted monkeys. J. Exper. Anal. Behav. 10; 235–42, 1967.

Goldstein, D. S. Stress and science. In O. G. Cameron (Ed.). Adrenergic dysfunction and psychobiology. Washington, D.C., American Psychiatric Press, Inc., 1994, pp. 179–236.

Goldstein, D. S., P. C. Chang, G. Eisenhofer, R. Miletich, R. Finn, J. Bacher, K. L Kirk, S. Bacharach, I. J. Kopin. Positron emission tomographic imaging of cardiac sympathetic innervation and function. Circulation 81; 1606–21, 1990.

Gonder-Frederick, L. A., D. J. Cox, S. A. Bobbitt, J. W. Pennebaker. Blood glucose symptom beliefs of diabetic patients: Accuracy and implications. Health Psychol. 5; 327–41, 1986.

Gonder-Frederick, L. A., D. J. Cox, S. A. Bobbitt, J. W. Pennebaker. Mood changes associated with blood glucose fluctuations in insulin-dependent diabetes mellitus. Health Psychol. 8; 45–59, 1989.

Gorelick, A. B., S. S. Koshy, F. G. Hooper, T. C. Bennett, W. D. Chey, W. L. Hasler. Differential effects of amitriptyline on perception of somatic and visceral stimulation in healthy humans. Am. J. Physiol. 275; G460–6, 1998.

Gorman, J. M., J. Askanazi, M. R. Liebowitz, A. J. Fyer, J. Stein, J. M. Kinney, D. F. Klein. Response to hyperventilation in a group of patients with panic disorder. Am. J. Psychiatry 141; 857–61, 1984.

Gorman, J. M., J. M. Kent, G. M. Sullivan, J. D. Coplan. Neuroanatomical hypothesis of panic disorder, revisited. Am. J. Psychiatry 157; 493–505, 2000.

Gorman, J. M., M. R. Liebowitz, A. J. Fyer, J. Stein. A neuroanatomical hypothesis for panic disorder. Am. J. Psychiatry 146; 148–61, 1989.

Graham, J. M., C. Desjardins. Classical conditioning: Induction of luteinizing hormone and testosterone secretion in anticipation of sexual activity. Science 210; 1039–41, 1980.

Gray, J. A. Elements of a two-process theory of learning. London, Academic Press, 1975a.

Gray, P. Effect of adrenocorticotropic hormone on conditioned avoidance in rats interpreted as state-dependent learning. J. Comp. Physiol. Psychol. 88; 281–4, 1975b.

Greenstadt, L., D. Shapiro, R. Whitehead. Blood pressure discrimination. Psychophysiology 23; 500–9, 1986.

Greenwald, A. G. Three cognitive markers of unconscious semantic activation. Science 273; 1699–1702, 1996.

Griez, E. J. L., H. Lousberg, M. A. van den Hout, G. M. van der Molen. CO_2 vulnerability in panic disorder. Psychiatry Res. 20; 87–95, 1987.

Grigg, L., R. Ashton. Heart rate discrimination viewed as a perceptual process: A replication and extension. Psychophysiology 19; 13–20, 1982.

Griggs, R. C., A. Stunkard. The interpretation of gastric motility: Sensitivity and bias in the perception of gastric motility. Arch. Gen. Psychiatry 11; 82–9; 1964.

Grilly, D. M., S. K. Johnson, R. Minardo, D. Jacoby, J. LaRiccia. How do tranquilizing agents selectively inhibit conditioned avoidance responding? Psychopharmacology 84; 262–7, 1984.

Grimaldi, A., F. Bosquet, P. Davidoff, J. P. Digy, C. Sachon, C. Landault, F. Thervet, F. Zoghbi, J. C. Legrand. Unawareness of hypoglycemia by insulin-dependent diabetics. Horm. Metab. Res. 22; 90–5, 1990.

Groenewegen, H. J., C. I. Wright, A. V. J. Beijer. The nucleus accumbens: Gateway for limbic structures to reach the motor system? In G. Holstege, R. Bandler, C. S. Saper (Eds.). The emotional motor system: Progress in brain research, vol. 107. Amsterdam, Elsevier, 1996, pp. 485–511.

Grossman, P. Respiration, stress, and cardiovascular function. Psychophysiology 20; 284–300, 1983.

Grossman, P., J. C. G. de Swart. Diagnosis of hyperventilation syndrome on the basis of reported complaints. J. Psychosom. Res. 28; 97–104, 1984.

Grossman, S. P. A textbook of physiological psychology. New York, John Wiley and Sons, 1967, pp. 564–95.

Gruber, R. P., D. R. Reed, J. D. Block. Transfer of the conditioned GSR from drug to nondrug state without awareness. J. Psychol. 70; 149–55, 1968.

Grunhaus, L. J., O. G. Cameron, A. C. Pande, R. F. Haskett, P. J. Hollingsworth, C. B. Smith. Comorbidity of panic disorder and major depressive disorder: Effects on platelet alpha2 adrenergic receptors. Acta Psychiatr. Scand. 81; 216–9, 1990.

Guglielmi, R. S., A. H. Roberts. Volitional vasomotor lability and vasomotor control. Biol. Psychol. 39; 29–44, 1994.

Haines, D. E. (Ed.). Fundamental neuroscience. New York, Churchill Livingstone, 1997.

Hale, F., S. Margen, D. Rabak. Postprandial hypoglycemia and 'psychological' symptoms. Biol. Psychiatry 17; 125–30, 1981.

Hammond, E. J., B. M. Uthman, S. A. Reid, B. J. Wilder, R. E. Ramsey. Vagus nerve stimulation in humans: Neurophysiological studies and electrophysiological monitoring. Epilepsia 31 (suppl. 2); S51–9, 1990.

Hamner, M. B., J. P. Lorberbaum, M. S. George. Potential role of the anterior cingulate cortex in PTSD: Review and hypothesis. Anxiety 9; 1–14, 1999.

Hanamori, T., T. Kunitake, K. Kato, H. Kannan. Responses of neurons in the insular cortex to gustatory, visceral, and nociceptive stimuli in rats. J. Neurophysiol. 79; 2535–45, 1998.

Hantas, M. N., E. S. Katkin, S. D. Reed. Cerebral lateralization and heartbeat discrimination. Psychophysiology 21; 274–8, 1984.

Hardy, S. G. P, J. P. Naftel. Viscerosensory pathways. In D. E. Haines (Ed.). Fundamental neuroscience. New York, Churchill Livingstone, 1997, pp. 255–63.

Harford, W. V. The syndrome of angina pectoris: Role of visceral pain perception. Am. J. Med. Sci. 307; 305–15, 1994.

Harley, C. W. A role for norepinephrine in arousal, emotion, and learning?: Limbic modulation by norepinephrine and the Kety hypothesis. Prog. Neuropharmacol. Biol. Psychiatr. 11; 419–58, 1987.

Haroutunian, V., B. A. Campbell. Emergence of interoceptive and exteroceptive control of behavior in rats. Science 205; 927–9, 1979.

Harver, A., E. S. Katkin, E. Bloch. Signal-detection outcomes on heartbeat and respiratory resistance detection tasks in male and female subjects. Psychophysiology 30; 223–30, 1993.

Harver, A., N. K. Squires, E. Block-Salisbury, E. S. Katkin. Event-related potentials to airway occlusion in young and old subjects. Psychophysiology 32; 121–9, 1995.

Hausken, T., S. Svebak, I. Wilhelmsen, T. T. Haug, K. Olafsen, E. Pettersen, K. Hveem, A. Berstad. Low vagal tone and antral dysmotility in patients with functional dyspepsia. Psychosom. Med. 55; 12–22, 1993.

Hefferline, R. F., B. Keenan, R. A. Harford. Escape and avoidance conditioning in human subjects without their observation of the response. Science 130; 1338–1339, 1959.

Hegel, M. T., T. A. Ahles. Behavioral analysis and treatment of reflexive vomiting associated with visceral sensations: A case study of interoceptive conditioning? J. Behav. Therap. Exper. Psychiatry 23; 237–42, 1992.

Heidbreder, E., A. Ziegler, K. Schafferhans, A. Heidland, W. Gruninger. Psychomental stress in tetraplegic man: Dissociation in autonomic variables and emotional responsiveness. J. Human Stress 10; 157–64, 1984.

Heller, S. R., I. A. MacDonald, M. Herbert, R. B. Tattersall. Influence of sympathetic nervous system on hypoglycaemia warning symptoms. Lancet 2; 359–63, 1987.

Hendry, S. H. C., S. S. Hsiao, M. D. Bushnell. Somatic sensation. In M. J. Zigmond, F. E. Bloom, S. C. Landis, J. L. Roberts, L. R. Squire (Eds.). Fundamental neuroscience. San Diego, Academic Press, 1999, pp. 761–89.

Henke, P. G., A. Ray, R. M. Sullivan. The amygdala. Emotions and gut functions. Digest. Dis. Sci. 36; 1633–43, 1991.

Henry, T. R., R. A. Bakay, J. R. Votaw, P. B. Pennell, C. M. Epstein, T. L. Faber, S. T. Grafton, J. M. Hoffman. Brain blood flow alterations induced by therapeutic vagus nerve stimulation in partial epilepsy: I. Acute effects at high and low levels of stimulation. Epilepsia 39; 983–90, 1998.

Herd, J. A. Cardiovascular response to stress in man. Annu. Rev. Physiol. 46; 177–85, 1984.

Herrnstein, R. J. Method and theory in the study of avoidance. Psychol. Rev. 76; 49–69, 1969.

Hibbert, G. A. Hyperventilation as a cause of panic attacks. Br. Med. J. 288; 263–4, 1984.

Hilgard, E. R., G. H. Bower. Theories of learning. New York, Appleton, Century, Crofts, 1966.

Hill, A. Phantom limb pain: A review of the literature on attributes and potential mechanisms. J. Pain Symptom Manage. 17; 125–42, 1999.

Hirschhorn, I. D., R. L. Hayes, J. A. Rosecrans. Discriminative control of behavior by electrical stimulation of the dorsal raphe nucleus: Generalization to lysergic acid diethylamide (LSD). Brain Res. 86; 134–8, 1975.

Ho, B. T., D. W. Richards, D. L. Chute (Eds.). Drug discrimination and state dependent learning. New York, Academic Press, 1978.

Hobson, A. R., S. Sarkar, P. L. Furlong, D. G. Thompson, O. Aziz. A cortical evoked potential study of afferents mediating human esophageal sensation. Am. J. Physiol. 279; G139–47, 2000.

Hoehn-Saric, R., D. R. McLeod, W. D. Zimmerli. Symptoms and treatment responses of generalized anxiety disorder patients with high versus low levels of cardiovascular complaints. Am. J. Psychiatry 146; 854–9, 1989a.

Hoehn-Saric, R., D. R. McLeod, W. D. Zimmerli. Somatic manifestations in women with generalized anxiety disorder. Arch. Gen. Psychiatry 46; 1113–9, 1989b.

Hoeldtke, R. D., G. Boden, C. R. Shuman, O. E. Owen. Reduced epinephrine secretion and hypoglycemia unawareness in diabetic autonomic neuropathy. Ann. Int. Med. 96; 459–62, 1982.

Hohmann, G. W. Some effects of spinal cord lesions on experienced emotional feelings. Psychophysiology 3; 143–56, 1966.

Hollerbach, S., D. Fitzpatrick, G. Shine, M. V. Kamath, A. R. Upton, G. Tougas. Cognitive evoked potentials to anticipated oesophageal stimulus in humans: Quantitative assessment of the cognitive aspects of visceral perception. Neurogastroenterology and Motility 11; 37–46, 1999.

Hollerbach, S., P. Hudoba, D. Fitzpatrick, R. Hunt, A. R. M. Upton, G. Tougas. Cortical evoked responses following esophageal balloon distention and electrical stimulation in healthy volunteers. Digest. Dis. Sci. 43; 2558–66, 1998.

Holstege, G. The emotional motor system. Europ. J. Morphol. 30; 67–79, 1992.

Holtzman, S. G. Discriminative stimulus properties of caffeine in the rat: Noradrenergic mediation. J. Pharmacol. Exper. Therap. 239; 706–14, 1986.

Holzer, H. H., H. E. Raybould. Vagal and splanchnic sensory pathways mediate inhibition of gastric motility induced by duodenal distention. Am. J. Physiol. 262; G603–8, 1992.

Holzl, R., L. P. Erasmus, A. Moltner. Detection, discrimination and sensation of visceral stimuli. Biol. Psychol. 42; 199–214, 1996.

Holzl, R., A. Moltner, C. W. Neidig. Somatovisceral interactions in visceral perception: Abdominal masking of colonic stimuli. Integr. Physiol. Behav. Sci. 34; 269–84, 1999.

Hornsveld, H., B. Garssen, M. F. Dop, P. van Spiegel. Symptom reporting during voluntary hyperventilation and mental load: Implications for diagnosing hyperventilation syndrome. J. Psychosom. Res. 34; 687–97, 1990.

Horwath, E., P. Adams, P. Wickramaratne, D. Pine, M. M. Weissman. Panic disorder with smothering symptoms: Evidence for increased risk in first-degree relatives. Depr. Anx. 6; 147–53, 1997.

Hothersall, D., J. Brener. Operant conditioning of changes in heart rate in curarized rats. J. Comp. Physiol. Psychol. 68; 338–42, 1969.

Howorka, K., G. Heger, A. Schabmann, P. Anderer, G. Tribl, J. Zeithofer. Severe hypoglycaemia unawareness is associated with an early decrease in vigilance during hypoglycaemia. Psychoneuroendocrinology 21; 295–312, 1996.

Howorka, K., G. Heger, A. Schabmann, F. Skrabal, J. Pumprla. Weak relationship between symptom perception and objective hypoglycaemia-induced changes in autonomic function in hypoglycaemia unawareness in diabetes. Acta Diabetologica 35; 1–8, 1998.

Howorka, K., J. Pumprla, B. Saletu, P. Anderer, M. Krieger, A. Schabmann. Decrease of vigilance assessed by EEG-mapping in Type I diabetic patients with history of recurrent severe hypoglycaemia. Psychoneuroendocrinology 25; 85–105, 2000.

Hu, W. H. C., N. J. Talley. Visceral perception in functional gastro-intestinal disorders: Disease markers or epiphenomenon? Digest. Dis. 14; 276–88, 1996.

Hubel, K. A. Voluntary control of gastrointestinal function: Operant conditioning and biofeedback. Gastroenterology 66; 1085–90, 1974.

Hugdahl, K. Cortical control of human classical conditioning: Autonomic and positron emission tomography data. Psychophysiology 35; 170–8, 1998.

Iovino, P., F. Azpiroz, E. Domingo, J. R. Malagelada. The sympathetic nervous system modulates perception and reflex responses to gut distention in humans. Gastroenterology 108; 680–6, 1995.

Jaeger, T. V., D. van der Kooy. Separate neural substrates mediate the motivating and discriminative properties of morphine. Behav. Neurosci. 100; 181–201, 1996.

James, W. What is an emotion? Mind 9; 188–205, 1884.

James, W. The physical basis of emotion. Psychol. Rev. 1; 516–29, 1894. (reprinted in Psychol. Rev. 101; 205–10, 1994.)

James, W. The principles of psychology, vols. one and two. New York, Dover Publications, Inc., 1950.

Janig, W. Neurobiology of visceral afferent neurons: Neuroanatomy, functions, organ regulations and sensations. Biol. Psychol. 42; 29–51, 1996.

Janssen, S. A., A. Arntz, S. Bouts. Anxiety and pain: Epinephrine-induced hyperalgesia and attentional influences. Pain 76; 309–16, 1998.

Jarbe, T. U. State-dependent learning and the discriminative control of behaviour: An overview. Acta Neurol. Scand. Suppl. 109; 37–59, 1986.

Jarbe, T. U. C. Cocaine as a discriminative cue in rats: Interactions with neuroleptics and other drugs. Psychopharmacology 59; 183–7, 1978.

Jasnos, T. M., K. L. Hakmiller. Some effects of lesion level and emotional cues on affective expression in spinal cord patients. Psychol. Rep. 37; 859–70, 1975.

Javanmard, M., J. Shlik, S. H. Kennedy, F. J. Vaccarino, S. Houle, J. Bradwejn. Neuroanatomic correlates of CCK-4-induced panic attacks in healthy humans: A comparison of two time points. Biol. Psychiatry 45; 872–82, 1999.

Jaynes, J. The origin of consciousness in the breakdown of the bicameral mind. Boston, Houghton Mifflin Co., 1976.

Jensen, H. H., B. Hutchings, J. C. Poulsen. Conditioned emotional responding under diazepam: A psychophysiological study of state-dependent learning. Psychopharmacology 98; 392–7, 1989.

Jensen, T. S., D. F. Smith. Selective association in conditioned stress-induced analgesia: Functional differences in interoceptive and exteroceptive sensory pathways. Behav. Neural Biol. 43; 218–21, 1985.

Johanson, C. E., K. Preston. The influence of an instruction on the stimulus effects of drugs in humans. Exper. Clin. Psychopharmacol. 6; 427–32, 1998.

Johansson, J. O., T. U. C. Jarbe. Physostigmine as a discriminative cue in rats. Arch. Int. Pharmacodyn. 219; 97–102, 1976.

Johnsen, B. H., K, Hugdahl. Hemispheric asymmetry in conditioning to facial emotional expressions. Psychophysiology 28; 154–62, 1991.

Johnsen, B. H., K, Hugdahl. Right hemisphere representation of autonomic conditioning to facial emotional expressions. Psychophysiology 30; 274–8, 1993.

Johnson, A. K., R. L. Thunhorst. The neuroendocrinology of thirst and salt appetite: Visceral sensory signals and mechanisms of central integration. Frontiers Neuroendocrinol. 18; 292–353, 1997.

Johnson, D. D., K. E. Dorr, W. M. Swenson, F. J. Service. Reactive hypoglycemia. JAMA 243; 1151–5, 1980.

Johnston, D. W., P. Anastasiades. The relationship between heart rate and mood in real life. J. Psychosom. Res. 34; 21–7, 1990.

Johnston, B. T., J. Shils, L. P. Leite, D. O. Castell. Effects of actreotide on esophageal visceral perception and cerebral evoked potentials induced by balloon distention. Am. J. Gastroenterol. 94; 65–70, 1999.

Jolley, R., R. B. Lydiard, M. E. Assey, B. W. Usher, W. H. Bardwell, J. C. Ballenger. Cardiovascular status of panic disorder patients with and without prominent cardiac symptoms. Psychosomatics 33; 81–4, 1992.

Jones, A. K. P., K. Friston, R. S. J. Frackowiak. Localization of response to pain in human cerebral cortex. Science 255; 215–6, 1992.

Jones, C. N., L. D. Grant, A. J. Prange. Stimulus properties of thyrotropin-releasing hormone. Psychopharmacology 59; 217–24, 1978.

Jones, C. N., H. F. Hill, R. T. Harris. Discriminative response control by d-amphetamine and related compounds in the rat. Psychopharmacologia 36; 347–56, 1974.

Jones, G. E. Perception of visceral sensations: A review of recent findings, methodologies, and future directions. In J. R. Jennings, P. K. Ackles, M. G. H. Coles (Eds.) Advances in psychophysiology, vol. 5. London, Jessica Kingsley Publishers, 1994, pp. 55–192.

Jones, G. E. Constitutional and physiological factors in heartbeat perception. In D. Vaitl, R. Schandry (Eds.). From the heart to the brain: The psychophysiology of circulation-brain interaction. Frankfurt, Peter Lang, 1995, pp. 173–92.

Jones, G., K. R. Jones, R. A. Cunningham, J. A. Caldwell. Cardiac awareness in infarct patients and normals. Psychophysiology 22; 480–7, 1985.

Jones, G. E., K. R. Jones, C. H. Rouse, D. M. Scott, J. A. Caldwell. The effect of body position on the perception of cardiac sensations: An experiment and theoretical implications. Psychophysiology 24; 300–11, 1987.

Jordan, S., H. C. Jackson, D. J. Nutt, S. L. Handley. Central alpha2 adrenoceptors are responsible for a clonidine-induced cue in the rat drug discrimination paradigm. Psychopharmacology 110; 209–12, 1993.

Joseph, R. The right cerebral hemisphere: Emotion, music, visual-spatial skills, body-image, dreams, and awareness. J. Clin. Psychol. 44; 630–73, 1988.

Justesen, D. R., E. W. Braun, R. G. Garrison, R. B. Pendleton. Pharmacological differentiation of allergic and classically conditioned asthma in the Guinea pig. Science 170; 864–6, 1970.

Kalaska, J. F. Central neural mechanisms of touch and proprioception. Can. J. Physiol. Pharmacol. 72; 542–5, 1994.

Kamath, M. V., S. Hollerbach, A. Bajwa, E. L. Fallen, A. R. M. Upton, G. Tougas. Neurocardiac and cerebral responses evoked by esophageal vago-afferent stimulation in humans: Effect of varying intensities. Cardiovasc. Res. 40; 591–9, 1998.

Kamien, J. B., W. K. Bickel, J. R. Hughes, S. T. Higins, B. J. Smith. Drug discrimination by humans compared to nonhumans: Current status and future directions. Psychopharmacology 111; 259–70, 1993.

Kanazawa, M., T. Nomura, S. Fukudo, M. Hongo. Abnormal visceral perception in patients with functional dyspepsia: Use of cerebral potentials evoked by electrical stimulation of the oesophagus. Neurogastroenterol. Motil. 12; 87–94, 2000.

Kandel, E. R., I. J. Kupfermann. Emotional states. In E. R. Kandel, J. H.

Schwartz, T. M. Jessell (Eds.). Essentials of neural science and behavior. Stamford, Connecticut, Appleton and Lange, 1995, pp. 595–612.

Kandel, E. R., J. H. Schwartz, T. M. Jessell. Principles of neural science, fourth edition. New York, McGraw-Hill, 2000.

Kannel, W. B., T. R. Dawber. The electrocardiogram in neurocirculatory asthenia (anxiety neurosis or neurasthenia): A study of 203 neurocirculatory asthenia patients and 757 healthy controls in the Framingham study. Ann. Int. Med. 49; 1351–60, 1958.

Kapp, B. S., J. S. Schawer, P. A. Driscoll. The organization of insular cortex projections to the amygdaloid central nucleus and autonomic regulatory nuclei of the dorsal medulla. Brain Res. 360; 355–60, 1985.

Karege, F., P. Bovier, H. Hilleret, J. M. Gaillard. Lack of effect of anxiety on total plasma MHPG in depressed patients. J. Affect. Disord. 28; 211–7, 1993.

Katerndahl, D. A. The sequence of panic symptoms. J. Family Pract. 26; 49–52, 1988.

Katkin, E. S. Blood, sweat, and tears: Individual differences in autonomic self-perception. Psychophysiology 22; 125–137, 1985.

Katkin, E. S., V. L. Cestaro, R. Weitkunat. Individual differences in cortical evoked potentials as a function of heartbeat detection ability. Internat. J. Neurosci. 61; 269–76, 1991.

Katkin, E. S., M. A. Morell, S. Goldband, G. L. Bernstein, J. A. Wise. Individual differences in heartbeat discrimination. Psychophysiology 19; 160–6, 1982.

Katkin, E. S., S. D. Reed. Cardiovascular asymmetries and cardiac perception. Internat. J. Neurosci. 39; 45–52, 1988.

Katner, S. N., J. G. Magalong, F. Weiss. Reinstatement of alcohol-seeking behavior by drug-associated discriminative stimuli after prolonged extinction in the rat. Neuropsychopharmacology 20; 471–9, 1999.

Katon, W. Panic disorder and somatization. Am. J. Med. 77; 101–106, 1984.

Katz, J., R. Melzack. Pain 'memories' in phantom limbs: Review and clinical observations. Pain 43; 319–36, 1990.

Kawachi, I., G. A. Colditz, A. Ascherio, E. R. Rimm, E. Giovannucci, M. J. Stampfer, W. C. Willett. Prospective study of phobic anxiety and risk of coronary heart disease in men. Circulation 89; 1992–7, 1994.

Kellner, R. Anxiety, somatic sensations and bodily complaints. In R. Noyes, M. Roth, G. D. Burrows (Eds.). Handbook of anxiety, vol. 2: Classification, etiological factors and associated disturbances. New York, Elsevier, 1988, pp. 213–37.

Kellner, R. Psychosomatic syndromes and somatic symptoms. Washington, D.C., American Psychiatric Press, 1991.

Kellner, R., B. F. Sheffield. The one-week prevalence of symptoms in neurotic patients and normals. Am. J. Psychiatry 130; 102–5, 1973.

Kennedy, S. H., M. Javanmard, F. J. Vaccarino. A review of functional neuroimaging in mood disorders: Positron emission tomography and depression. Can. J. Psychiatry 42; 467–75, 1997.

Keshner, E. A., H. Cohen. Current concepts of the vestibular system reviewed:

1. The role of the vestibulospinal system in postural control. Am. J. Occup. Therap. 43; 320–30, 1989.

Ketter, T. A., P. J. Andreasen, M. S. George, C. Lee, D. S. Gill, P. I. Parekh, M. W. Willis, P. Herscovitch, R. M. Post. Anterior paralimbic mediation of procaine-induced emotional and psychosensory experiences. Arch. Gen. Psychiatry 53; 59–69, 1996.

Kihlstrom, J. F. The cognitive unconscious. Science 237; 1445–55, 1987.

Kilbey, M. M., E. H. Ellinwood. Discriminative stimulus properties of psychomotor stimulants in the cat. Psychopharmacology 63; 151–3, 1979.

Kimble, G. A. Conditioning and learning. New York, Appleton, Century, Crofts, 1961.

Kimmel, H. D. Instrumental conditioning of autonomically mediated responses in human beings. Am. Psychol. 29; 325–35, 1974.

King, R. J., E. P. Bayon, D. B. Clark, C. B. Taylor. Tonic arousal and activity: Relationships to personality and personality disorder traits in panic patients. Psychiatry Res. 25; 65–72, 1988.

King, A. B., R. S. Menon, V. Hachinski, D. F. Cechetto. Human forebrain activation by visceral stimuli. J. Comp. Neurol. 41; 572–82, 1999.

Kinomura, S., R. Kawashima, K. Yamada, S. Ono, M. Itoh, S. Yoshioka, T. Yamaguchi, H. Matsui, H. Miyazawa, H. Itoh, R. Goto, T. Fujiwara, K. Satoh, H. Fukuda. Functional anatomy of taste perception in the human brain studied with positron emission tomography. Brain Res. 659; 263–6, 1994.

Kinomura S., J. Larsson, B. Gulyas, P. E. Roland. Activation by attention of the human reticular formation and thalamic intralaminar nuclei. Science 271; 512–5, 1996.

Klein, D. F. False suffocation alarms, spontaneous panics, and related conditions. Arch. Gen. Psychiatry 50; 306–17, 1993.

Klein, D. F. Testing the suffocation false alarm theory of panic disorder. Anxiety 1; 1–7, 1994.

Klein, S. B. Adrenal-pituitary influence in reactivation of avoidance-learning memory in the rat after intermediate intervals. J. Comp. Physiol. Psychol. 79; 341–59, 1972.

Knight, M. L., R. J. Borden. Autonomic and affective reactions of high and low socially-anxious individuals awaiting public performance. Psychophysiology 16; 209–13, 1979.

Knott, V., D. Bakish, S. Lusk, J. Barkely. Relaxation-induced EEG alterations in panic disorder patients. J. Anx. Disord. 11; 365–76, 1997.

Knott, V., D. Bakish, S. Lusk, J. Barkely, M. Perugini Quantitative EEG correlates of panic disorder. Psychiatry Res. 68; 31–9, 1996.

Knuepfer, M. M., A. Eismann, I. Schutze, H. Stumpf, G. Stock. Responses of single neurons in amygdala to interoceptive and exteroceptive stimuli in conscious cats. Am. J. Physiol. 268 (3 Pt 2); R666–75, 1995.

Ko, D., C. Heck, S. Grafton, M. L. Apuzzo, W. T. Couldwell, T. Chen, J. D. Day, V. Zelman, T. Smith, C. M. DeGiorgio. Vagus nerve stimulation activates central nervous system structures in epileptic patients during PET $H_2^{(15)}$ O blood flow imaging." Neurosurgery 39; 426–30, 1996.

Koehler, K., D. Vartzopoulos, H. Ebel. The relationship of panic attacks to autonomically labile generalized anxiety. Compr. Psychiatry 29; 91–7, 1988.

Kollenbaum, V. E., B. Dahme, G. Kirchner. 'Interoception' of heart rate, blood pressure, and myocardial metabolism during ergometric work load in healthy young subjects. Biol. Psychol. 42; 183–97, 1996.

Koriath, J. J. A view of cardio-cortical connections. J. Am. Coll. Cardiol. 14; 528–9, 1989.

Koriath, J. J., E. Lindholm. Cardiac-related cortical inhibition during a fixed foreperiod reaction time. Internat. J. Psychophysiol. 4; 183–95, 1986.

Koriath, J. J., E. Lindholm, D. M. Landers. Cardiac-related cortical activity during variations in mean heart rate. Internat. J. Psychophysiol. 5; 289–99, 1987.

Koslovskaya, I. B., R. P. Vertes, N. E. Miller. Instrumental learning without proprioceptive feedback. Physiol. Behav. 10; 101–7, 1973.

Kosslyn, S. M., L. M. Shin, W. L. Thompson, R. J. McNally, S. L. Rauch, R. K. Pitman, N. M. Alpert. Neural effects of visualizing and perceiving aversive stimuli: A PET investigation. NeuroReport 7; 1569–76, 1996.

Kroenke, K. Symptoms in medical patients: An untended field. Am. J. Med. 92 (suppl. 1A); 3S–6S, 1992.

Kroeze, S., M. van den Hout, M. A. Haenen, A. Schmidt. Symptom reporting and interoceptive attention in panic patients. Percept. Motor Skills 83 (3 Pt 1); 1019–26, 1996.

Krushel, L. A., D. van der Kooy. Visceral cortex: Integration of the mucosal senses with limbic information in the rat agranular insular cortex. J. Comp. Neurol. 270; 39–54, 1988.

Kujala, T., K. Alho, R. Naatanen. Cross-modal reorganization of human cortical functions. Trends Neurosci. 23; 115–20, 2000.

Kunst-Wilson, W. R., R. B. Zajonc. Affective discrimination of stimuli that cannot be recognized. Science 207; 557–8, 1980.

Kupfermann, I., J. Schwartz. Motivation." In E. R. Kandel, E. R., Schwartz, T. M. Jessell (Eds.). Essentials of neural science and behavior. Stamford, Connecticut, Appleton and Lange, 1995, pp. 613–28.

Kurtz, P., T. Palfai. State-dependent learning produced by Metrazol. Physiol. Behav. 10; 91– 5, 1973.

Lacey, J. H., S. A. Birtchnell. Body image and its disturbances. J. Psychosom. Res. 30; 623–31, 1986.

Lacroix, J. M. The acquisition of autonomic control through biofeedback: The case against an afferent process and a two-process alternative. Psychophysiology 18; 573–87, 1981.

Ladabaum, U., S. Minoshima, C. Owyang. Pathobiology of visceral pain: Molecular mechanisms and therapeutic implications V. Central nervous system processing of somatic and visceral sensory signals. Am. J. Physiol. 279; G1–6, 2000.

Lader, M. H. Palmer skin conductance measures in anxiety and phobic states. J. Psychosom. Res. 11; 271–81, 1967.

Lader, M. The psychophysiology of mental illness. London, Routledge and Kegen Paul, 1975.

Lader, M., A. Mathews. Physiological changes during spontaneous panic attacks. J. Psychosom. Res. 14; 377–82, 1970.

Lader, M., I. Marks. Clinical anxiety. London, William Heinemann, 1971.

Lakoff, G., M. Johnson. Philosophy in the flesh: The embodied mind and its challenge to western thought. New York, Basic Books, 1999.

Lal, H. (Ed.). Discriminative stimulus properties of drugs. New York, Plenum Press, 1977.

Lal, H., S. Yaden. Discriminative stimuli produced by clonidine in spontaneously hypertensive rats: Generalization to antihypertensive drugs with different mechanisms of action. J. Pharmacol. Exper. Therap. 232; 33–9, 1985.

Lambert, J. D. C. GABA (gamma aminobutyric acid). In G. Fink (Ed.). The encyclopedia of stress, vol. 2. New York, Academic Press, 2000, pp. 177–91.

Landis, C., W. A. Hunt. Adrenalin and emotion. Psychol. Rev. 39; 467–85, 1932.

Lane, R. D., E. M. Reiman, G. L. Ahern, G. E. Schwartz, R. J. Davidson. Neuroanatomical correlates of happiness, sadness, and disgust. Am. J. Psychiatry 154; 926–33, 1997.

Lang, P. J. The varieties of emotional experience: A meditation on James–Lange theory. Psychol. Rev. 101; 211–21, 1994.

Lang, P. J., M. K. Greenwald, M. M. Bradley, A. O. Hamm. Looking at pictures: Affective, facial, visceral, and behavioral reactions. Psychophysiology 30; 261–73, 1993.

Lapides, J., R. B. Sweet, L. W. Lewis. Role of striated muscle in urination. J. Urology 77; 247–50, 1957.

Lasiter, P. S. Gastrointestinal reactivity in rats lacking anterior insular neocortex. Behav. Neural Biol. 39; 149–54, 1983.

Laurent, B., R. Peyron, L. Garcia-Larrea, F. Mauguiere. Positron emission tomography to study central pain integration. Revue Neurol. 156; 341–51, 2000.

Lautenbacher, S., J. C. Krieg. Pain perception in psychiatric disorders: A review of the literature. J. Psychiatr. Res. 28; 109–22, 1994.

Lautenbacher, S., S. Roscher, D. Strain, K. Fassbender, K. Krumrey, J. C. Kreig. Pain perception in depression: Relationships to symptomatology and naloxone-sensitive mechanisms. Psychosom. Med. 56; 345–52, 1994.

Le, A. D., C. X. Poulos, H. Cappell. Conditioned tolerance to the hypothermic effect of ethyl alcohol. Science 206; 1109–10, 1979.

LeDoux, J. The emotional brain. New York, Simon and Schuster, 1996.

LeDoux, J. Fear and the brain: Where have we been and where are we going? Biol. Psychiatry 44; 1229–38, 1998.

LeDoux, J. E., M. E. Thompson, C. Iadecola, L. W. Tucker, D. J. Reis. Local cerebral blood flow increases during auditory and emotional processing in the conscious rat. Science 221; 576–8, 1983.

Lee, Y. J., G. C. Curtis, J. G. Weg, J. L. Abelson, J. G. Modell, K. M. Campbell. Panic attacks induced by doxapram. Biol. Psychiatry 33; 295–7, 1993.

Lehofer, M., M. Moser, R. Hoehn-Saric, D. McLeod, P. Liebman, B. Drnovsek, S. Egner, G. Hildebrandt, H. G. Zapotoczky. Major depression and cardiac autonomic control. Biol. Psychiatry 42; 914–9, 1997.

Leibrecht, B. C., A. J. Lloyd, S. Pounder Auditory feedback and conditioning of the single motor unit. Psychophysiology 10; 1–7, 1973.

Lembo, T., B. Naliboff, J. Munakata, S. Fullerton, L. Saba, S. Tung, M. Schmulson, E. A. Mayer. Symptoms and visceral perception in patients with pain-predominant irritable bowel syndrome. Am. J. Gastroenterol. 94; 1320–6, 1999.

Lenz, F. A., R. H. Gracely, A. J. Romanoski, E. J. Hope, L. H. Rowland, P. M. Dougherty. Stimulation in the human somatosensory thalamus can reproduce both the affective and sensory dimensions of previously experienced pain. Nature Med. 1; 910–3, 1995.

Lenz, F. A., R. H. Gracely, T. A. Zirh, D. A. Leopold, L. H. Rowland, P. M. Dougherty. Human thalamic nucleus mediating taste and multiple other sensations related to ingestive behavior. J. Neurophysiol. 77; 3406–9, 1997.

Leuchter, A. F., I. A. Cook, S. H. Uijtdehaage, J. Dunkin, R. B. Lufkin, C. Anderson-Hanley, M. Abrams, S. Rosenberg-Thompson, R. O'Hara, S. L. Simon, S. Osato, A. Babaie. Brain structure and function and the outcomes of treatment for depression. J. Clin. Psychiatry 58 (suppl. 16); 22–31, 1997.

Levander-Lindgren, M., S. Ek. Studies in neurocirculatory asthenia (DaCosta's Syndrome) II. Statistical analysis of the associations between symptoms and signs in the work test and the orthostatic test, and the correlation between these factors and immobilization and bodily constitution. Acta Med. Scand. 172; 677–83, 1962.

Levin, E. R., D. G. Gardner, W. K. Willis. Mechanisms of disease: Natiuretic peptides. N. Engl. J. Med. 339; 321–8, 1998.

Levine, J. D., N. C. Gordon, H. L. Fields. The mechanisms of placebo analgesia. The Lancet 2; 654–7, 1978.

Levis, D. J. The case for a return to a two-factor theory of avoidance: The failure of non-fear interpretation. In S. B. Klein, R. R. Mowrer (Eds.). Contemporary learning theories: Pavlovian conditioning and the status of traditional learning theory. Hillsdale, N.J., Lawrence Erlbaum Associates, Inc., 1989, pp. 227–77.

Lewis, S., N. Higgins (Eds.). Brain imaging in psychiatry. Oxford, England, Blackwell Science, 1996.

Leyton, M., C. Belanger, J. Martial, S. Beaulieu, E. Corin, J. Pecknold, N. M. K. Ng Ying Kin, M. Meaney, J. Thavundayil, S. Larue, N. P. Vasavan Nair. Cardiovascular, neuroendocrine, and monoaminergic responses to psychological stressors: Possible differences between remitted panic disorder patients and healthy controls. Biol. Psychiatry 40; 353–60, 1996.

Li, M., D. E. McMillan. The effects of drug discrimination history on drug discrimination and on punished and unpunished responding. Pharmacol. Biochem. Behav. 61; 93–105, 1998.

Libet, B. The neural time factor in conscious and unconscious events. In Ciba Foundation symposium 174: Experimental and theoretical studies of consciousness. Chichester, England, Wiley Publishers, 1993, pp. 123–46.

Linton, J. C., M. Hirt. A comparison of predictions from peripheral and central theories of emotion. Br. J. Med. Psychol. 52; 11–5, 1979.

Liotti, M., H. S. Mayberg, S. K. Brannan, S. McGinnis, P. Jerabek, P. T. Fox. Differential limbic-cortical correlates of sadness and anxiety in healthy subjects: Implications for affective disorders. Biol. Psychiatry 48; 30–42, 2000.

Locke, K. W., S. G. Holtzman. Characterization of the discriminative stimulus effects of centrally administered morphine in the rat. Psychopharmacology 87; 1–6, 1985.

Loewy, A. D. Central autonomic pathways. In A. D. Loewy, K. M. Spyer (Eds.). Central regulation of autonomic functions. New York, Oxford University press, 1990, pp. 88–103.

Logothetis, N. K. Vision: A window on consciousness. Scien. Amer. 281; 68–75, 1999.

Lovand, D. G., J. J. Kim, R. E. Thompson. Mammalian brain substrates of aversive classical conditioning. Annu. Rev. Psychol. 44; 317–42, 1993.

Lown, B. Mental stress, arrhythmias and sudden death. Am. J. Med. 72; 177–80, 1982.

Lown, B. Sudden cardiac death: Biobehavioral perspective. Circulation 76 (suppl. I); I186–96, 1987.

Lucey, J. V., D. C. Costa, G. Adshead, M. Deahl, G. Busatto, S. Gacinovic, M. Travis, L. Pilowsky, P. J. Ell, I. M. Marks, R. W. Kerwin. Brain blood flow in anxiety disorders. OCD, panic disorder with agoraphobia, and posttraumatic stress disorder on 99mTcHMPAO single photon emission tomography (SPET). Br. J. Psychiatry 171; 346–50, 1997.

Ludwick-Rosenthal, R., R. W. J. Neufeld. Heart beat perception: A study of individual differences. Internat. J. Psychophysiol. 3; 57–65, 1985.

Lum, L. C. Hyperventilation: The tip and the iceberg. J. Psychosom. Res. 19; 375–83, 1975.

Lundqvist, C., A. Siosteen, C. Blomstrand, B. Lind, M. Sullivan. Spinal cord injuries. Classical, functional, and emotional status. Spine 16; 78–83, 1991.

Luperallo, T., H. A. Lyons, E. R. Bleecker, E. R. McFadden. Influences of suggestion of airway reactivity in asthmatic subjects. Psychosom. Med. 30; 819–25, 1968.

Lynn, R. B., L. S. Friedman. Irritable bowel syndrome. N. Engl. J. Med. 329; 1940–5, 1993.

Macht, M. L., N. E. Spear, D. J. Levin. State-dependent retention in humans induced by alterations in affective state. Bull. Psychonomic Soc. 10; 415–8, 1977.

Mackey, W. B., J. Keller, D. van der Kooy. Visceral cortex lesions block conditioned taste aversions induced by morphine. Pharmacol. Biochem. Behav. 24; 71–8, 1986.

MacLeod, A. K., C. Haynes, T. Sensky. Attributions about common bodily sensations: Their associations with hypochondriasis and anxiety. Psychol. Med. 28; 225–8, 1998.

Maddock, R. J. The retrosplenial cortex and emotion: New insights from functional neuroimaging of the human brain. Trends Neurosci. 22; 310–6, 1999.

Maddock, R. J., C. S. Carter, L. Tavano-Hall, E. A. Amsterdam. Hypocapnia associated with cardiac stress scintigraphy in chest pain patients with panic disorder. Psychosom. Med. 60; 52–5, 1998.

Maes, M., H. Y. Meltzer, E. Suy, B. Minner, J. Calabrese, P. Cosyns. Sleep disorders and anxiety as symptom profiles of sympathoadrenal system hyperactivity in major depression. J. Affect. Disord. 27; 197–207, 1993.

Magarian, G. J. Hyperventilation syndromes: Infrequently recognized common expressions of anxiety and stress. Medicine 61; 219–36, 1982.

Malagelada, J. R. Conscious perception of gut activity. Digest. Dis. Sci. 39; 51S–3S, 1994.

Manji, H. K., M. V. Rudorfer, W. Z. Potter. Affective disorders and adrenergic function. In O. G. Cameron (Ed). Adrenergic dysfunction and psychobiology. Washington, D.C., American Psychiatric Press, 1994, pp. 365–401.

Maranon, G. Contribution a l'étude de l'action émotive de l'adrénaline. Rev. Francaise Endocrinol. 2; 301–25, 1924.

Marchand, W. E., B. Sarota, H. C. Marble, T. M. Leary, C. B. Burbank, M. J. Bellinger. Occurrence of painless acute surgical disorders in psychotic patients. N. Engl. J. Med 260; 580–5, 1959.

Maren, S. Long-term potentiation in the amygdala: A mechanism for emotional learning and memory. Trends Neurosci. 22; 561–7, 1999.

Maren, S., S. G. Anagnostaras, M. S. Fanselow. The startled seahorse: Is the hippocampus necessary for contextual fear conditioning? Trends Cogn. Sci. 2; 39–42, 1998.

Margraf, J., C. B. Taylor, A. Ehlers, W. T. Roth, W. S. Agras. Panic attacks in the natural environment. J. Nerv. Ment. Dis. 175; 558–65, 1987.

Martinez, J. M., L. A. Papp, J. D. Coplan, D. E. Anderson, C. M. Mueller, D. F. Klein, J. M. Gorman. Ambulatory monitoring of respiration in anxiety. Anxiety 2; 296–302, 1996.

Masand, P. S., D. S. Kaplan, S. Gupta, A. N. Bhandary. Irritable bowel syndrome and dysthymia. Is there a relationship? Psychosomatics 38; 63–9, 1997.

Mason, S. T. Noradrenaline and selective attention: A review of the model and the evidence. Life Sci. 27; 617–31, 1980.

Mason, S. T. Catecholamines and behavior. Cambridge, Cambridge University Press, 1984.

Masterson, F. A., M. Crawford. The defense motivation system: A theory of avoidance behavior. Behav. Brain Sci. 5; 661–96, 1982.

Matthews, P. B. C. Proprioceptors and their contribution to somatosensory mapping: Complex messages require complex processing. Can. J. Physiol. Pharmacol. 66; 430–8, 1988.

Mayberg, H. S., M. Liotti, S. K. Brannan, S. McGinnis, R. K. Mahurin, P. A. Jerabek, J. A., Silva, J. L. Tekell, C. C. Martin, J. L. Lancaster, P. T. Fox. Reciprocal limbic-cortical function and negative mood: Converging PET findings in depression and normal sadness. Am. J. Psychiatry 156; 675–82, 1999.

Mayer, E. A., G. F. Gebhart. Basic and clinical aspects of visceral hyperalgesia. Gastroenterology 107; 271–93, 1994.

Mayer, E. A., H. E. Raybould. Role of visceral afferent mechanisms in functional bowel disorders. Gastroenterology 99; 1688–1704, 1990.

Mayse, J. F., T. L. DeVietti. A comparison of state dependent learning induced by electroconvulsive shock and pentobarbital. Physiol. Behav. 7; 717–21, 1971.

Mazziota, J. C., A. W. Toga, R. S. J. Frackowiak (Eds.). Brain mapping: The disorders. San Diego, Academic Press, 2000.

McCaughey, S. A., T. R. Scott. The taste of sodium. Neurosci. Biobehav. Rev. 22; 663–76, 1998.

McCleary, R. A., R. Y. Moore. Subcortical mechanisms of behavior. New York, Basic Books, 1965.

McCormack, D. A., H. C. Pape, A. Williamson. Actions of norepinephrine in the cerebral cortex and thalamus: Implications for function of the central noradrenergic system." In C. D. Barnes, O. Pompeiano (Eds.). Neurobiology of the locus coeruleus: Progress in brain research, vol. 88. Amsterdam, Elsevier, 1991, pp. 293–306.

McCrea, C. W., A. B. Summerfield, B. Rosen. Body image: A selective review of existing measurement techniques." Br. J. Med. Psychol. 55; 225–33, 1982.

McFarland, R. A. Heart rate perception and heart rate control. Psychophysiology 12; 402–6, 1975.

McFarland, R. A., R. Kennison. Asymmetry in the relationship between finger temperature changes and emotional state in males. Biofeedback Self Regul. 14; 281–90, 1989.

McIntyre, D. C., J. L. Gunter. State-dependent learning induced by low intensity electrical stimulation of the caudate or amygdala nuclei in rats. Physiol. Behav. 23; 449–54, 1979.

McIntyre, D. C., R. J. Stenstrom, D. Taylor, K. A. Stokes, N. Edson. State-dependent learning following electrical stimulation of the hippocampus: Intact and split-brain rats. Physio. Behav. 34; 133–9, 1985.

McLachlan, R. S. Vagus nerve stimulation for intractable epilepsy: A review." J. Clin. Neurophysiol. 14; 358–68, 1997.

McLeod, D. R., R. Hoehn-Saric. Perception of physiological changes in normal and pathological anxiety. In R. Hoehn-Saric, D. R. McLeod (Eds.). Biology of anxiety disorders. Washington, D.C., American Psychiatric Press, 1993, pp. 223–43.

McLeod, D. R., R. Hoehn-Saric, R. L. Stefan. Somatic symptoms of anxiety:

Comparison of self-report and physiological measures. Biol. Psychiatry 21; 301–10, 1986.

McMahon, S. B. Are there fundamental differences in the peripheral mechanisms of visceral and somatic pain? Behav. Brain Sci. 20; 381–91, 1997.

McNaughton, N. Biology and emotion. Cambridge, Cambridge University Press, 1989.

Meadows, M. E., R. F. Kaplan. Dissociation of autonomic and subjective responses to emotional slides in right hemisphere damaged patients. Neuropsychologia 32; 847–856, 1994.

Mearin, F., M. Cucala, F. Azpiroz, J. R. Malagelada. The origin of symptoms on the brain–gut axis in functional dyspepsia. Gastroenterology 101; 999–1006, 1991.

Mega, M. S., J. L. Cummings, S. Salloway, P. Malloy. The limbic system. In S. Salloway, P. Malloy, J. L. Cummings (Eds.). The neuropsychiatry of limbic and subcortical disorders. Washington, D.C., American Psychiatric Press, Inc., 1997, pp. 3–18.

Melzack, R. Phantom limbs. Scien. Amer. 266; 120–6, 1992.

Mertz, H., R. Fass, A. Kodner, F. Yan-Go, S. Fullerton, E. A. Mayer. Effect of amitriptyline on symptoms, sleep, and visceral perception in patients with functional dyspepsia. Am. J. Gastroenterol. 93; 160–5, 1998a.

Mertz, H., S. Fullerton, B. Naliboff, E. A. Mayer. Symptoms and visceral perception in severe functional and organic dyspepsia. Gut 42; 814–22, 1998b.

Mertz, H., B. Naliboff, J. Munakata, N. Naizi, E. A. Mayer. Altered rectal perception is a biological marker of patients with irritable bowel syndrome. Gastroenterology 109; 40–52, 1995.

Mesulam, M. M., E. J. Mufson. The insula of Reil in man and monkey: Architectonics, connectivity, and function." In A. Peters, E. G. Jones (Eds.). Cerebral cortex, vol. 4: Association and auditory cortices. New York, Plenum Press, 1985, pp. 179–226.

Meyer, C. R. Grossmann, A. Mitrakou, R. Mahler, T. Veneman, J. Gerich, R. G. Bretzel. Effects of autonomic neuropathy on counterregulation and awareness of hypoglycemia in type I diabetic patients. Diabetes Care 21; 1960–6, 1998.

Meyer, R. B. Kroner-Herwig, H. Sporkel. The effect of exercise and induced expectations on visceral perception in asthmatic patients. J. Psychosom. Res. 34; 455–60, 1990.

Millan, M. J. The induction of pain: An integrative review. Prog. Neurobiol. 57; 1–164, 1999.

Miller, N. E. Effects of drugs on motivation: The value of using a variety of measures. Ann. N.Y. Acad. Sci. 65; 318–33, 1956.

Miller, N. E. Learning of visceral and glandular responses. Science 163; 434–45, 1969.

Miller, N. E. Some examples of psychophysiology and the unconscious. Biofeedback Self Regul. 17; 3–16, 1992.

Miller, N. E., A. Banuazizi. Instrumental learning by curarized rats of a specific visceral response, intestinal or cardiac. J. Comp. Physiol. Psychol. 65; 1–7, 1968.

Miller, N. E., A. Carmona. Modification of a visceral response, salivation in thirsty dogs, by instrumental training with water reward. J. Comp. Physiol. Psychol. 63; 1–6, 1967.

Miller, N. E., L. V. DiCara. Instrumental learning of urine formation in rats: Changes in renal blood flow." Am. J. Physiol. 215; 677–83, 1968.

Miller, N. E., L. V. DiCara. Instrumental learning of heart rate changes in curarized rats: Shaping, and specificity to discriminative stimulus. J. Comp. Physiol. Psychol. 63; 12–9, 1967.

Miller, N. E., L. V. DiCara, G. Wolf. Homeostasis and reward: T-maze learning induced by manipulating antidiuretic hormone. Am. J. Physiol. 215; 684–6, 1968.

Miller, N. E., B. Dworkin. Visceral learning: Recent difficulties with curarized rats and significant problems for human research. In L. V. DiCara, T. X. Barber, J. Kamiya, N. E. Miller, D. Shapiro, J. Stoyva (Eds.). Biofeedback and self-control, 1974 annual. Chicago, Aldine Publishing Company, 1975, pp. 83–103.

Milner, A. D. Streams and consciousness: Visual awareness and the brain. Trends Cogn. Sci. 2; 25–30, 1998.

Milner, P. M. Brain-stimulation reward: A review. Can. J. Psychol. 14; 1–36, 1991.

Missri, J. C., S. Alexander. Hyperventilation syndrome. A brief review. JAMA 240; 2093–6, 1978.

Mitrakou, A., C. Ryan, T. Veneman, M. Moken, T. Jenssen, I. Kiss, J. Durrant, P. Cryer, J. Gerich. Hierarchy of glycemic thresholds for counterregulatory hormone secretion, symptoms, and cerebral dysfunction. Am. J. Physiol. 260; E67–74, 1991.

Modrow, H. E., D. K. Bliss. Electrophysiological correlates of state-dependent learning. Physiol. Psychol. 7; 259–62, 1979.

Mokan, M., A. Mitrakou, T. Veneman, C. Ryan, M. Korytkowski, P. Cryer, J. Gerich. Hypoglycemia unawareness in IDDM. Diabetes Care 17; 1397–1403, 1994.

Montgomery, W. A., G. E. Jones. Laterality, emotionality, and heartbeat perception." Psychophysiology 21; 459–465, 1984.

Montoya, P., K. Ritter, E. Huse, W. Larbig, C. Braun, S. Topfner, W. Lutzenberger, W. Grodd, H. Flor, N. Birbaumer. The cortical somatotopic map and phantom phenomena in subjects with congenital limb atrophy and traumatic amputees with phantom limb pain. Europ. J. Neurosci. 10; 1095–102, 1998.

Montoya, P., R. Schandry, A. Muller. Heartbeat evoked potentials (HEP): Topography and influence of cardiac awareness and focus of attention. Electroencephal. Clin. Neurophysiol. 88; 163–72, 1993.

Morecraft, R. J., C. Geula, M. M. Mesulam. Cytoarchitecture and neural afferents of orbitofrontal cortex in the brain of the monkey. J. Comp. Neurol. 323; 341–58, 1992.

Morgane, P. J., J. Panksepp (Eds.). Handbook of the hypothalamus, vol 1. Anatomy of the hypothalamus. New York, Marcel Dekker, 1979.

Morgane, P. J., J. Panksepp (Eds.). Handbook of the hypothalamus, vol 2. Physiology of the hypothalamus. New York, Marcel Dekker, 1980a.

Morgane, P. J., J. Panksepp (Eds.). Handbook of the hypothalamus, vol 3, part A: Behavioral studies of the hypothalamus. New York, Marcel Dekker, 1980b.

Morgane, P. J., J. Panksepp (Eds.). Handbook of the hypothalamus, vol 3, part B: Behavioral studies of the hypothalamus. New York, Marcel Dekker, 1981.

Morris, J. S., C. D. Frith, D. I. Perrett, D. Rowland, A. W. Young, A. J. Calder, R. J. Dolan. A differential neural response in the human amygdala to fearful and happy facial expressions. Nature 383; 812–5, 1996.

Morris, J. S., A. Ohman, R. J. Dolan. Conscious and unconscious emotional learning in the human amygdala. Nature 393; 467–70, 1998.

Morris, P., S. I. Rapaport. Neuroimaging and affective disorder in late life: A review. Can. J. Psychiatry 35; 347–54, 1990.

Morrow, G. R., A. H. Labrum. The relationship between psychological and physiological measures of anxiety. Psychol. Med. 8; 95–101, 1978.

Morrow, G. R., C. Morrell. Behavioral treatment for the anticipatory nausea and vomiting induced by cancer chemotherapy. N. Engl. J. Med. 307; 1476–80, 1982.

Moses, J. L., C. Bradley. Accuracy of subjective blood glucose estimation by patients with insulin-dependent diabetes. Biofeedback Self Regul. 10; 301–14, 1985.

Mountz, J. M., J. G. Modell, M. W. Wilson, G. C. Curtis, M. A. Lee, S. Schmaltz, D. E. Kuhl. Positron emission tomographic evaluation of cerebral blood flow during state anxiety in simple phobia. Arch. Gen. Psychiatry 46; 501–4, 1989.

Mowrer, O. H. On the dual nature of learning: A reinterpretation of 'conditioning' and 'problem solving.' Harvard Educ. Rev. 17; 102–48, 1947.

Mowrer, O. H. Learning theory and behavior. New York, John Wiley and Sons, 1960.

Mraovitch, S., C. Iadecola, D. A. Ruggiero, D. J. Reis. Widespread reductions in cerebral blood flow and metabolism elicited by electrical stimulation of the parabrachial nucleus in rat. Brain Res. 341; 283–96, 1985.

Muller, A., P. Montoya, R. Schandry, L. Hartl. Changes in physical symptoms, blood pressure and quality of life over 30 days. Behav. Res. Therap. 32; 593–603, 1994.

Munjack, D. J., B. Crocker, D. Cabe, R. Brown, R. Usigli, A. Zulueta, M. McManus, D. McDowell, R. Palmer, M. Leonard. Alprazolam, propranolol, and placebo in the treatment of panic disorder and agoraphobia with panic attacks. J. Clin. Psychopharmacol. 9; 22–7, 1989.

Myrtek, M., G. Brugner. Perception of emotions in everyday life: Studies with patients and normals. Biol. Psychol. 42; 147–64, 1996.

Myrtek, M., W. Stiels, J. M. Herrmann, G. Brugner, W. Muller, V. Hoppner, A. Fichtler. Emotional arousal, pain, and ECG changes during ambulatory monitoring in patients with cardiac neurosis and controls: Methodological considerations and first results. In D. Vaitl, R. Schandry (Eds.). From the heart to the brain: The psychophysiology of circulation-brain interaction. Frankfurt, Peter Lang, 1995, pp. 319–34.

Nader, K., J. E. LeDoux. Is it time to invoke multiple fear learning systems in the amygdala? Trends Cogn. Sci. 1; 241–4, 1997.

Nakagawa, Y., T, Iwasaki, T. Ishima, K. Kimura. Interaction between benzodiazepine and GABA-A receptors in state-dependent learning." Life Sci. 52; 1935–1945, 1993.

Naliboff, B. D., J. Munakata, L. Chang, E. A. Mayer. Toward a biobehavioral model of visceral hypersensitivity in irritable bowel syndrome. J. Psychosom. Res. 45; 485–92, 1998.

Naritoku, D. K., A. Morales, T. L. Pencek, D. Winkler. Chronic vagus nerve stimulation increases the latency of the thalamocortical somatosensory evoked potential. Pacing Clin. Electrophysiol. 15 (10 Pt 2); 1572–8, 1992.

Natelson, B. H. Neurocardiology: An interdisciplinary area for the 80s. Arch. Neurol. 42; 178–84, 1985.

Natelson, B. H., E. Grover, N. A. Cagin, J. E. Ottenweller, W. N. Tapp. Learned fear: A cause of arrhythmia onset in the presence of digitalis. Pharmacol. Biochem. Behav. 33; 431–4, 1989.

Natelson, B. H., R. McCarty. Conditioned catecholamine changes in Rhesus monkeys. Pav. J. Biol. Sci. 15; 188–96, 1980.

Ness, T. J., G. F. Gebhart. Interactions between visceral and cutaneous nociception in the rat. II. Noxious visceral stimuli inhibit cutaneous nociceptive neurons and reflexes. J. Neurophysiol. 66; 29–39, 1991.

Nesse, R. M., O. G. Cameron, G. C. Curtis, D. S. McCann, M. J. Huber-Smith. Adrenergic function in patients with panic anxiety. Arch. Gen. Psychiatry 41; 771–6, 1984.

Nestoros, J. N., L. A. Demers-Desrosiers, L. A. Dalicandro. Levels of anxiety and depression in spinal cord–injured patients. Psychosomatics 23; 823–30, 1982.

Newman, F., M. B. Stein, J. R. Trettau, R. Coppola, T. W. Uhde. Quantitative electroencephalographic effects of caffeine in panic disorder. Psychiatry Res. 45; 105–13, 1992.

Newman, J. Thalamic contributions to attention and consciousness. Consciousness Cogn. 4; 172–93, 1995.

Newman, P. P. Visceral afferent functions of the nervous system. London, Edward Arnold Publishers, 1974.

Nicolelis, M. A., D. Katz, D. J. Krupa. Potential circuit mechanisms underlying concurrent thalamic and cortical plasticity. Rev. Neurosci. 9; 213–224, 1998.

Nordahl, T. E., W. E. Semple, M. Gross, T. A. Mellman, M. B. Stein, P. Goyer, A. C. King, T. W. Uhde, R. M. Cohen. Cerebral glucose metabolic differences in patients with panic disorder. Neuropsychopharmacology 3; 261–72, 1990.

Nordahl, T. E., M. B. Stein, C. Benkelfat, W. E. Semple, P. Andreason, A. Zametkin, T. W. Uhde, R. M. Cohen. Regional cerebral metabolic asymmetries replicated in an independent group of patients with panic disorder. Biol. Psychiatry 44; 998–1006, 1998.

Nowlis, D. P., Kamiya, J. The control of electroencephalographic alpha rhythms through auditory feedback and the associated mental activity. Psychophysiology 6; 476–84, 1970.

O'Brien, W. H., G. J. Reid, K. R. Jones. Differences in heartbeat awareness among males with higher and lower levels of systolic blood pressure. Internat. J. Psychophysiol. 29; 53–63, 1998.

O'Brien, C. P., T. Testa, T. J. O'Brien, J. P. Brady, B. Wells. Conditioned narcotic withdrawal in humans. Science 195; 1000–2, 1977.

O'Carroll, R. E., A. P. Moffoot, M. Van Beck, N. Dougall, C. Murray, K. P. Ebmeier, G. M. Goodwin. The effect of anxiety induction on the regional uptake of 99mTc-exametazine in simple phobia as shown by single photon emission tomography (SPET). J. Affect. Disord. 28; 203–10, 1993.

O'Connor, P. J., J. C. Smith, W. P. Morgan. Physical activity does not provoke panic attacks in patients with panic disorder: A review of the evidence. Anx. Stress Coping 13; 333–53, 2000.

Ohman, A. Nonconscious control of autonomic responses: A role for Pavlovian conditioning? Biol. Psychol. 27; 113–35, 1988.

Ongur, D., J. L. Price. The organization of networks within the orbital and medial prefrontal cortex of rats, monkeys and humans. Cereb. Cortex 10; 206–19, 2000.

Ono, T., H. Nishijo, T. Uwano. Amygdala role in conditioned associative learning. Prog. Neurobiol. 46; 401–22, 1995.

Oppenheimer, S. M., D. F. Cechetto, V. C. Hachinski. Cerebrogenic cardiac arrhythmias: Cerebral electrocardiographic influences and their role in sudden death. Arch Neurol 47; 513–9, 1990.

Oppenheimer, S. M., T. M. Saleh, J. X. Wilson, D. F. Cechetto. Plasma and organ catecholamine levels following stimulation of the rat insular cortex. Brain Res. 569; 221–8, 1992.

Orne, M. T., D. A. Paskewitz. Aversive situational effects on alpha feedback training. Science 186; 458–60, 1974.

Ostrowsky, K., J. Isnard, P. Ryvlin, M. Guenot, C. Fischer, F. Mauguiere. Functional mapping of the insular cortex: Clinical implications in temporal lobe epilepsy. Epilepsia 41; 681–6, 2000.

Overton, D. A., J. C. Winter. Discriminable properties of drugs and state-dependent learning. Fed. Proc. 33; 1785–6, 1974.

Palardy, J., J. Havrankova, R. Lepage, R. Matte, R. Belanger, P. D'Amour, L. G. Ste. Marie. Blood glucose measurements during symptomatic episodes in patients with suspected postprandial hypoglycemia. N. Engl. J. Med 321; 1421–5, 1989.

Pappas, B. A., P. Gray. Cue value of dexamethasone for fear-motivated behavior. Physiol. Behav. 6; 127–30, 1971.

Paradiso, S., R. G. Robinson, N. D. Andreasen, J. E. Downhill, R. J. Davidson, P. T. Kirchner, G. L. Watkins, L. L. Boles Ponto, R. D. Hichwa. Emotional activation of limbic circuitry in elderly normal subjects in a PET study. Am. J. Psychiatry 154; 384–9, 1997.

Pardo, J. V., P. J. Pardo, M. E. Raichle. Neural correlates of self-induced dysphoria." Am. J. Psychiatry 150; 713–9, 1993.

Partiot, A., J. Grafman, N. Sadato, J. Wachs, M. Hallett. Brain activation during the generation of non-emotional and emotional plans. NeuroReport 6; 1269–72, 1995.

Paterson, W. G., H. Wang, S. J. Vanner. Increasing pain sensation to repeated esophageal balloon distention in patients with chest pain of undetermined etiology. Digest. Dis. Sci. 40; 1325–31, 1995.

Pauli, P., L. Hartl, C. Marquardt, H. Stalmann, F. Strain. Heartbeat and arrhythmia perception in diabetic autonomic neuropathy. Psychol. Med. 21; 413–21, 1991.

Pauli, P., C. Marquardt, L. Hartl, D. O. Nutzinger, R. Hotzl, F. Strain. Anxiety induced by cardiac perception in patients with panic attacks: A field study. Behav. Res. Therap. 29; 137–45, 1991.

Pavlov, I. P. Conditioned reflexes. New York, Dover Publications, 1960.

Pearce, S. A., S. Isherwood, D. Hrouda, P. H. Richardson, A. Erskine, J. Skinner. Memory and pain: Tests of mood congruity and state dependent learning in experimentally induced and clinical pain. Pain 43; 187–93, 1990.

Pearlson, G. D., P. J. Jeffrey, G. J. Harris, C. A. Ross, M. W. Fischman, E. E. Camargo. Correlation of acute cocaine-induced changes in local cerebral blood flow with subjective effects. Am. J. Psychiatry 150; 495–7, 1993.

Peghini, P. L., P. O. Katz, D. O. Castell. Imipramine decreases oesophageal pain perception in human male volunteers. Gut 42; 807–13, 1998.

Pennebaker, J. W. Stimulus characteristics influencing estimation of heart rate. Psychophysiology 18; 540–8, 1981.

Pennebaker, J. W. The psychology of physical symptoms. New York, Springer-Verlag, 1982.

Pennebaker, J. W., D. J. Cox, L. Gonder-Frederick, M. G. Wunsch, W. S. Evans, S. Pohl. Physical symptoms related to blood glucose in insulin-dependent diabetics. Psychosom. Med. 43; 489–500, 1981.

Pennebaker, J. W., L. Gonder-Frederick, H. Stewart, L. Elfman, J. A. Skelton. Physical symptoms associated with blood pressure. Psychophysiology 19; 201–210, 1982.

Pennebaker, J. W., C. W. Hoover. Visceral perception versus visceral detection: Disentangling methods and assumptions. Biofeedback Self Regul. 9; 339–52, 1984.

Pennebaker, J. W., J. M. Lightner. Competition of internal and external information in an exercise setting. J. Personality Social Psychol. 39; 165–74, 1980.

Pennebaker, J. W., D. Watson. Blood pressure estimation and beliefs among normotensives and hypertensives. Health Psychol. 7; 309–28, 1988.

Peper, E. Localized EEG alpha feedback training: A possible technique for mapping subjective, conscious, and behavioral experiences. Kybernetik 11; 166–9, 1972.

Perkins, K. A., M. Sanders, C. Fonte, A. S. Wilson, W. White, R. Stiller, D. McNamara. Effects of central and peripheral nicotinic blockade on human nicotinic discrimination. Psychopharmacology 142; 158–64, 1999.

Permutt, M. A. Postprandial hypoglycemia. Diabetes 25; 719–33, 1976.

Perna, G., M. Battaglia, A. Garberi, C. Arancio, A. Bertani, L. Bellodi. Carbon dioxide/oxygen challenge test in panic disorder. Psychiatry Res. 52; 159–71, 1994.

Perna, G., A. Bertani, E. Politi, G. Colombo, L. Bellodi. Asthma and panic attacks. Biol. Psychiatry 42; 625–30, 1997.

Perna, G., S. Cocchi, A. Bertani, C. Arancio, L. Bellodi. Sensitivity to 35% CO_2 in healthy first-degree relatives of patients with panic disorder. Am. J. Psychiatry 152; 623–5, 1995.

Perry, E., M. Walker, J. Grace, R. Perry. Acetylcholine in mind: A neurotransmitter correlate of consciousness? Trends Neurosci. 22; 273–80, 1999.

Persson, L. O., L. Sjoberg. Mood and somatic symptoms. J. Psychosom. Res. 31; 499–511, 1987.

Peyron, R., L. Garcia-Larrea, M. C. Gregoire, N. Costes, P. Convers, F. Lavenne, F. Mauguiere, D. Michel, B. Laurent. Haemodynamic brain responses to acute pain in humans: Sensory and attentional networks. Brain 122; 1765–80, 1999.

Phelps, M. E., J. C. Mazziotta, H. R. Schelbert (Eds.). Positron emission tomography and autoradiography. New York, Raven Press, 1986.

Phillips, A. G., F. G. LePiane. Electrical stimulation of the amygdala as a conditioned stimulus in a bait-shyness paradigm. Science 201; 536–8, 1978.

Phillips, A. G., F. G. LePiane. Differential effects of electrical stimulation of amygdala or caudate on inhibitory shock avoidance: A role for state-dependent learning. Behav. Brain Res. 2; 103–11, 1981.

Phillips, G. C., G. E. Jones, E. J. Rieger, J. B. Snell. Effects of the presentation of false heart-rate feedback on the performance of two common heartbeat-detection tasks. Psychophysiology 36; 504–10, 1999.

Phillips, K. A., G. E. Vaillant, P. Schnurr. Some physiologic antecedents of adult mental health. Am. J. Psychiatry 144; 1009–13, 1987.

Phillips, M. L., C. Senior, T. Fahy, A. S. David. Disgust—the forgotten emotion in psychiatry. Br. J. Psychiatry 172; 373–5, 1998.

Phillips, M. L., A. W. Young, C. Senior, M. Brammer, C. Andrews, A. J. Calder, E. T. Bullmore, D. I. Perrett, D. Rowland, S. C. R. Williams, J. A. Gray, A. S. David. A specific neural substrate for perceiving facial expressions of disgust. Nature 389; 495–8, 1997.

Pianka, M. J. Cortical spreading depression: A case of state-dependent learning. Physiol. Behav. 17; 565–70, 1976.

Piccinelli, M., P. Rucci, B. Ustun, G. Simon. Topologies of anxiety, depression and somatization symptoms among primary care attenders with no formal mental disorder. Psychol. Med. 29; 677–88, 1999.

Pick, J. The autonomic nervous system. Philadelphia, J. B. Lippincott, 1970.

Ploghaus, A., I. Tracey, J. S. Gati, S. Clare, R. S. Menon, P. M. Matthews, J. Nicholas, P. Rawlins. Dissociating pain from its anticipation in the human brain. Science 284; 1979–81, 1999.

Plotkin, W. B., R. Cohen. Occipital alpha and the attributes of the alpha experience. Psychophysiology 13; 16–21, 1976.

Pohl, R., V. Yeragani, R. Balon, A. Ortiz, A. Aleem. Isoproterenol-induced panic: A beta-adrenergic model of panic disorder. In J. C. Ballenger (Ed.). Neurobiology of panic disorder. New York, Wiley-Liss, 1990., pp. 107–20.

Pons, T. P., P. E. Garraghty, D. P. Friedman, M. Mishkin. Physiological evidence for serial processing in somatosensory cortex. Science 237; 417–20, 1987.

Porges, S. W. Emotion: An evolutionary by-product of the neural regulation of the autonomic nervous system. Ann. N.Y. Acad. Sci. 807; 62–77, 1997.

Porsolt, R. D., C. Pawelec, S. Roux, M. Jalfre. Discrimination of the amphetamine cue: Effects of A, B, and mixed type inhibitors of monoamine oxidase. Neuropharmacology 23; 569–73, 1984.

Posner, M. I., M. E. Raichle. Images of mind. New York, Scientific American Library, 1994.

Postone, N. Phantom limbs: A review. Internal. J. Psychiatry Med. 17; 57–70, 1987.

Poulos, C. X., R. Hinson. Pavlovian conditional tolerance to haloperidol catalepsy: Evidence of dynamic adaptation in the dopaminergic system. Science 218; 491–2, 1982.

Pratt, J. A., I. P. Stolerman, H. S. Garcha, V. Giardini, C. Feyerabend. Discriminative stimulus properties of nicotine: Further evidence for mediation at a cholinergic receptor. Psychopharmacology 81; 54–60, 1983.

Price, D. D. Psychological and neural mechanisms of the affective dimension of pain. Science 288; 1769–72, 2000.

Rakover, S. S. One or two different sets of laws of learning—Is this an empirical question? Bull. Psychonomic Soc. 13; 41–3, 1979.

Ramachandran, V. S., W. Hirstein. The perception of phantom limbs. Brain 121; 1603–30, 1998.

Randell, W. C. Nervous control of cardiovascular function. New York, Oxford University Press, 1984.

Randich, A., G. F. Gebhart. Vagal afferent modulation of nociception. Brain Res. Rev. 17; 77–99, 1992.

Rao, S. S. C. Visceral hyperalgesia: The key for unraveling functional gastrointestinal disorders. Digest. Dis. 14; 271–5, 1996.

Raskin, M. Decreased skin conductance response habituation in chronically anxious patients. Biol. Psychol. 2; 309–19, 1975.

Rauch, S. L., C. R. Savage, N. M. Alpert, A. J. Fischman, M. A. Jenike. The functional neuroanatomy of anxiety: A study of three disorders using positron emission tomography and symptom provocation. Biol. Psychiatry 42; 446–52, 1997.

Rauch, S. L., C. R. Savage, N. M. Alpert, E. C. Miguel, L. Baer, H. C. Breiter, A. J. Fischman, P. A. Manzo, C. Moretti, M. A. Jenike. A positron emission tomographic study of simple phobic symptom provocation. Arch. Gen. Psychiatry 52; 20–8, 1995.

Rauch, S. L., B. van der Kolk, R. E. Fisher, N. M. Alpert, S. P. Orr, C. R. Savage, A. J. Fischman, M. A. Jenike, R. K. Pitman. A symptom provocation study of posttraumatic stress disorder using positron emission tomography and script-driven imagery. Arch. Gen. Psychiatry 53; 380–7, 1996.

Ray, W. J. The relationship of locus of control, self-report measures, and feedback to the voluntary control of heart rate. Psychophysiology 11; 527–34, 1974.

Razran, G. The observable unconscious and the inferable conscious in current Soviet psychophysiology. Psychol. Rev. 68; 81–147, 1961.

Reed, S. D., A. Harver, E. S. Katkin. Interoception. In J. T. Cacioppo, L. G. Tassinary (Eds.). Principles of psychophysiology: Physical, social, and inferential elements. New York, Cambridge University Press, 1990, pp. 253–91.

Reilly, S. The role of gustatory thalamus in taste-guided behavior. Neurosci. Biobehav. Rev. 22; 883–901, 1998.

Reiman, E. M. The application of positron emission tomography to the study of normal and pathologic emotions. J. Clin. Psychiatry 58 (suppl. 16); 4–12, 1997.

Reiman, E. M., M. J. Fusselman, P. T. Fox, M. E. Raichle. Neuroanatomical correlates of anticipatory anxiety. Science 243; 1071–4, 1989a.

Reiman, E. M., R. D. Lane, G. L. Ahern, G. E. Schwartz, R. J. Davidson, K. J. Friston, L. S. Yun, K. Chen. Neuroanatomical correlates of externally and internally generated human emotion." Am. J. Psychiatry 154; 918–925, 1997.

Reiman, E. M., M. E. Raichle, F. K. Butler, P. Herscovitch, E. Robins. A focal brain abnormality in panic disorder, a severe form of anxiety. Nature 310; 683–5, 1984.

Reiman, E. M., M. E. Raichle, E. Robins, F. K. Butler, P. Herscovitch, P. Fox, J. Perlmutter. The application of positron emission tomography to the study of panic disorder." Am. J. Psychiatry 143; 469–77, 1986.

Reiman, E. M, M. E. Raichle, E. Robins, M. A. Mintun, M. J. Fusselman, P. T. Fox, J. L. Price, K. A. Hackman. Neuroanatomical correlates of a lactate-induced anxiety attack. Arch. Gen. Psychiatry 46; 493–500, 1989b.

Reitveld, S., W. Everaerd, I. Vanbeest. Can biased symptom perception explain false-alarm choking sensations? Psychol. Med. 29; 121–6, 1999.

Rescorla, R. A., R. L. Solomon. Two-process learning theory: Relationships between Pavlovian conditioning and instrumental learning." Psychol. Rev. 74; 151–82, 1967.

Reus, V. I. A neuroanatomic perspective on state-dependent learning: The role of the striatum. Acta Neurol. Scand. Suppl. 109; 31–6, 1986.

Reus, V. I., H. Weingartner, R. M. Post. Clinical implications of state-dependent learning. Am. J. Psychiatry 136; 927–31, 1979.

Riddle, E. E., L. L. Hernandez. Adrenocorticotropic hormone fragment ACTH/MSH(4–10) can act as a discriminative stimulus in rats. Peptides 10; 1101–3, 1989.

Ring, C., J. Brener. Influence of beliefs about heart rate and actual heart rate on heartbeat counting. Psychophysiology 33; 541–6, 1996.

Ring, C., X. Liu, J. Brener. Cardiac stimulus intensity and heartbeat detection: Effects of tilt-induced changes in stroke volume. Psychophysiology 31; 553–64, 1994.

Rizzo, P. A., F. Pierelli, G. Pozzessere, F. Fattapposta, L. Sanarelli, C. Morocutti. Pain, anxiety, and contingent negative variation: A clinical and pharmacological study. Biol. Psychiatry 20; 1297–1302, 1985.

Robbins, T. W. Arousal systems and attentional processes. Biol. Psychol. 45; 57–71, 1997.

Roffman, M., H. Lal. Role of brain amines in learning associated with 'amphetamine state'. Psychopharmacologia 25; 195–204, 1972.

Rosecrans, J. A., W. T. Chance. Cholinergic and non-cholinergic aspects of the discriminative stimulus properties of nicotine. In Lal, H. (Ed.). Discriminative stimulus properties of drugs. New York, Plenum Press, 155–185 1977.

Rosen, S. D., E. Paulesu, C. D. Frith, R. S, Frackowiak, G. J. Davies, T. Jones, P. G. Camici. Central nervous pathways mediating angina pectoris. Lancet 344; 147–50, 1994.

Rosen, S., E. Paulesu, P. Nihoyannopoulos, D. Tousoulis, R. S. J. Frackowiak, C. D. Frith, T. Jones, P. Camici. Silent ischemia as a central problem: Regional brain activation compared in silent and painful myocardial ischemia. Ann. Int. Med. 124; 939–49, 1996.

Rosenfeld, J. P., B. E. Hetzler. Operant-controlled evoked responses: Discrimination of conditioned and normally occurring components. Science 181; 767–9, 1973.

Rosenfeld, J. P., A. P. Rudell, S. S. Fox. Operant control of neural events in humans. Science 165; 821–3, 1969.

Rosenthal, S. H., K. A. Porter, B. Coffey. Pain insensitivity in schizophrenia: Case report and review of the literature. Gen. Hosp. Psychiatry 12; 319–22, 1990.

Ross, E. D. Hemispheric specialization for emotions, affective aspects of language and communication and the cognitive control of display behaviors in humans. In G. Holstege, R. Bandler, C. S. Saper (Eds.). The emotional motor system: Progress in brain research, vol. 107. Amsterdam, Elsevier, 1996, pp. 583–94.

Rossel, P., A. M. Drewes, P. Petersen, J. Nielsen, L. Arendt-Nielsen. Pain produced by electric stimulation of the rectum in patients with irritable bowel syndrome: Further evidence of visceral hyperalgesia. Scand. J. Gastroenterol. 34; 1001–6, 1999.

Rostrup, M., O. Ekeberg. Awareness of high blood pressure influences on psychological and sympathetic responses. J. Psychosom. Res. 36; 117–23, 1992.

Roth, W. T., M. J. Telch, C. B. Taylor, J. A. Sachitano, C. C. Gallen, M. L. Kopell, K. L. McClenahan, W. S. Agras, A. Pfefferbaum. Autonomic characteristics of agoraphobia with panic attacks. Biol. Psychiatry 1133–54, 1986.

Roth, W. T., J. R. Tinklenberg, C. M. Doyle, T. B. Horvath, B. S. Kopell. Mood states and 24-hour cardiac monitoring. J. Psychosom. Res. 20; 179–86, 1976.

Roth, W. T., F. H. Wilhelm, W. Trabert. Autonomic instability during relaxation in panic disorder. Psychiatry Res. 80; 155–64, 1998.

Roth, W. T., F. H. Wilhelm, W. Trabert. Voluntary breath holding in panic and generalized anxiety disorder. Psychosom. Med. 60; 671–9, 1998.

Rothstein, R. D., M. Stecker, M. Reivich, A. Alavi, X. S. Ding, J. Jaggi, J. Greenberg, A. Ouyang. Use of positron emission tomography and evoked potentials in the detection of cortical afferents from the gastrointestinal tract. Am. J. Gastroenterology 91; 2372–6, 1996.

Rouse, C. H, G. E. Jones, K. R. Jones. The effect of body composition and gender on cardiac awareness. Psychophysiology 25; 400–7, 1988.

Roy-Byrne, P., T. W. Uhde, R. M. Post, A. C. King, M. S. Buchsbaum. Normal pain sensitivity in patients with panic disorder. Psychiatry Res. 14; 75–82, 1985.

Rozin, P. Disgust faces, basal ganglia and obsessive-compulsive disorder: Some strange bedfellows. Trends Cogn. Sci. 1; 321–2, 1997.

Rozin, P., J. W. Kalat. Specific hungers and poison avoidance as adaptive specializations of learning. Psychol. Rev. 78; 459–86, 1971.

Rozanski, A., N. Bairey, D. S. Krantz, J. Friedman, K. J. Resser, M. Morell, S. Hilton-Chalfen, L. Hestrin, J. Bietendorf, D. S. Berman. Mental stress and the induction of silent myocardial ischemia in patients with coronary artery disease. N. Engl. J. Med. 318; 1005–12, 1988.

Ruch, T. C. Neurophysiology of emotion. In T. C. Ruch, H. D. Patton (Eds.). Physiology and biophysics. Philadelphia, W. B. Saunders Co., 1965, pp. 508–22.

Ruggiero, D. A., S. Anwar, J. Kim, S. B. Glickstein. Visceral afferent pathways to the thalamus and olfactory tubercle: Behavioral implications. Brain Res. 799; 159–71, 1998.

Ruggiero, D. A., S. Mraovitch, A. R. Granata, M. Anwar, D. J. Reis. A role of insular cortex in cardiovascular function. J. Comp. Neurol. 257, 189–207, 1987.

Rush, A. J., M. S. George, H. A. Sackeim, L. B. Marangell, M. M. Husain, C. Giller, Z. Nahas, S. Haines, R. K. Simpson, R. Goodman. Vagus nerve stimulation (VNS) for treatment-resistant depressions: A multicenter study. Biol. Psychiatry 47; 276–86, 2000.

Russ, M. J., S. D. Roth, A. Lerman, T. Kakuma, K. Harrison, R. D. Shindledecker, J. Hull, S. Mattis. Pain perception in self-injurious patients with borderline personality disorder. Biol. Psychiatry 32; 501–11, 1992.

Russ, M. J., S. D. Roth, T. Kakuma, K. Harrison, J. W. Hull. Pain perception in self-injurious borderline patients: Naloxone effects. Biol. Psychiatry 35; 207–9, 1994.

Sainsbury, P., J. G. Gibson. Symptoms of anxiety and tension and the accompanying physiological changes in the muscular system. J. Neurol. Neurosurg. Psychiatr. 17; 216–24, 1954.

Saleh, T. M., B. J. Connell. The parabrachial nucleus mediates the decreased cardiac baroreflex sensitivity observed following short-term visceral afferent activation. Neurosci. 87; 135–46, 1998.

Salet, G. A., M. Samson, J. M. Roelofs, G. P. van Berge Henegouwen, A. J. Smout, L. M. Akkermans. Responses to gastric distention in functional dyspepsia. Gut 42; 823–9, 1998.

Salinsky, M. C., K. J. Burchiel. Vagus nerve stimulation has no effect on awake EEG rhythms in humans. Epilepsia 34; 299–304, 1993.

Samuels, M. A. Neurogenic heart disease: A unifying hypothesis. Am. J. Cardiol. 60; 15J-9J, 1987.

Sananes, C. B., B. A. Campbell. Role of the central nucleus of the amygdala in olfactory heart rate conditioning. Behav. Neurosci. 103; 519–25, 1989.

Saper, C. B. Role of the cerebral cortex and striatum in emotional motor response. In G. Holstege, R. Bandler, C. S. Saper (Eds.). The emotional motor system: Progress in brain research, vol. 107. Amsterdam, Elsevier, 1996, pp. 538–50.

Sato, A. Neural mechanisms of autonomic responses elicited by somatic sensory stimulation. Neurosci. Behav. Physiol. 27; 610–21, 1997.

Sawchenko, P. E., E. R. Brown, R. K. W. Chan, A. Ericsson, H. Y. Li, B. L. Roland, K. J. Kovacs. The paraventricular nucleus of the hypothalamus and the functional neuroanatomy of visceromotor responses to stress. In G. Holstege, R. Bandler, C. S. Saper (Eds.). The emotional motor system: Progress in brain research, vol. 107. Amsterdam, Elsevier, 1996, pp. 201–22.

Saxena, S., A. L. Brodu, J. M. Schwartz, L. R. Baxter. Neuroimaging and frontal-subcortical circuitry in obsessive-compulsive disorder. Br. J. Psychiatry (suppl.) 35; 26–37, 1998.

Schacter, D. L. Memory and awareness. Science 280; 59–60, 1998.

Schacter, D. L. (section Ed.). Consciousness. In Gazzaniga, M. S. The new cognitive neurosciences (second ed.). Cambridge, Mass., The MIT Press, 2000, pp. 1271–1363.

Schachter, S., J. E. Singer. Cognitive, social and physiological determinants of emotional state. Psychol. Rev. 69; 379–99, 1962.

Schandry, R. Heart beat perception and emotional experience. Psychophysiology 18; 483–8, 1981.

Schandry, R., M. Bestler. The association between parameters of cardiovascular function and heartbeat perception. In D. Vaitl, R. Schandry (Eds.). From the heart to the brain: The psychophysiology of circulation-brain interaction. Frankfurt, Peter Lang, 1995, pp. 223–50.

Schandry, R., M. Bestler, P. Montoya. On the relation between cardiodynamics and heartbeat perception. Psychophysiology 30; 467–74, 1993.

Schandry, R., C. Leopold, M. Vogt. Symptom reporting in asthma patients and insulin-dependent diabetics. Biol. Psychol. 42; 231–44, 1996.

Schandry, R, P. Montoya. Event-related brain potnetials and the processing of cardiac activity. Biol. Psychol. 42; 75–85, 1996.

Schandry, R., B. Sparrer, R. Weitkunat. From the heart to the brain: A study of heartbeat contingent scalp potentials. Internat. J. Neurosci. 30; 261–75, 1986.

Schapiro, H., L. G. Britt, C. W. Gross, K. J. Gaines. Sensory deprivation on visceral activity. III. The effect of olfactory deprivation of canine gastric secretion. Psychosom. Med. 33; 429–35, 1971.

Schapiro, H., C. W. Gross, T. Nakamura, L. D. Wruble, L. G. Britt. Sensory deprivation on visceral activity. II. The effect of auditory and vestibular deprivation on canine gastric secretion. Psychosom. Med. 32; 515–21, 1970a.

Schapiro, H., L. D. Wruble, L. G. Britt, T. A. Bell. Sensory deprivation on visceral activity. I. The effect of visual deprivation on canine gastric secretion. Psychosom. Med. 32; 379–96, 1970b.

Schechter, M. D., J. T. Concannon, R. E. Maloney, L. Bellush. Can ACTH analogs support discriminative learning in rats? Peptides 4; 11–4, 1983.

Schechter, M. D., P. G. Cook. Dopaminergic mediation of the interoceptive cue produced by d-amphetamine in rats. Psychopharmacologia 42; 185–93, 1975.

Schechter, M. D., J. A. Rosecrans. Nicotine as a discriminative stimulus in rats depleted of norepinephrine or 5-hydroxytryptamine. Psychopharmacologia 24; 417–29, 1972.

Schmahmann, J. D. An emerging concept: The cerebellar contribution to higher function. Arch. Neurol. 48; 1178–87, 1991.

Schmulson, M. J., E. A. Mayer. Gastrointestinal sensory abnormalities in functional dyspepsia. Bailliere's Clin. Gastroenterol. 12; 545–56, 1998.

Schneider, F., W. Grodd, U. Weiss, U. Klose, K. R. Mayer, T. Nagele, R. C. Gur. Functional MRI reveals left amygdala activation during emotion. Psychiatry Res.: Neuroimaging 76; 75–82, 1997.

Schneider, F., U. Weiss, C. Kessler, H. W. Muller-Gartner, S. Posse, J. B. Salloum, W. Grodd, F. Himmelmann, W. Gaebel, N. Birbaumer. Subcortical correlates of differential classical conditioning of aversive emotional reactions in social phobia. Biol. Psychiatry 45; 863–71, 1999.

Schnitzler, A., J. Volkmann, P. Enck, T. Frieling, O. W. Witte, H. J. Freund. Different cortical organization of visceral and somatic sensation in humans. Europ. J. Neurosci. 11; 305–15, 1999.

Schonecke, O. W. Functional cardiac disorder and cardiac perception: Attempts of quantification. In D. Vaitl, R. Schandry (Eds.). From the heart to the brain: The psychophysiology of circulation-brain interaction. Frankfurt, Peter Lang, 1995, pp. 335–50.

Schott, G. D. Visceral afferents: Their contribution to 'sympathetic dependent' pain. Brain 117; 397–413, 1994.

Schramke, C. J., R. M. Bauer. State-dependent learning in older and younger adults. Psychol. Aging 12; 255–62, 1997.

Schuster, C. R., J. V. Brady. The discriminative control of a food-reinforced operant by interoceptive stimulation. Pavl. J. High. Nerv. Act. 14; 448–58, 1964.

Schwartz, N. S., W. E. Clutter, S. D. Shah, P. E. Cryer. Glycemia thresholds for activation of glucose counterregulatory systems are higher than the threshold for symptoms. J. Clin. Invest. 79; 777–81, 1987.

Seiden, L. S., L. A. Dykstra (Eds.). Psychopharmacology: A biochemical and behavioral approach. New York, Van Nostrand Reinhold Company, 1977.

Servan-Schreiber, D., W. M. Perlstein, J. D. Cohen, M. Mintun. Selective pharmacological activation of limbic structures in human volunteers: A positron emission tomography study." J. Neuropsychiatr. Clin. Neurosci. 10; 148–59, 1998.

Shapiro, D., G. E. Schwartz, B. Tursky. Control of diastolic blood pressure in man by feedback and reinforcement. Psychophysiology 9; 296–304, 1972.

Shapiro, D., B. Tursky, E. Gershon, M. Stein. Effects of feedback and reinforcement on the control of human systolic blood pressure. Science 163; 588–9, 1969.

Shapiro, P. A., R. P. Sloan, E. Bagiella, J. T. Bigger, J. M. Gorman. Heart rate reactivity and heart period variability throughout the first year after heart transplantation. Psychophysiology 33; 54–62, 1996.

Shear, M. K., J. J. Polan, G. Harshfield, T. Pickering, J. J. Mann, A. Frances, G. James. Ambulatory monitoring of blood pressure and heart rate in panic patients. J. Anx. Disord. 6; 213–21, 1992.

Shin, L. M., S. M. Kosslyn, R. J. McNally, N. M. Alpert, W. L. Thompson, S. L. Rauch, M. L. Macklin, R. K. Pitman. Visual imagery and perception in posttraumatic stress disorder: A positron emission tomographic investigation. Arch. Gen. Psychiatry 54; 233–41, 1997.

Shipley, M. T., Y. Geinisman. Anatomical evidence for convergence of olfactory, gustatory, and visceral afferent pathways in mouse cerebral cortex. Brain Res. Bull. 12; 221–26, 1984.

Shioiri, T., T. Kato, J. Murashita, H. Hamakawa, T. Inubushi, S. Takahashi. High-energy phosphate metabolism in the frontal lobes of patients with panic disorder detected by phase-encoded 31P-MRS. Biol. Psychiatry 40; 785–93, 1996.

Silverman, D. H. S., J. A. Munakata, H. Ennes, M. A. Mandelkern, C. K. Hoh, E. A. Mayer. Regional cerebral activity in normal and pathological perception of visceral pain. Gastroenterology 112; 64–72, 1997.

Skerritt, P. W. Anxiety and the heart—a historical review. Psychol. Med. 13; 17–25, 1983.

Slucki, H., G. Adam, R. W. Porter. Operant discrimination of an interoceptive stimulus in Rhesus monkeys. J. Exper. Anal. Behav. 8; 405–14, 1965.

Sluckin, W. (Ed.). Fear in animals and man. New York, Van Nostrand Reinhold Company, 1979.

Smith, H. M., J. V. Basmajian, S. F. Vanderstoep. Inhibition of neighboring motoneurons in conscious control of single spinal motoneurons. Science 183; 975–6, 1974.

Smythies, J. The functional neuroanatomy of awareness: With a focus on the role of various anatomical systems in the control of intermodal attention. Consciousness Cogn. 6; 455–81, 1997.

Smythies, J. R. Neurophilosophy. Psychol. Med. 22; 547–9, 1992.

Soares, J. C., J. J. Mann. The anatomy of mood disorders—review of structural neuroimaging studies. Biol. Psychiatry 41; 86–106, 1997.

Sobotka, P. A., J. H. Mayer, R. A. Bauernfeind, C. Kanakis, K. M. Rosen. Arrhythmias documented by 24-hour continuous ambulatory electrocardiographic monitoring in young women without apparent heart disease. Am. Heart J. 101; 753–9, 1981.

Soldoff, S., H. Slucki. Operant discrimination of interoceptive urinary bladder stimulation in the monkey. Physiol. Behav. 12; 583–7, 1974.

Southwick, S. M., J. D. Bremner, A. Rasmusson, C. A. Morgan, A. Arnsten, D. S. Charney. Role of norepinephrine in the pathophysiology and treatment of posttraumatic stress disorder. Biol. Psychiatry 46; 1192–1204, 1999.

Spence, S., D. Shapiro, E. Zaidel. The role of the right hemisphere in the physiological and cognitive components of emotional processing. Psychophysiology 33; 112–22, 1996.

Spencer, D. G., S. Yaden, H. Lal. Behavioral and physiological detection of classically-conditioned blood pressure reduction. Psychopharmacology 95; 25–8, 1988.

Spinhoven, P., A. J. van der Does. Somatization and somatosensory amplification in psychiatric outpatients: An explorative study. Compr. Psychiatry 38; 93–7, 1997.

Stark, R. P., A. L. McGinn, R. F. Wilson. Chest pain in cardiac-transplant recipients: Evidence of sensory reinnervation after cardiac transplantation. N. Engl. J. Med. 324; 1791–4, 1991.

Starkman, M. N., O. G. Cameron, R. M. Nesse, T. Zelnik. Peripheral catecholamine levels and the symptoms of anxiety: Studies in patients with and without pheochromocytoma. Psychosom. Med. 52; 129–42, 1990.

Starkman, M. N., T. C. Zelnik, R. M. Nesse, O. G. Cameron. Anxiety in patients with pheochromocytoma. Arch. Int. Med. 145; 248–52, 1985.

Stein, M. B., W. D. Leslie. A brain single photon-emission computed tomographic (SPECT) study of generalized social phobia. Biol. Psychiatry 39; 825–8, 1996.

Steriade, M. Arousal: Revisiting the reticular activating system. Science 272; 225–6, 1996.

Stern, R. M., N. L. Lewis. Ability of actors to control their GSRs and express emotions. Psychophysiology 4; 294–9, 1968.

Stewart, R. S., M. D. Devous, A. J. Rush, L. Lane, F. J. Fonte. Cerebral blood flow changes during sodium-lactate panic attacks. Am. J. Psychiatry 145; 442–9, 1988.

Stewart, J., W. H. Krebs, E. Kaczender. State-dependent learning produced with steroids. Nature 216; 1223–4, 1967.

Stockhorst, U., E. Gritzmann, K. Klopp, Y. Schottenfeld-Noar, A. Hubinger, H. W. Berresheim, H. J. Steingruber, F. A. Gries. Classical conditioning of insulin effects in healthy humans. Psychosom. Med. 61; 424–35, 1999.

Stokes, K. A., D. C. McIntyre. Lateralized asymmetrical state-dependent learning produced by kindled convulsions from the rat hippocampus. Physiol. Behav. 26; 163–9, 1981.

Stokes, K. A., D. C. McIntyre. Lateralized state-dependent learning produced by hippocampal kindled convulsions: Effect of split-brain. Physiol. Behav. 34; 217–24, 1985.

Stolerman, I. P., F. Rasul, P. J. Shine. Trends in drug discrimination research analyzed with a cross-indexed bibliography, 1984–1987. Psychopharmacology 98; 1–19, 1989.

Stone, H. L., K. J. Dormer, R. D. Foremena, R. Thies, R. W. Blair. Neural regulation of the cardiovascular system during exercise. Fed Proc. 44; 2271–8, 1985.

Stunkard, A., C. Koch. The interpretation of gastric motility: Apparent bias in the reports of hunger by obese persons. Arch. Gen. Psychiatry 11; 74–82; 1964.

Sturges, L. V., V. L. Goetsch. Psychophysiological reactivity and heartbeat awareness in anxiety sensitivity. J. Anx. Disord. 10; 283–94, 1996.

Sturges, L. V., V. L. Goetsch, J. Ridley, M. Whittal. Anxiety sensitivity and response to hyperventilation challenge: Physiologic arousal, interoceptive acuity, and subjective distress. J. Anx. Disord. 12; 103–15, 1998.

Sullivan, G. M., J. D. Coplan, J. M. Kent, J. M. Gorman. The noradrenergic system in pathological anxiety: A focus on panic with relevance to generalized anxiety and phobias. Biol. Psychiatry 46; 1205–18, 1999.

Svensson, T. H. Peripheral, autonomic regulation of locus coeruleus noradrenergic neurons in brain: Putative implications for psychiatry and psychopharmacology. Psychopharmacology 92; 1–7, 1987.

Sylven, C. Mechanisms of pain in angina pectoris—a critical review of the adenosine hypothesis. Cardiovasc. Drugs Therap. 7; 745–59, 1993.

Talbot, J. D., S. Marrett, A. C. Evans, E. Meyer, M. C. Bushnell, G. H. Duncan. Multiple representations of pain in human cerebral cortex. Science 251; 1355–8, 1991.

Tallman, J. F., J. V. Cassella, G. White, D. W. Gallager. GABA-A receptors: Diversity and its implications for CNS disease. The Neuroscientist 5; 351–61, 1999.

Tan, B. K. Physiological correlates of anxiety: A preliminary investigation of the orienting reflex. Can. Psychiatr. J. 9; 63–70, 1964.

Tanaka, M. Emotional stress and characteristics of brain noradrenaline release in the rat. Industrial Health 37; 143–56, 1999.

Taylor, C. B., C. Hayward, R. King, A. Ehlers, J. Margraf, R. Maddock, D. Clark,

W. T. Roth, W. S Agras. Cardiovascular and symptomatic reduction effects of alprazolam and imipramine in patients with panic disorder: Results of a double-blind, placebo-controlled trial. J. Clin. Psychopharmacol. 10; 112–8, 1990.

Taylor, C. B., M. J. Telch, D. Havvik. Ambulatory heart rate changes during panic attacks. J. Psychiatr. Res. 17; 261–6, 1983.

Taylor, L. A., S. J. Rachman. The effects of blood sugar level changes on cognitive function, affective state, and somatic symptoms. J. Behav. Med. 11; 279–91, 1988.

Thayer, J. F., B. H. Friedman, T. D. Borkovec. Autonomic characteristics of generalized anxiety disorder and worry. Biol. Psychiatry 39; 255–66, 1996.

Thompson, T., R. Pickens (Eds.). Stimulus properties of drugs. New York, Appleton-Century-Crofts, 1971.

Thompson, T., R. Pickens, R. A. Meisch (Eds.). Readings in behavioral pharmacology. New York, Appleton-Century-Crofts, 1970.

Thornton, E. W., C. Van Toller. Operant conditioning of heart rate changes in the functionally decorticate curarized rat. Physiol. Behav. 10; 983–9, 1973a.

Thornton, E. W., C. Van Toller. Effect of immunosympathectomy on operant heart rate conditioning in the curarized rat. Physiol. Behav. 10; 197–201, 1973b.

Thyer, B. A., J. D. Papsdorf, P. Wright. Physiological and psychological effects of acute intentional hyperventilation. Behav. Res. Therap. 22; 587–90, 1984.

Tiihonen, J., J. Kuikka, K. Bergstrom, U. Lepola, H. Koponen, E. Leinonen. Dopamine reuptake site densities in patients with social phobia. Am. J. Psychiatry 154; 239–42, 1997.

Toga, A. W., J. C. Mazziotta (Eds.). Brain mapping: The methods. New York, Academic Press, 1996.

Toga, A. W., J. C. Mazziotta (Eds.). Brain mapping: The systems. San Diego, Academic Press, 2000.

Tongjaroenbuangam, W., D. Meksuriyen, P. Goyitrapong, N. Kotchabhakdi, B. A. Baldwin. Drug discrimination analysis of pseudoephedrine in rats. Pharmacol. Biochem. Behav. 59; 505–10, 1998.

Tougas, G. The autonomic nervous system in functional bowel disorders. Canad. J. Gastroenterol. 13 (suppl. A); 15A-7A, 1999.

Towler, D. A., C. E. Havlin, S. Craft, P. Cryer. Mechanism of awareness of hypoglycemia: Perception of neurogenic (predominantly cholinergic) rather than neuroglycopenic symptoms. Diabetes 42; 1791–8, 1993.

Townsend, M. H., N. B. Bologna, J. G. Barbee. Heart rate and blood pressure in panic disorder, major depression, and comorbid panic disorder with major depression. Psychiatry Res. 79; 187–90, 1998.

Travis, T. A., C. Y. Kondo, J. R. Knott. Subjective aspects of alpha enhancement. Br. J. Psychiatry 127; 122–6, 1975.

Trimble, K. C., R. Farouk, A. Pryde, S. Douglas, R. C. Heading. Heightened visceral sensation in functional gastrointestinal disease is not site-specific: Evidence for a generalized disorder of gut sensitivity. Digest. Dis. 40; 1607–13, 1995.

Trimble, M. R., M. F. Mendez, J. L. Cummings. Neuropsychiatric symptoms from the temporolimbic lobes. In S. Salloway, P. Malloy, J. L. Cummings (Eds.). The neuropsychiatry of limbic and subcortical disorders. Washington, D.C., American Psychiatric Press, Inc., 1997, pp. 123–32.

Trost, R. C. Differential classical conditioning of abstinence syndrome in morphine-dependent rats. Psychopharmacologia 30; 153–61, 1973.

Trowill, J. A. Instrumental conditioning of the heart rate in the curarized rat. J. Comp. Physiol. Psychol. 63; 7–11, 1967.

Tuma, A. H., J. D. Maser (Eds.). Anxiety and the anxiety disorders. Hillsdale, N.J., Lawrence Erlbaum Associates Publishers, 1985.

Turner, E. G., H. L. Altshuler. Conditioned suppression of an operant response using d-amphetamine as the conditioned stimulus. Psychopharmacology 50; 139–43, 1976.

Tursky, B., J. F. Papillo, R. Friedman. The perception and discrimination of arterial pulsations: Implications for the behavioral treatment of hypertension. J. Psychosom. Res. 26; 485–93, 1982.

Tyrer, P. Current status of beta-blocking drugs in the treatment of anxiety disorders. Drugs 36; 773–83, 1988.

Tyrer, P. J., M. H. Lader. Response to propranolol and diazepam in somatic and psychic anxiety. Br. Med. J. 2; 14–6, 1974.

Tyrer, P. J., M. H. Lader. Central and peripheral correlates of anxiety: A comparative study. J. Nerv. Ment. Dis. 162; 99–104, 1976.

Tyrer, P., I. Lee, J. Alexander. Awareness of cardiac function in anxious, phobic and hypochondriacal patients. Psychol. Med. 10; 171–4, 1980.

Umachandran, V., K. Ranjadayalan, G. Ambepityia, B. Marchant, P. G. Kopelman, A. D. Timmis. Aging, autonomic function, and the perception of angina. Br. Heart J. 66; 15–8, 1991.

Uno, T. The effects of awareness and successive inhibition on interoceptive and exteroceptive conditioning of the galvanic skin response. Psychophysiology 7; 27–43, 1970.

Vaitl, D., H. Gruppe. CNS modulation induced by hemodynamic change. In D. Vaitl, R. Schandry (Eds.). From the heart to the brain: The psychophysiology of circulation-brain interaction. Frankfurt, Peter Lang, 1995, pp. 89–104.

Vaitl, D., H. Gruppe, R. Stark, P. Possel. Simulated micro-gravity and cortical inhibition: A study of the hemodynamic–brain interaction. Biol. Psychol. 42; 87–103, 1996.

Valentino, R. J., G. Drolet, G. Aston-Jones. Central nervous system noradrenergic–peptide interactions. In O. G. Cameron (Ed.). Adrenergic dysfunction and psychobiology. Washington, D.C., American Psychiatric Press, 1994, pp. 33–72.

Van Beek, N., E. Griez. Reactivity to a 35% CO_2 challenge in healthy first-degree relatives of patients with panic disorder. Biol. Psychiatry 47; 830–5, 2000.

Van den Bergh, O., K. Stegen, K. P. van de Woestijne. Learning to have psychosomatic complaints: Conditioning of respiratory behavior and somatic complaints in psychosomatic patients. Psychosom. Med. 59; 13–23, 1997.

Van den Hout, M. A., E. Griez. Peripheral panic symptoms occur during changes in alveolar carbon dioxide. Compr. Psychiatry 26; 381–7, 1985.

Van den Hout, M. A., G. M. van der Molen, E. Griez, H. Lousberg, A. Nansen. Reduction of CO_2-induced anxiety in patients with panic attacks after repeated CO_2 exposure. Am. J. Psychiatry 144; 788–91, 1987.

Van der Does, A. J. W., M. M. Antony, A. Ehlers, A. J. Barsky. Heartbeat perception in panic disorder: A reanalysis. Behav. Res. Therap. 38; 47–62, 2000.

Van der Does, A. J. W., R Van Dyck, P. Spinhoven. Accurate heartbeat perception in panic disorder: Fact and artefact. J. Affect. Disord. 43; 121–30, 1997.

Vanderwolf, C. H. Brain, behavior, and mind: What do we know and what can we know? Neurosci. Biobehav. Rev. 22; 125–42, 1998.

Van Toller, C. The nervous body. New York, John Wiley and Sons, 1979.

Van Zijderveld, G. A., B. J. TenVoorde, D. J. Veltman, L. J. P. van Doornen, J. F. Orlebeke, R. van Dyck, F. J. H. Tilders. Cardiovascular, respiratory, and panic reactions to epinephrine in panic disorder patients. Biol. Psychiatry 41; 249–51, 1997.

Velden, M. C. Wolk. Cardiac cycle effects: A key to the understanding of neural heart-brain interactions." In D. Vaitl, R. Schandry (Eds.). From the heart to the brain: The psychophysiology of circulation–Brain interaction. Frankfurt, Peter Lang, 1995, pp. 121–31.

Veltman, D. J., G. A. Van Zijderveld, R. van Dyck, A. Bakker. Predictability, controllability, and fear of symptoms of anxiety in epinephrine-induced panic. Biol. Psychiatry 44; 1017–26, 1998.

Verfaellie, M., M. M. Keane. The neural basis of aware and unaware forms of memory. Sem. Neurol. 17; 153–61, 1997.

Verrier, R. L. Mechanisms of behaviorally induced arrhythmias. Circulation 76 (suppl. I); I48–56, 1987.

Verrier, R. L., B. Lown. Behavioral stress and cardiac arrhythmias. Annu. Rev. Physiol. 46; 155–76, 1984.

Vidergar, L. J., R. M. Lee, M. S. Goldman. Discrimination of systolic blood pressure. Biofeedback Self Regul. 8; 45–61, 1983.

Violani, C., C. Lombardo, L. DeGennaro, A. Devoto. Left movers' advantage in heartbeat discrimination: A replication and extension. Psychophysiology 33; 234–8, 1996.

Vogt, B. A., S. Derbyshire, A. K. Jones. Pain processing in four regions of human cingulate cortex localized with co-registered PET and MR imaging. Europ. J. Neurosci. 8; 1461–73, 1996.

Vogt, B. A., D. M. Finch, C. R. Olsen. Functional heterogeneity in cingulate cortex: The anterior executive and posterior evaluative regions. Cereb. Cortex 2; 435–43, 1992.

Vonck, K., P. Boon, K. Van Laere, M. D'Have, T. Vandekerckhove, S. O'Connor,

B. Brans, R. Dierckx, J. DeReuck. Acute single photon emission computed tomographic study of vagus nerve stimulation in refractory epilepsy. Epilepsia 41; 601–9, 2000.

Wagman, W. D., G. C. Maxey. The effects of scopolamine hydrobromide and methyl scopolamine hydrobromide upon the discrimination of interoceptive and exteroceptive stimuli. Psychopharmacologia 15; 280–8, 1969.

Walker, B. B., C. A. Sandman. Human visual evoked responses are related to heart rate. J. Comp. Physiol. Psychol. 93; 18–25, 1979.

Walker, B. B., C. A. Sandman. Visual evoked potentials change as heart rate and carotid pressure change. Psychophysiology 19; 520–7, 1982.

Walsh, D. H. Interactive effects of alpha feedback and instructional set on subjective state. Psychophysiology 11; 428–35, 1974.

Waterhouse, B. D., F. M. Sessler, W. Liu, C. S. Lin. Second messenger–mediated actions of norepinephrine on target neurons in central circuits: A new perspective on intracellular mechanisms and functional consequences." In C. D. Barnes, O. Pompeiano (Eds.). Neurobiology of the locus coeruleus: Progress in brain research, vol. 88. Amsterdam, Elsevier, 1991, pp. 351–62.

Watkins, L. L., P. Grossman, R. Krishnan, A. Sherwood. Anxiety and vagal control of heart rate. Psychosom. Med. 60; 498–502, 1998.

Watts, J. M. J. Anxiety and the habituation of the skin conductance response. Psychophysiology 12; 596–601, 1975.

Weinberger, N. M., P. E. Gold, D. B. Sternberg. Epinephrine enables Pavlovian fear conditioning under anesthesia. Science 223; 605–7, 1984.

Weingarten, H. P., T. L. Powley. Pavlovian conditioning of the cephalic phase of gastric acid secretion in the rat. Physiol. Behav. 27; 217–21, 1981.

Weingartner, H., H. Miller, D. L. Murphy. Mood-state–dependent retrieval of verbal associations. J. Abnorm. Psychol. 86; 276–84, 1977.

Weiss, J. M. Effects of coping behavior in different warning signal conditions on stress pathology in rats. J. Comp. Physiol. Psychol. 77; 1–13, 1971a.

Weiss, J. M. Effects of coping behavior with and without a feedback signal on stress pathology in rats. J. Comp. Physiol. Psychol. 77; 22–30, 1971b.

Weiss, T., B. T. Engel. Operant conditioning of heart rate in patients with premature ventricular contractions. Psychosom. Med. 33; 301–21, 1971.

Weisz, J., L. Balazs, G. Adam. The influence of self-focused attention on heartbeat perception." Psychophysiology 25; 193–9, 1988.

Weisz, J., L. Balazs, G. Adam. The effect of monocular viewing on heartbeat discrimination. Psychophysiology 31; 370–4, 1994.

Weisz, J., L. Balazs, E. Lang, G. Adam. The effect of lateral visual fixation and the direction of eye movements on heartbeat discrimination. Psychophysiology 27; 523–7, 1990.

Welgan, P. R. Learned control of gastric acid secretions in ulcer patients. Psychosom. Med. 36; 411–9, 1974.

Weitkunat, R., R. Schandry. Motivation and heartbeat evoked potentials. J. Psychophysiol. 4; 33–40, 1990.

Wells, D. T., B. W. Feather, M. W. Headrick. The effects of immediate feedback upon voluntary control of salivary rate. Psychophysiology 10; 501–9, 1973.

Wenning, A. Sensing effectors make sense. Trends Neurosci. 22; 550–5, 1999.

Westlund, K. N., D. Zhang, S. M. Carlton, L. S. Sorkin, W. D. Willis. Noradrenergic innervation of somatosensory thalamus and spinal cord. In C. D. Barnes, O. Pompeiano (Eds.). Neurobiology of the locus coeruleus: Progress in brain research, vol. 88. Amsterdam, Elsevier, 1991, pp. 77–88.

Whalen, P. J., G. Bush, R. J. McNally, S. Wilhelm, S. C. McInerney, M. A. Jenike, S. L. Rauch. The emotional counting Stroop paradigm: A functional magnetic resonance imaging probe of the anterior cingulate affective division. Biol. Psychiatry 44; 1219–28, 1998.

Wheeler, E. O., P. D. White, E. W. Reed, M. E. Cohen. Neurocirculatory asthenia (anxiety neurosis, effort syndrome, neurasthenia). JAMA 142; 878–89, 1950.

Whitehead, W. E. Biofeedback treatment of gastrointestinal disorders. Biofeedback Self Regul. 17; 59–76, 1992.

Whitehead, W. E., M. D. Crowell, A. L. Davidoff, O. S. Palsson, M. M. Schuster. Pain from rectal distention in women with irritable bowel syndrome: Relationship to sexual abuse. Digest. Dis. Sci. 42; 796–804, 1997.

Whitehead, W. E., V. M. Drescher. Perception of gastric contractions and self-control of gastric motility. Psychophysiology 17; 552–8, 1980.

Whitehead, W. E., O. S. Palsson. Is rectal pain sensitivity a biological marker for irritable bowel syndrome: Psychological influences on pain perception. Gastroenterology 115; 1263–71, 1998.

Whitehead, W. E., P. F. Renault, I. Goldiamond. Modification of human acid secretion with operant-conditioning procedures. J. Appl. Behav. Anal. 8; 147–56, 1975.

Wiedemann, G., A. Stevens, P. Pauli, W. Dengler. Decreased duration and altered topography of electroencephalographic microstates in patients with panic disorder. Psychiatry Res. 84; 37–48, 1998.

Wientjes, C. J. E., P. Grossman. Overreactivity of the psyche or the soma? Interindividual associations between psychosomatic symptoms, anxiety, heart rate, and end-tidal partial carbon dioxide pressure. Psychosom. Med. 56; 533–40, 1994.

Wik, G., M. Fredrikson, K. Ericson, L. Eriksson, S. Stone-Elander, T. Greitz. A functional cerebral response to frightening visual stimulation. Psychiatry Res. 50; 15–24, 1993.

Wik, G., M. Fredrikson, H. Fischer. Cerebral correlates of anticipated fear: A PET study of specific phobia. Internal. J. Neurosci. 87; 267–76, 1996.

Wik, G., M. Fredrikson, H. Fischer. Evidence of altered cerebral blood-flow relationships in acute phobia. Internat. J. Neurosci. 91; 253–63, 1997.

Wilkinson, D. J. C., J. M. Thompson, G. W. Lambert, G. L. Jennings, R. G. Schwarz, D. Jeffreys, A. G. Turner, M. D. Esler. Sympathetic activity in patients with panic disorder at rest, under laboratory mental stress, and during panic attacks. Arch. Gen. Psychiatry 55; 511–20, 1998.

Willis, W. D., K. N. Westlund. Neuroanatomy of the pain system and the pathways that modulate pain. J. Clin. Neurophysiol. 14; 2–31, 1997.

Windmann, S., Schonecke, O. W., G. Frohlig, G. Maldener. Dissociating beliefs about heart rates and actual heart rates in patients with cardiac pacemakers. Psychophysiology 36; 339–42, 1999.

Winters, M., G. V. Padilla. Autonomic arousal and fear in the spinal cord injured. Rehabilitation Nursing 11; 13–7, 1986.

Wolf, S. Behavioral aspects of cardiac arrhythmia and sudden death. Circulation 76 (suppl. I); I174–6, 1987.

Wolf, S. Forebrain involvement in fatal cardiac arrhythmias. Integr. Physiol. Behav. Sci. 30; 215–25, 1995.

Woods, S. W., K. Koster, J. K. Krystal, E. O. Smith, I. G. Zubal, P. B. Hoffer, D. S. Charney. Yohimbine alters regional cerebral blood flow in panic disorder. Lancet 2; 678, 1988.

Wu, J. C., M. S. Buchsbaum, T. G. Hershey, E. Hazlett, N. Sicotte, J. C. Johnson. PET in generalized anxiety disorder. Biol. Psychiatry 29; 1181–99, 1991.

Wurthmann, C., J. Gregor, B. Baumann, O. Effenberger, W. Dohring, B. Bogerts. Qualitative evaluation of brain structure in CT in panic disorder. Nervenarzt 69; 763–8, 1998.

Yates, B. J., S. D. Stocker. Integration of somatic and visceral inputs by the brainstem. Exper. Brain Res. 119;269–75, 1998.

Yeragani, V. K., R. Balon, R. Pohl, A. Ortiz, P. Weinberg, J. M. Rainey. Do higher preinfusion heart rates predict laboratory-induced panic attacks? Biol. Psychiatry 22; 554–8, 1987a.

Yeragani, V. K., R. Pohl, R. Balon, P. Weinberg, R. Berchou, J. M. Rainey. Preinfusion anxiety predicts lactate-induced panic attacks in normal controls. Psychosom. Med. 49; 383–9, 1987b.

Yoon, B. W., C. A. Morillo, D. Cechetto, V. Hackinski. Cerebral hemispheric lateralization in cardiac autonomic control. Arch. Neurol. 54; 741–4, 1997.

Young, L. D., E. B. Blanchard. Awareness of heart activity and self-control of heart rate: A failure to replicate. Psychophysiology 21; 361–2, 1984.

Young, R., R. A. Glennon. Discriminative stimulus properties of (-) ephedrine." Pharmacol. Biochem. Behav. 60; 771–5, 1998.

Zajonc, R. B. Emotion and facial efference: A theory reclaimed. Science 228; 15–21, 1985.

Zamrini E Y, Meador, K J, Loring D W, Nichols F T, Lee G P, Figueroa R E, Thompson W O. Unilateral cerebral inactivation produces differential left/right heart rate responses. Neurology 1990;40:1408–11.

Zandbergen, J., V. van Aalst, C. de Loof, H. Pols, E. Griez. No chronic hyperventilation in panic disorder patients. Psychiatry Res. 47; 1–6, 1993.

Zandbergen, J., M. Strahm, H. Pols, E. J. L. Griez. Breath-holding in panic disorder. Compr. Psychiatry 33; 47–51, 1992.

Zener, K. The significance of behavior accompanying conditioned salivary secretion for theories of the conditioned response. Am. J. Psychol. 50; 384–403, 1937.

Zighelboim, J., N.J. Talley, S.F. Phillips, W.S. Harmsen, A.R. Zinsmeister. Visceral perception in irritable bowel syndrome: Rectal and gastric responses to distention and serotonin type 3 antagonism. Digest. Dis. Sci. 40; 819–27, 1995.

Zigmond, M.J., F.E. Bloom, S.C. Landis, J.L. Roberts, L.R. Squire (Eds.). Fundamental neuroscience. San Diego, Academic Press, 1999.

Zipes, D.P. Influence of myocardial ischemia and infarction on autonomic innervation of heart. Circulation 82; 1095–1105, 1990.

Zoellner, L.A., M.G. Craske. Interoceptive accuracy and panic. Behav. Res. Therap. 37; 1141–58, 1999.

Zubieta, J.K., J.A. Chinitz, U. Lombardo, L.M. Fig, O.G. Cameron, I. Liberzon. Medial frontal cortex involvement in PTSD symptoms: A SPECT study. J. Psychiatr Res. 33; 259–64, 1999.

NAME INDEX

Page numbers in italics refer to illustrations.

SUBJECT INDEX

Pages in italics refer to illustrations.